S0-ABD-849

Innate and Adaptive Immunity in the Tumor Microenvironment

Eitan Yefenof
Editor

Innate and Adaptive Immunity in the Tumor Microenvironment

Eitan Yefenof
Lautenberg Center for General
and Tumor Immunology

ISBN: 978-1-4020-6749-5 e-ISBN: 978-1-4020-6750-1

Library of Congress Control Number: 2007939400

Printed on acid-free paper.

9 8 7 6 5 4 3 2 1

springer.com

Foreword

Tumor Microenvironment under the Magnifying Glass: A New Challenge to Cancer Immunologists

The theory of immunological surveillance against cancer has been introduced half a century ago by McFarlane Burnet and Lewis Thomas. Despite extensive research to validate and generalize this concept, approaches to intervene in cancer by immunological means has led to more disappointments than successes, more doubts than convictions. Is this concept still valid? If the question addresses the possibility that cancer cells induce protective immunity sufficient to enact tumor rejection, the answer is no. But what about manipulating innate and adaptive, cellular and soluble immune components to fight tumor progression and metastasis? A pessimist's answer would allude to the ample evidence demonstrating that most malignant cells are non-immunogenic and do not become "non-self" targets. The optimist will highlight cases of melanoma and non-Hodgkin's lymphoma that can be managed by active and passive immunotherapy. Realistically, however, one has to admit that approaches to induce specific immunity to cancerous cells may be useful in the treatment of tumors with a well-defined etiological factor (viral or carcinogen), leaving most carcinomas and sarcomas outside the scope of potential targets for adaptive immune attack. Nevertheless, when it comes to innate immune responses and factors, the prospects of harnessing inflammatory and growth restricting cytokines, NK cell death receptors and ligands has not been exploited in full. We may be surprised, after all, by research endeavors focusing on these immune components in the future.

In recent years, the study of cancer immunology has been benefited by the emerging concept of the tumor microenvironment.

> It is not in heaven …neither it is beyond the sea. …It is very nigh unto thee that you mayest do it. (Deuteronomy 30:11)

The message is clear and straight forward. To discern immune responses pertinent to tumor regression (or progression), one should look nigh to the tumor. Studying of systemic immunity in cancer patients or experimental animals is not always relevant to the antitumor effects operating at the tumor site. Both innate and adaptive immune responses taking place within the tumor and its immediate vicinity are unique and their study should yield novel insights into the role that immunity may play in protecting the host against an emerging or expanding tumor.

In the present issue we undertook the challenge to review and discuss recent studies focusing on the uniqueness of loco-regional anti-(or pro) cancer immunity, manifested at the tumor microenvironment. In Chapter 1, Whiteside reviews evidence suggesting that components necessary for mounting an antitumor immune response are present in cancer patients. Yet, such responses fail to arrest tumor growth due to a loco-regional immune dysfunction within and around the tumor. Multiple mechanisms are involved in creating an immunosuppressive microenvironment at the tumor site, which call for the development of therapies aimed to avoid tumor induced suppression and sustain the function of immune effector cells.

Mechanisms leading to immunosuppression at the tumor microenvironment are discussed in Chapter 2 by Ferrone. It appears that immunosuppressive cytokines such as IL10 and TGF-β are abundantly produced by tumor, stromal and immune cells at the microenvironment. In addition to their inhibitory effect, such cytokines act directly on cancerous cells by impairing their antigen presenting capacity. Another inhibitory activity is mediated by soluble HLA molecules generated at the tumor microenvironment that induce apoptosis of infiltrating CTL and NK cells with residual cytotoxic activity.

In Chapter 3, Schirrmacher describes and discusses malignant diseases in which tumor growth occurs despite the existence of functional tumor specific memory T cells in the bone marrow. This apparent paradoxical situation reflects the disparity between peripheral and loco-regional immune responses during the progression of cancer. The unique loco-regional characteristics of host tumor relationship require novel intervention modalities which counteract immune and metabolic deregulation in the tumor microenvironment.

Modern technologies enabling gene expression profiling can be applied in the study of cancer microenvironmental immunity. This approach is discussed by Gajewski in Chapter 4 using melanoma microenvironment as a model. An oligonucleotide array platform preformed on melanoma biopsies and cell lines identified numerous genes that are upregulated in tumor versus stromal cells. Many such gene transcripts are involved in regulating T and B cell immunity and in crating immunosuppressive microenvironment. Such throughput analysis may help in discerning new targets for manipulating the tumor microenvironment in favor of immunoreactivity.

How can tumors manipulate their microenvironment to escape immune regulation and attack? Multiple mechanisms exist, one of which involves tumor-associated antigens (TAA). Such antigens, as discussed by Engelmann and Finn in Chapter 5, play a Dr. Jekyll and Mr. Hyde game. While in experimental systems and some clinical circumstances they can be targeted by immune surveillance, at the tumor microenvironment they promote tumorigenesis by acting as adhesion or signaling molecules. The Mr. Hyde type of action becomes even more deleterious when TAA are expressed on cancer stem cells, by virtue of their capacity to prevent apoptosis, increase proliferation and induce chemotaxis of immune cells.

Under most circumstances, cancer cells shape their microenvironment in their favor. However, at the same time they are affected by microenvironmental factors that confer upon them heightened status of immunoresistant phenotype. This outcome is discussed by Bonavida in Chapter 6. While immuno-competent cytotoxic lymphocytes infiltrate the growing tumor, they fail to kill tumor cells, since the latter develops a high degree of apoptotic resistance. Both extrinsic and intrinsic mechanisms generated inside and outside the tumor contribute to the gradual acquisition of a resistant phenotype and their identification is required in order to re-sensitize the cells for cytotoxicity.

The cancer microenvironment may be envisaged as a site of chronic inflammation. This view is presented by Selleri et al. in Chapter 7. Evidence has been generated in recent years to indicate that the evolution and progression of certain cancers is promoted by a chronic inflammatory response to viral or bacterial infection. Hence, cellular and soluble components of inflammation at the tumor microenvironment are likely to create a niche favoring tumor growth. Immune modulation and suppression is one outcome of chronic inflammation, manifested *in situ*, at close proximity to the growing tumor.

The most important effector cells which are directly engaged in tumor cell destruction are Cytotoxic T lymphocytes (CTLs) and natural killer (NK) cells. In Chapter 8, Lotem et al. discuss the events taking place during CTL activation, migration, target recognition, immune synapse formation and killing. While all these steps occur and advance during viral infection or graft rejection, they fail to operate within the tumor microenvironment. This deficiency may be restored using gene transfer technologies by which CTLs are equipped with high affinity T cell receptors (TCR) or TCR replacing molecules, such as chimeric monoclonal antibodies. These manipulations, together with *ex vivo* expansion and selection of highly reactive CTLs should boost the destructive capacity of the cells, beyond the limiting threshold of the immunosuppressive tumor microenvironment.

Whereas CTLs are a cellular component of the specific/adaptive immune system, NK cells belong to the innate arm of the immune response. As reviewed by Gazit et al. in Chapter 9, the potency of tumor cell killing by NK cells has been demonstrated in various experimental systems. Yet, tumor rejection by NK cell killing is not evident due to microenvironmental factors that inhibit NK activity. Since NK cells are equipped with an array of inhibitory receptors (NKIRs) that interact with MHC and non-MHC ligands, their killing activity at the tumor microenvironment may be inhibited by direct or indirect engagement of such receptors. In addition the immunosuppressive microenvironment of the tumor disarms NK cells by inducing multiple phenotypic changes which should be studied in detail before modalities for harnessing of NK cells to fight cancer are developed.

Conceptually and traditionally, the tumor microenvironment is envisaged as a consequence of malignant transformation in that a growing tumor creates a shield that protects its cells from destruction by immune and non-immune mechanisms. Klein et al. in Chapter 10 provide a different look at the tumor microenvironment. In certain hemopoietic cancers it may play a causative role and promote malignant transformation. This is the case in Hodgkin's and nasal NK/T lymphomas which emerge following infection by EBV. The transforming capacity of this virus is usually offset by robust immunity in the host, leading to persistent latent infection. However, when a certain pattern of viral protein expression ensues, a complex interactive microenvironment comes to exist with unique composition of cellular and soluble components. Such microenvironmental factors promote, rather than inhibit, cellular transformation and tumor progression, and are responsible for the formation of the Hodgkin's lymphoma Granuloma.

This publication is first in a series of Springer Science books dealing with various aspects of the tumor microenvironment. To the best of my knowledge it is a first compilation of chapters discussing the interrelationship between the immune response, the tumor and its microenvironment. Evidently, a focus on the tumor microenvironment provides new and exciting insight into the complex interaction between innate and adaptive, cellular and humoral immune responses and cancer cells. It is a young and evolving concept doing its first steps on a long path of scientific inquiry. The ultimate goal is to design local and systemic immune strategies that will affect the tumor and its environment in a therapeutic manner. Hopefully, the ongoing investment in studying cancer microenvironment will ripe to yield practical dividends in the near future.

Eitan Yefenof

Contents

List of Contributors

Philipp Beckhove
Division of Cellular Immunology, German Cancer Research Center, Heidelberg, Germany

Benjamin Bonavida
Department of Microbiology, Immunology, and Molecular Genetics, Jonsson Comprehensive Cancer, David Geffen School of Medicine, University of California, Los Angeles, CA, USA

Sara Deola
Immunogenetics Section, Department of Transfusion Medicine, Clinical Center, National Institutes of Health, Bethesda, MD, USA

Katja Engelmann
Department of Immunology, University of Pittsburgh School of Medicine, Pittsburgh, PA, USA

Olivera J. Finn
Department of Immunology, University of Pittsburgh School of Medicine, Pittsburgh, PA, USA

Soldano Ferrone
Hillman Cancer Center, University of Pittsburgh, Cancer Institute, Pittsburgh, PA, USA

Shoshana Frankenburg
Sharett Institute of Oncology, Hadassah Hebrew University Hospital, Jerusalem, Israel

Thomas F. Gajewski
Department of Pathology and Department of Medicine, Section of Hematology/Oncology, University of Chicago, Chicago, IL, USA

Roi Gazit
Lautenberg Center for General and Tumor Immunology, the Hebrew University–Hadassah Medical School, Jerusalem, Israel

L. L. Kis
Department of Microbiology, Tumor and Cell Biology (MTC),
Karolinska Institutet Stockholm, Sweden

Eva Klein
Department of Microbiology, Tumor and Cell Biology (MTC),
Karolinska Institutet, Stockholm, Sweden

Michal Lotem
Sharett Institute of Oncology, Hadassah Hebrew University Hospital, Jerusalem, Israel

Arthur Machlenkin
Sharett Institute of Oncology, Hadassah Hebrew University Hospital, Jerusalem, Israel

Ofer Mandelboim
Lautenberg Center for General and Tumor Immunology, the Hebrew University–Hadassah Medical School, Jerusalem, Israel

Francesco M. Marincola
Immunogenetics Section, Department of Transfusion Medicine Clinical Center National Institutes of Health, Bethesda, MD, USA

Tamar Peretz
Sharett Institute of Oncology, Hadassah Hebrew University Hospital, Jerusalem, Israel

Cristiano Rumio
Universita degli Studi di Milano, Milan, Italy

Volker Schirrmacher
Division of Cellular Immunology, German Cancer Research Center, Heidelberg, Germany

Silvia Selleri
Immunogenetics Section, Department of Transfusion Medicine, Clinical Center, National Institutes of Health, Bethesda, MD, USA

M. Takahara
Department of Otolaryngology, Asahikawa Medical College, Asahikawa, Japan

Theresa L. Whiteside
Departments of Pathology, Immunology and Otolaryngology, University of Pittsburgh, School of Medicine, University of Pittsburgh Cancer Institute, Pittsburgh, PA, USA

Eitan Yefenof
Lautenberg Center for General and Tumor Immunology, the Hebrew University–Hadassah Medical School, Jerusalem, Israel

Chapter 1
Immune Effector Cells in the Tumor Microenvironment

Their Role in Regulation of Tumor Progression

Theresa L. Whiteside

Abstract Immune effector cells have the ability to eliminate malignant cells and thus regulate tumor progression. However, immune cells in the tumor microenvironment are dysfunctional, generally fail to control tumor growth and may even promote its progression. Molecular mechanisms responsible for tumor-induced local and systemic immune suppression are currently under intense scrutiny. It appears that tumors can deregulate recruitment, effector functions and survival of immune cells, interfering with all stages of antitumor response. Suppressive mechanisms targeting key signaling pathways in immune cells have been identified. Strategies for reversal of tumor-mediated immunsuppression are being developed. Understanding of multiple and varied mechanisms used by tumors to escape immune surveillance is crucial for the future design of more effective antitumor therapies.

Keywords Antitumor immunity, immune suppression, mechanisms of suppression, tumor escape, tumor microenvironment

1.1 Introduction

Evidence supports the concept that the host immune system interacts with the developing tumor and, in some cases, may be responsible for the arrest of tumor growth and tumor regression. The presence of antibodies (Abs) to tumor-associated antigens (TAA) and of specific (i.e. tetramer+) as well as nonspecific effector cells in the peripheral circulation and at the tumor site of cancer patients has been often reported. This implies that immune cells and Abs are potentially capable of mediating tumor rejection in these patients. *In vitro*, isolated effector cells, including T cells, natural killer (NK) cells, Antibody-armed NK cells or macrophages, have each been shown to mediate the death of tumor targets in short-term assays by mechanisms such as osmotic lysis (cytotoxicity) or apoptosis and/or antibody-dependent

---

Departments of Pathology, Immunology and Otolaryngology, University of Pittsburgh School of Medicine and University of Pittsburgh Cancer Institute Pittsburgh, PA 15213, USA

E. Yefenof (ed.), *Innate and Adaptive Immunity in the Tumor Microenvironment.*

cellular cytotoxicity (ADCC). *In vivo*, however, tumors seem to have evolved multiple means to evade these immune effector cells, a paradigm known as "immunologic escape." Tumors can engineer such immune escape via a wide range of mechanisms, and the ability to interfere with numbers of antitumor immune cells and/or to subvert their function is considered to be a principal reason underlying the failure of the host to ultimately control tumor progression.

Immunotherapies, including antitumor vaccines, are designed to activate and mobilize the host adaptive immune response directed at the tumor. However, despite the current availability of immunogenic TAA and peptides [133], vaccination-based clinical trials using these antigens in patients with advanced malignancies have failed to demonstrate a conclusive relationship between clinical responses and the presence or frequency of antitumor effector cells [63, 137]. In many cases, immune responses to vaccines have been weak or undetectable [62, 96]. In other cases, the frequency of peptide- or TAA-specific CD8+ effector cells (cytolytic T lymphocytes or CTL) increases following vaccine administration, without any measurable impact on the patient's clinical course [e.g. 117, 125]. This lack of convincing and reproducible associations of immunologic with objective clinical responses to immunotherapies in cancer patients has been both unexpected and disconcerting, and it has cast doubts regarding the utility of biologic therapies in cancer. These findings do not fit with the evidence accumulated from a wide range of animal tumor models, where tumor-specific T cells have been shown to play an active role in eliminating tumor/metastasis and inducing antitumor memory responses [99, 213]. Although human PBMC isolated from patients with cancer contain tumor-specific CD8+ and CD4+ T cell precursors that are capable of killing tumor cell targets after appropriate *in vitro* activation [187, 191, 197, 198], attempts at the expansion and sustaining of antitumor functions of these effector cells in the clinical setting remain a major challenge.

Current evidence suggest that the components necessary for mounting the antitumor immune response are present in cancer patients, including TAA, antigen presenting cells (APC), immune T cell subsets and cytokines. Nevertheless, antitumor immune responses generally fail to arrest tumor growth. Protective antitumor responses are not promoted and sustained in tumor-bearing hosts. Efforts to discern what happens to immune cells in the tumor microenvironment that prevents the development of clinically relevant TAA-specific immune responses are of central importance. Two possibilities have been considered: (a) tumor-specific immune responses are generated both *in situ* and systemically, but they are inadequate, inappropriate or otherwise skewed toward tolerance; and (b) tumor-specific immune responses are effectively generated and are not biased, but the tumor cleverly avoids immune recognition. The mechanisms underlying the capability of human tumors to cause an imbalance of the host immune system and to successfully avoid recognition by immune cells are being identified. The role of tumor microenvironment in development and progression of cancer in man has been recognized, and interactions between immune cells and other cellular or molecular components of the tumor milieu are known to significantly influence the disease outcome. Design of novel therapies to restore or balance antitumor immune responses *in situ* and systemically, thereby achieving therapeutic benefits, remains a major incentive of tumor immunology today.

1.2 Evidence for Loco-regional Immune Dysfunction in Cancer

The commonly observed presence of leukocytes in human tumors has led to speculations that these cells contribute to the control of tumor growth. Despite many attempts to correlate the degree or type of cellular infiltrates with prognosis or patient survival, no consensus exists [reviewed in ref. 198], and the significance of leukocyte infiltrates in tumors remains equivocal. However, it seems reasonable to predict that it is not the presence per se, but the functional potential of the infiltrating cells confronting the tumor that determines their utility in host defense. Most tumor-infiltrating lymphocytes (TIL) are activated T cells containing variable proportions of CD8+ and CD4+ T cell subsets, which are almost exclusively CD45RO+ memory T cells [reviewed in 187, 191]. In comparison to autologous peripheral blood lymphocytes or those isolated from tissues distant from the tumor, TIL-T cells have been consistently found to be poorly responsive or unresponsive to traditional T-cell activating stimuli [197, 198]. Table 1.1 summarizes evidence accumulated to date for functional deficiencies observed in fresh TIL isolated from a wide variety of human tumors. In aggregate, these data indicate that activated T-cells derived from tumor sites or tumor-involved human lymph nodes are functionally compromised and do not behave like normal activated T cells.

Nevertheless, TIL freshly isolated from certain tumors, especially melanoma and renal cell carcinoma (RCC), are “responsive” to immunologic intervention *in vitro* [24]. This suggests that even in the face of tumor-induced immune deviation, these lymphocytes can be salvaged and may be of clinical benefit to the patient, if an appropriate treatment is administered. Functional impairments have also been

Table 1.1 Immune deviation in T cells present in the tumor microenvironment

1. Activation of proteolytic enzymes in tumor-infiltrating leukocytes: rapid degradation of cellular proteins [129].
2. Signaling defects in TIL and PBL-T:
   (a) NFκB abnormalities [82, 172]
   (b) ζ chain defects: either low expression or absence [43, 72, 98, 104, 132, 211]
   (c) Ca+ + flux alterations [130]
3. Cytokine expression: absent/decreased Th1-type cytokines [86, 128, 160].
4. Inhibition of lymphocyte proliferation, cytotoxic activity or cytokine production [29, 130, 170, 197].
5. Inhibition of leukocyte migration [121].
6. Induction of T-cell apoptosis [56, 129, 130, 190].
7. Expansion of immunosuppressive macrophages [8, 94, 95].
8. Accumulation of regulatory CD4+ $CD25^{high}$ T cells [6, 31, 84, 142, 146, 158].

noted for alternate effector cells that accumulate within tumor sites. Several recent reports indicate that tumor-associated DC (TADCs) are functionally defective, especially in their antigen-presenting capacity [46]. Tumor-associated macrophages (TAMs) also exhibit functional defects relative to their counterparts obtained from tumor-uninvolved inflammatory sites in the same patient [94, 95].

1.2.1 *Biologic Significance of Loco-regional Immune Suppression*

Functional defects in TIL appear to be biologically significant. For example, we have documented the presence of signaling defects in T cells isolated from tumor biopsies of 138 patients with oral carcinoma [132]. We evaluated the expression of the CD3ζ chain in TIL, using semiquantitative analysis of immunostained sections of paraffin-embedded tumors [132]. TIL were scored as negative/weakly stained (0 or 1) or as positive (2) for ζ expression. The results showed that 32% of tumors had absent or low expression of ζ in TIL, and that this alteration was significantly associated with an advanced stage (T3 or T4) as well as nodal involvement [132]. In oral carcinoma patients with advanced disease, normal expression of ζ in TIL was predictive of a significantly better 5-year survival (Fig. 1.1) and was independent of other established prognostic parameters [132]. This report of a direct association between reduced ζ expression in TIL-T cells, disease progression and overall patient survival suggested that functional abnormalities observed in TIL have important biologic consequences. It has been also reported that decreased expression of ζ occurs in patient peripheral blood mononuclear cells (PBMC) [72, 131]. These observations suggest that in patients with cancer, ζ may be a marker of

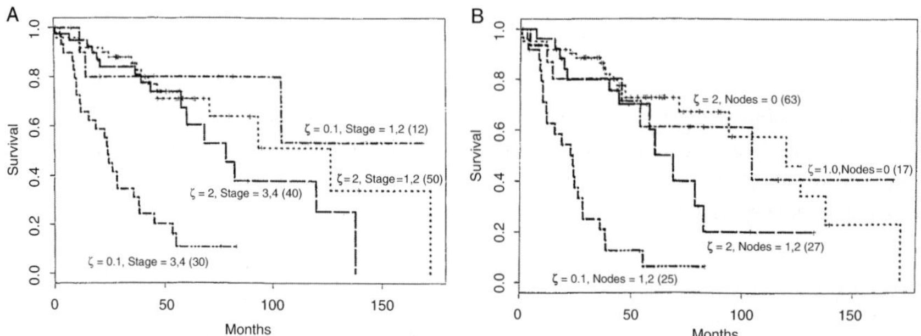

Fig. 1.1 (A), Kaplan-Meier plots presenting survival of patients with oral SCC by ζ chain expression in TIL and by tumor stage. Numbers in parentheses, numbers of patients in each category (low ζ, stage I–II; normal ζ, stage I–II, low ζ, stage III–IV; and normal ζ, stage III–IV). (B), Kaplan-Meier plots presenting survival of patients with oral SCC by lymph node involvement and low or normal ζ chain expression in TIL

immune competence, and individuals who have normal ζ expression are most likely to respond favorably to biotherapy [195].

1.3 Evidence for Systemic Immune Deviation in Cancer

In patients with cancer, alterations also occur in systemic antitumor immunity [56, 131, 190, 192]. In a tumor-bearing host, dissemination of antigens released from the growing (shed or secreted antigens) or dying (apoptotic/necrotic debris) tumor cells assures their access to APC and allows for the cross-priming of immune effector cells in lymphoid tissues draining the site of the tumor. It is likely that the composition, number and the state of activation/differentiation of the immune infiltrate at the local site will vary over time, depending on the evolving (constitutive or therapy induced) systemic immune response of the host to TAA and the ability of these cells to be recruited into and survive/function within the tumor microenvironment. Circulating T cells obtained from patients with cancer are either biased in their cytokine profile or otherwise functionally compromised [17, 131, 171]. Furthermore, dysfunctions in circulating lymphocytes was linked to the extent of dysfunction seen in paired TIL [131] and to the disease stage and/or activity [24, 131]. Thus, patients with advanced stage carcinomas had more lymphocyte dysfunction (low ζ expression, low proliferative index, greater NFκB dysfunction) than patients with early or inactive disease [24, 131]. The presence of such systemic alterations may explain, in part, why vaccines and other immunotherapies yield objective clinical responses in only a small minority (at best 20%) of patients with cancer.

1.4 Mechanisms Responsible for Dysfunction of Immune Cells in Cancer Patients

To date, many different mechanisms that are responsible for immune dysfunction have been identified. Some are directly mediated by factors produced by tumors, while others result from alterations of normal tissue homeostasis occurring in the presence of cancer. Until recently, little was known about molecular alterations in tumor cells *in situ* as they progressed from the pre-malignant to metastatic phenotype. Genetic instability, now recognized as a principal characteristic of all tumors, may result in changes in their epitope profile. Molecular changes, already detectable during early stages of tumorigenesis, become more pronounced as the tumor progresses. The net result of these changes is increased resistance of tumor cells to immune surveillance. Remarkably, tumors appear to be able to interfere with immune cell development, differentiation, function and even their elimination. Thus, all stages of antitumor immune response are subject to adverse intervention in the tumor microenvironment as indicated in Table 1.2 and discussed below.

Table 1.2 Mechanisms responsible for tumor immune escape[a]

Interference with distinct stages of TAA-specific immune responses	Mechanisms	References
I. Induction of antitumor immune responses	1. Molecular signaling in the tumor micro-environment	
	a. Paucity of co-stimulatory signals due to decreased expression of co-stimulatory molecules on tumor or APC surface	[75, 176]
	b. Death receptor/ligand signaling and "tumor counterattack"	[112, 113]
	c. NFκB signaling	[83, 122]
	2. Dysfunction of DC and inadequate cross-presentation of TAA to T cells	[1, 65]
	3. DC apoptosis in the tumor	[38, 123]
II. Compromised function/survival of effector T lymphocytes	1. Suppression of T-cell responses by Treg	[31, 183, 206]
	2. Suppression by myeloid suppressor cells (MSC)	[23, 144]
	3. Apoptosis of effector T cells in the tumor and in the periphery	[56, 190]
	4. MV (exosomes secreted by human tumors)	[2, 12, 70]
III. Loss of tumor cell recognition by immune cells	1. Changes in surface expression of HLA molecules	[25, 109]
	2. Down-regulation of surface TAA displayed by tumor cells: antigen loss variants	[92, 135]
	3. Alterations in APM component expression in tumor cells or APC	[42, 184]
	4. Suppression of NK activity in the tumor microenvironment	[77, 126]
IV. Resistance to tumor cells to immune intervention	1. Lack of susceptibility to immune effector cells	[87, 92, 203]
	2. Immunoselection of resistant variants	[30, 92]
	3. Tumor stem cells	[60]

[a] The mechanisms listed are selected from among others known to illustrate the diversity of tumor escape strategies. See text for details.

1.4.1 Interference with the Induction of TAA-specific Responses

Tumors are known to interfere with the early stages of the immune response by targeting cellular and molecular mechanisms responsible for its initiation and subsequent progression.

1.4.1.1 Molecular Signaling in the Tumor Microenvironment

Cells initiating immune responses that are driven by activating and co-stimulatory signals communicate with one another. However, normal signal transmission is disrupted or diverted in the tumor presence. Both murine and human tumors are known to interfere with different molecular pathways which mediate cellular cross talk [104, 129]. In addition to tumor cells, the tumor microenvironment contains stromal fibroblasts, infiltrating hematopoietic cells and blood vessel elements [193]. These normal cells were viewed as a scaffold necessary for tumor cell expansion (the stroma) or a manifestation of host antitumor activity (infiltrating leukocytes). Only recently molecular interactions between these cells and the tumor came to be recognized as regulatory elements of cell growth vs. cell death. Among them, the death receptor/ligand and NFkB activation pathways also play a key role in immune cell differentiation and functions [201]. However, the signaling immune cells encounter in the tumor milieu is distinct from that in normal tissues, and it does not favor the induction of TA-specific responses.

(a) Paucity and diversion of co-stimulatory signals

Immunohistology studies show that human tumors lack or down-regulate co-stimulatory molecules [75]. Although tumor cells are not "professional" APC, insufficient expression of molecules such as B7.1or B7.2 may result in tolerance, because the second signal necessary for T-cell activation is lacking [176]. If DC conditioned by the tumor have reduced levels of co-stimulatory molecules, their ability to fully activate T cells is diminished. Thus, tumors are inefficient in supporting T-cell differentiation and proliferation. In effect, DC in the tumor are locked in an immature state. Further, malignant cells express molecules such as programmed death ligands 1 and 2 (PD-L1 and PD-L2), which can interact with corresponding death receptors (PD-1 or PD-2) present on activated T cells and induce their apoptosis [118]. Also, PD-L1 and PD-L2 facilitate preferential development of CD25+CD152 (CTLA-4+) T suppressor cells and promote functional anergy [118]. Signaling processed via PD-L1 (B7-H1) has been shown to promote tumor progression in both murine and human SCID models [61], and its expression on tumors has been correlated with poor prognosis in humans [167]. CTLA-4, a homologue of CD28, is constitutively expressed on activated T cells and Treg. In the presence of DC cross-presenting TA and rich in B7.1 and B7.2 in the tumor microenvironment, activated T cells up-regulate CTLA-4. Because CTLA-4 molecule functions as a key negative regulator of CD28-dependent T-cell responses, such CTLA-4 upregulation on activated T cells results in the inhibition of Th1-type immunity and polarization to Th-2-type responses [164]. Evidence from animal models of tumor growth indicates that negative CTLA-4 signaling is in a large part responsible for functional paralysis of T cells in the tumor microenvironment [115].

(b) Death receptor/ligand signaling

Death ligands and their receptors, especially the Fas/FasL or TRAILR/TRAIL, are expressed on tumor cells and on activated immune cells [49, 200, reviewed in 201]. Immune cells expressing membrane-associated FasL and/or TRAIL are thus "armed" to eliminate death receptor-positive tumor cells and mediate immune surveillance [51, 100]. Death ligands, especially TRAIL, expressed on NK cells infiltrating tissues such as liver, contribute to protection from metastasis [51, 71, 101]. On the other hand, death ligands expressed on tissue cells maintain the immune privilege of the anterior chamber of the eye or testis, protecting them from immune interference [97, 162]. Death ligands present on the surface of tumor cells are used as means of achieving immune privilege. Best known as the "Fas-counterattack hypothesis," this concept is based on the demonstration that FasL+ colon cancer cells could induce Fas-mediated apoptosis of activated T cells *in vitro* [112, 113]. Expression of FasL in human tumors *in situ* has been confirmed by many studies [201, 209] and linked to Fas/FasL-mediated apoptosis of tumor-infiltrating lymphocytes (TIL) as well as to a poor prognosis for breast, ovarian, colon and liver cancers [9, 13, 28, 48, 89, 157]. While FasL+ tumor cells can counterattack immune cells, they have evolved tactics for protection from apoptosis by blocking signals delivered via Fas via intracellular inhibitors of apoptosis (IAPs) [203]. The overall results favor the tumor and interfere with antitumor immunity.

Surprisingly, FasL+ tumors are not protected from rejection *in vivo*. When implanted in experimental animals, FasL+ tumor cells are rapidly rejected due to a massive infiltration of the graft by granulocytes [103]. This suggests that FasL is a pro-inflammatory molecule, which promotes inflammatory cell infiltration into tissues [88, 93]. FasL has the ability to induce inflammatory gene expression in tissue-resident immune cells, especially macrophages, which leads to the release of TNF-α and IL-8, cytokines with potent pro-inflammatory activities [4, 22, 40, 181]. Pro-inflammatory cytokines, acting as autocrine or paracrine mediators, up-regulate death receptor (Fas, TRAILR) expression and death ligand (FasL, TRAIL) secretion from tumor/tissue cells, shifting signaling from the pro-apoptotic to pro-inflammatory mode. Death ligands can play diametrically opposing roles in tumor growth. On the one hand, their expression on the tumor confers immune privilege and promotes metastasis. However, they can also induce potent inflammatory responses leading to tumor eradication. The tumor takes advantage of this dual functionality of death ligands by consistently shifting the balance toward pro-tumorigenic effects. In the tumor microenvironment, immune cells are at a disadvantage, having been corrupted to produce factors favoring tumor growth.

(c) NFκB signaling

The developing tumor is a site of chronic inflammation [15]. It has been suggested that progression to malignancy is regulated at the level of NFκB signaling and pro-inflammatory cytokines [83, 122]. NFκB, a ubiquitous transcription factor, may be the link between cancer and inflammation [122]. NFκB, is regulated differently in normal vs. malignant tissues [83, 122]. In the former, NFκB regulates expression of various cytokines and the process of inflammation. In the latter, NFκB stimulates proliferation of tumor cells and inhibits their apoptosis [122]. In most normal cells, NFκB complexes are present in the cytoplasm but remain inactive due to their interaction with inhibitor of NFκB (IκB). Upon cell activation, IκB kinases phosphorylate IκB, which is degraded and frees NFκB to translocate to the nucleus, where it regulates gene transcription. During inflammation, activation of NFκB is initiated by, e.g. binding of TNF-α to its receptor (TNFR1) expressed on inflammatory cells, and it induces regulated expression of cytokine genes, which control cell migration, proliferation and death. In chronic inflammation, the microenvironment favors overproduction of pro-inflammatory cytokines by infiltrating as well as local tissue cells. Such deregulated cytokine secretion is driven by the NFκB activation in tumor cells.

Many neoplastic cells show constitutive NFκB activation, which contributes to abnormal proliferation, resistance to apoptosis and disease progression [83, 120]. NFκB activation is already present in pre-malignant tissue cells, leading to establishment of the pro-inflammatory environment [120, 199]. As the tumor develops, tumor cells which depend on cytokines and growth factors for survival, either produce and secrete them (autocrine regulation) or reprogram leukocytes found in the tumor microenvironment to produce them (paracrine regulation). Responding to this cytokine storm, tumor and stromal cells produce a panoply of soluble factors with biologic effects ranging from enhancement of cell proliferation, matrix remodeling, vessel growth, inhibition of apoptosis or cellular differentiation to sustained release of pro-inflammatory mediators [15]. NFκB is also implicated in the promotion of metastasis by regulating cell adhesion and migration [83]. This microenvironment is assiduously maintained by the tumor, promotes tumor growth and invasiveness but is suppressive to immune cells and, in concert with tumor progression, assumes the features of chronic inflammation.

Tumors have the ability to usurp normal biologic process of inflammation to promote tumor progression. The prominent role of TNF-α in this process has been emphasized [102]. The NFκB pathway can either promote survival of malignant cells by inhibiting apoptosis or sustain the production of cytokines necessary for tumor growth and tissue restructuring, thus mediating opposite biologic effects. Signals available in the milieu, including FasL and/or TRAIL, undoubtedly contribute to this molecular diversion, promoting tumor escape. The tumor, manipulating molecular circuits to its advantage, "never heals," and the overall anti-apoptotic character of the milieu promotes tumor progression. Whether NFκB is a friend or foe of cancer cells is currently a central question of tumor biology [119]. It is becoming clear, however, that not only NFκB but other signaling molecules have dual functions, and the tumor can capitalize on this biologic redundancy.

1.4.1.2 Dysfunction in TAA Cross-presentation by Dendritic Cells

The key role of DC in cross-presentation of TAA to T cells is well known [14, 156]. Dysfunction of DC cross-presenting TAA to T cells could lead to an inadequate or biased antitumor immune response. DC not only process and present TAA to T cells but are important sources of IL-1, IL-12, IL-15, IL-18, IL-23 and Interferons, among other cytokines and chemokines. They are also rich in co-stimulatory molecules necessary as second signals or in growth factors for T-cell differentiation, proliferation and memory development [46, 186]. Therefore, if DC are depleted or unable to perform normally, the induction of TAA-specific immunity is likely to be impaired. Indeed, inhibition of DC in the presence of tumors has been reported [1, 65]. Defective maturation of DC in the tumor microenvironment may be mediated by tumor-derived vascular endothelial growth factor (VEGF) [44]. Others report that tumor-derived exosomes interfere ex vivo with differentiation of human DC from peripheral blood monocytes [149]. Tumor-derived gangliosides interfere with expression of inducible proteasomal components of antigen processing machinery (APM) in DC [149, 168]. Expression of APM components in DC is likely to determine the functional potential of this cellular mechanism necessary for presentation of cell surface-bound HLA-peptide-β2 microglobulin complexes to cognate T cells. Downregulation of APM component expression in DC co-incubated with apoptotic tumor cells (HNC, melanoma) has been observed, and it may contribute to poor induction of antitumor immune responses in cancer patients [189].

1.4.1.3 Apoptosis of DC in the Tumor Milieu

It has been recently shown that elimination of DC or DC precursors in the tumor microenvironment is an important element of tumor-induced immune suppression and may, in part, contribute to a deficient antitumor immune response in cancer patients [37, 150, 151]. The molecular pathways that may be involved include: (1) downregulation in DC of the anti-apoptotic Bcl-2 family proteins [38, 124]; (2) accumulation of ceramides which could interfere with PI3K-mediated survival signals [38, 123]; (3) production by the tumor of NO, which suppresses expression of cellular inhibitors of apoptosis proteins (cIAPs) [38] or cFLIP. Analysis of gene and protein expression in DC and DC precursors in the tumor microenvironment has demonstrated that expression of several intracellular signaling molecules is reproducibly altered in DC co-incubated with tumor cells, including IRF2, IL-2Rγ, Mcl-1, and small Rho GTPases among others. It appears that both intrinsic and extrinsic apoptotic pathways are involved in tumor-induced apoptosis of DC, as determined by an increased resistance to apoptosis of DC genetically modified to overexpress XIAP, caspase-8, Bcl-xL or FLIP [123]. Finally, it has been shown that DC genetically engineered to overexpress Bcl-xL-induced strong antitumor immune responses and inhibited tumor growth in murine tumor models *in vivo* [123]. Taken together, these data suggest that protection of endogenous DC in cancer patients should significantly augment the generation of host antitumor immune responses and consequently inhibit tumor growth.

1.4.2 *Interference with Functions/Survival of Effector T Lymphocytes*

In addition to defective APC functions described in cancer patients, which interfere with the induction of antitumor responses, other functional aberrations result in effector T cell paralysis or their untimely elimination [56, 190, 195]. In this respect, the presence of death ligands and receptors, including the Fas/FasL system, in human lymphocytes has been of special interest, because of their role in immune homeostasis. Activated T cells are especially sensitive to receptor-mediated apoptosis through increased sensitivity to FasL [177, 194], leading to the elimination of CD95/Fas+ activated T cells, a process known as activation-induced cell death (AICD). This mechanism is used for downregulation of the immune response, when it is no longer necessary, to control unlimited lymphocyte expansion and to maintain homeostasis. TAA-specific T cells may be especially sensitive to apoptosis because most TAA are self-antigens, which are subject to control by peripheral tolerance [59]. Thus, suppressed tumor-specific immunity may reflect tolerance to self [59]. The underlying mechanisms are still poorly defined and difficult to study *in vivo*; however, several pathways used by tumors to counteract immune surveillance have been described, including the role played by regulatory T cells (Treg) and myeloid suppressor cells (MSC).

1.4.2.1 Suppression of T-cell Responses by Treg

Treg are currently considered to be responsible for maintaining peripheral tolerance [141, 148, 165, 166, 179, 207], including transplantation tolerance and the prevention of autoimmune diseases [141, 148]. Treg have a beneficial role in preventing autoimmunity but are the most potent opponents of antitumor immune cells in cancer and play an important role in suppressing TA-specific immunity [141, 147]. Studies indicate that CD4+CD25^{bright} FoxP3+T cells are present in blood or lymph nodes of subjects with cancer and accumulate at tumor sites [31, 146, 147, 206]. To date, at least three types of CD4+ Treg have been described in humans: (i) naturally occurring CD4^{+}CD25^{bright}Foxp3+T cells (nTreg), which arise in the thymus and can suppress responses of both CD4+CD25− and CD8+CD25^{-} T cells in a contact-dependent, cytokine-independent, Ag-nonspecific manner [81, 110, 183]; (ii) CD4+CD25^{negative} Foxp3^{low} cells known as Type-1 regulatory (Tr1) T cells, which arise in the periphery upon encountering Ag in a tolerogenic environment via an IL-10-dependent process [79, 80]; and (iii) Th3 suppressor cells [39], which are dependent on IL-4 for functional differentiation. The nTreg are a heterogenous population endowed with regulatory functions and differentially expressing CTLA-4, glucocorticoid-induced tumor necrosis factor receptor (GITR) and CD62L [35, 159]. Tr1 cells can be generated *in vitro* by activating naïve CD4^{+}T cells by CD2 co-stimulation, IL-10 or IL-4 plus IL-10 [50, 64, 182]. The Tr1 cells appear to be modulated by IL-10 and TGF-beta and may play a prominent role in cancer. Tr1 cells are induced in the tumor microenvironment, and tumor-derived factors such as, e.g. PGE_2, have been shown to promote this process [18]. We have recently established an *in vitro* model simulating the tumor microenvironment, in which COX-2+ , PGE_2-producing carcinoma co-cultured with CD4+CD25$^-$ T cells, autologous DC and low doses of IL-2, IL-15 and IL-10 induces CD4+CD25$^-$CD122+CD132+Foxp3+IL-10+TGFβ+Tr1 cells [19]. Apart from CD4+ cells, CD8+ regulatory T cells have been recently reported. CD8+ CD28$^-$ T cells can be generated *in vitro* after stimulation of human PBMC with either allogeneic or xenogeneic APC. CD8+ CD28$^-$ T cells induce Ag-specific tolerance by increasing the expression of inhibitory receptor ILT3 and ILT4 on APC rather than by IL-10 production [26]. Today, the nature of human Treg is only partially defined. The phenotype, functions (Ag specificity, stability, trafficking or survival), lineage, differentiation and the relationship between the various Treg subsets are under investigation. No single specific marker is sufficient for distinguishing Treg subpopulations. Given the expansion of these populations in the circulation and tumor tissues of cancer patients [6, 84, 142, 158, 205], it is important to perform Treg phenotypic and functional evaluations to better define their role in the regulation of tumor-specific responses. From a practical point of view, it is important to distinguish Treg from activated CD4+CD25+ T cells which mediate helper functions and are sensitive to AICD. In contrast, Treg appear to be resistant to apoptosis [159]. Recent reports suggest that human Treg preferentially expand and survive in the presence of Rapamycin [16, 159], providing a novel approach to the large-scale culture of these cells for possible therapeutic use.

1.4.2.2 Suppression of T-cell Responses by Myeloid Suppressor Cells

Most tumors secrete TGF-β or induce TGF-β secretion from immature myeloid cells (MSC) that tend to accumulate in the tumor microenvironment [145, 152]. Young et al. first reported accumulations of CD34+ cell-derived myeloid cells with immunosuppressive ability the peripheral blood of HNC patients [116]. These cells correspond to CD11b+ /Gr-1+ myeloid progenitor cells in mice [144]. In tumor-bearing mice, MSC accumulate in the spleen and peripheral circulation, reaching very high proportions and exerting potent immunosuppression, thus favoring tumor growth. MSC also control the availability of essential amino acids such as L-arginine and produce high levels of reactive oxygen species (ROS). MSC present in tumors constitutively express iNOS and arginase1, an enzyme involved in metabolism of L-arginine, which also synergizes with iNOS to increase superoxide and NO production, blunting lymphocyte responses [33]. MSC with the phenotype CD34+CD33+CD13+and CD15− and suppressive functions were found to be increased in the peripheral blood of patients with various cancers [11]. Further, maturation defects in DC of patients with cancer have been described [10, 57] and are attributable, in part, to vascular endothelial growth factor (VEGF) production by human tumors [44, 45]. GM-CSF, which is also a frequently secreted product of tumor cells, recruits MSC and induces dose-dependent *in vivo* immune suppression and tumor promotion [144]. At the same time, GM-CSF is widely used as immune adjuvant in antitumor vaccines [36]. This dual role of GM-CSF (stimulatory and suppressive) suggests that GM-CSF and MSC are involved in maintaining immune homeostasis under normal physiologic conditions but in the tumor presence are subverted to promote its escape.

1.4.2.3 Apoptosis of T cells in Patients with Cancer

It has been observed that when lymphocytes are co-incubated with autologous tumor cells, DNA fragmentation occurs in a proportion of activated T lymphocytes, presumably by the mechanism similar or identical to AICD. When TUNEL assays, which detect DNA fragmentation, were performed using human tumor biopsies and tumor-involved LN [130], it was not tumor cells, but TIL and DC that were TUNEL positive (Fig. 1.2). Control normal tissues or tumor-uninvolved tissues obtained from patients with cancer contained infrequent or no apoptotic lymphocytes [130]. This unexpected finding was subsequently confirmed, using tumor tissues isolated from a variety of patients with cancer [52, 66, 111, 129]. Further, TUNEL staining showed that CD8+ rather than CD4+ T cells were primarily undergoing apoptosis at the tumor site, suggesting that the fate of these two T-cell subsets *in situ* may differ due to their divergent sensitivity to apoptosis.

Apoptosis of immune cells is not limited to the tumor site. Apoptosis of circulating CD8+ T cells in subjects with cancer has been described in melanoma, breast, ovarian, and head and neck cancers [74, 139, 140, 190]. Studies involving TUNEL staining of TIL and Annexin V (ANX) binding to circulating T cells suggest that CD8+ rather than CD4+ T cells selectively undergo apoptosis at the tumor site and in the peripheral circulation of cancer patients [74, 190]. The proportion of

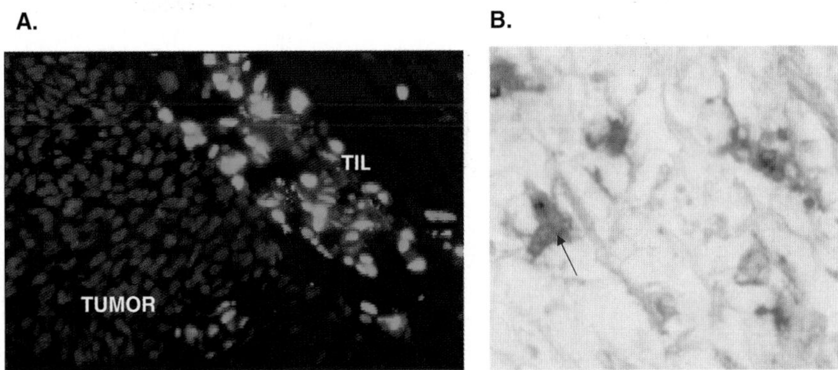

Fig. 1.2 Apoptosis of lymphocytes (A) or TADC (B) in the tumor microenvironment: TUNEL assay in sections of human oral CA (A) and caspase activity in prostate carcinoma *in situ* (B). TU = tumor; L = lymphocytes. The arrow points to a caspase+ (red) DC (blue = CD83+). (The photograph shown in B was generously contributed by Dr. Michael Shurin, University of Pittsburgh.)

CD8+Fas+ T cells that bind ANX is significantly increased in the patients' circulation relative to age-matched normal controls [56, 190]. Thus, the fate of CD8+ and CD4+ T-cell subsets may differ due to their divergent sensitivity to apoptosis [69]. Also, the effector subpopulations of CD8+ T cells (e.g. CD8+ CD45RO+ CD27⁻ and CD8+ CD28⁻) appear to be preferentially targeted for apoptosis in cancer patients [169]. The CD8+ CCR7+ subset of effector cells, which is resistant to apoptosis, is replaced by apoptosis sensitive CD8+ CCR7⁻ lymphocytes in patients with cancer [68 and Fig. 1.3]. Absolute numbers of circulating T-cell subsets are low in these patients [73]. Examination of the proliferative history of T-cells subsets using the T-cell receptor excision circle (TREC) PCR-based analysis confirms aberrant lymphocyte homeostasis characterized by a rapid turnover of T cells in cancer patients [73, 212]. Circulating Vβ-restricted CD8+ T cells are especially sensitive to apoptosis [7] and so are tetramer+ CD8+ T cells [5]. Tumor epitope-specific T cells (tetramer+) appear to preferentially bind ANX and are targeted for apoptosis [5]. Taken together, these findings suggest that a loss of effector T cell function through targeted apoptosis might compromise antitumor functions of the host immune system and contribute to tumor progression [73].

Recent studies of apoptosis in immune cells suggest that in T lymphocytes, sensitivity to Fas-mediated death is a regulated phenomenon, in which both IL-2 and antigenic stimulation play a crucial regulatory role [78, 177]. In AICD, which is an

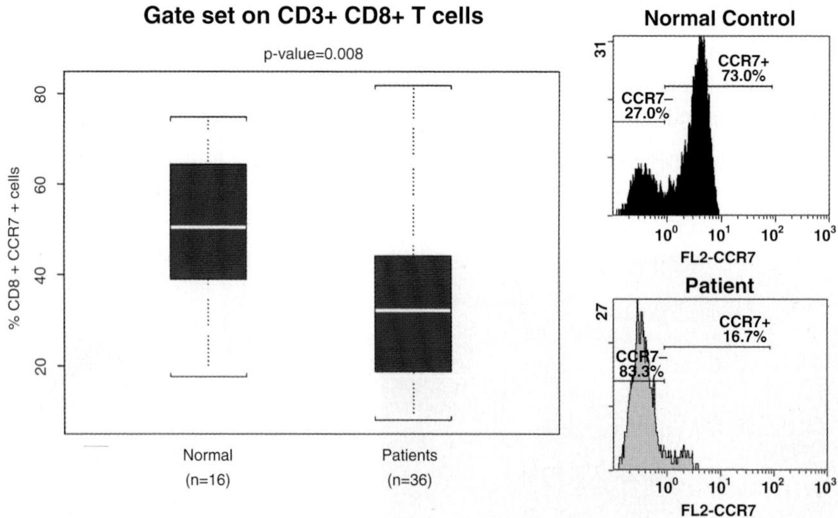

Fig. 1.3 Flow cytometry for expression of CCR7 on circulating CD8+ T cells in normal donors and patients with head and neck cancer (HNC). On the left, the boxplots show that the frequency of CCR7+ CD8+ T cells in HNC patients is significantly lower than that in normal donors. On the right, the data from a representative patient and normal donor indicate that in the patient CCR7+ CD8+ T cells are replaced by CCR7–CD8+ cells, which are sensitive to apoptosis. (Reproduced from Kim et al., 2005. With permission.)

essential part of any normal immune response, IL-2 is a potentiating cytokine. At appropriate concentrations and in the presence of a relevant antigen, it enhances the Fas/FasL pathway in activated T cells leading to expression of CD95 [128, 160, 177]. AICD is induced by repeated or chronic antigenic stimulation, and neither co-stimulatory molecules nor the Bcl-2 family members can rescue T cells from AICD. Furthermore, Th1 cells appear to be more sensitive to AICD than Th2 cells [78]. T cells in the tumor, LN or peripheral circulation of patients with cancer experience chronic or repeated antigenic stimulation with TAA, express CD95 on the cell surface [129, 130] and might be particularly sensitive to AICD. However, expression of IL-2 in the tumor has been shown to be low or absent both at the message and protein levels [86, 128]. TIL *in situ* do not appear to produce IL-2 or express IL-2R [86, 128]. Translation of IL-2 mRNA is defective in TIL isolated from human breast carcinomas [86]. Therefore, if IL-2 is required for the assembly or function of the Fas death complex in T lymphocytes, then AICD may not be the only mechanism responsible for the demise of these cells in the tumor microenvironment.

1.4.2.4 Microvesicles (MV) Released by Tumors as Intercellular Harbingers of T-cell Apoptosis

Sera and body fluids of patients with melanoma, ovarian carcinoma or HNC contain membranous 50–100 nm microvesicles (MV) presumably originating from the tumor, which contain biologically active 42 kDa FasL and MHC class I molecules and mediate apoptosis of Fas+ lymphocytes at sites distant from the tumor [2, 12, 70, 163]. Upon isolation, these MV induce TCRζ degradation and DNA fragmentation in activated T cells [reviewed in 196]. MV are prominent in tumor cell supernatants, but activated normal cells also produce MV [196]. However, tumor-derived MV have a molecular profile that that distinguishes them from MV produced by normal cells such as DC [196, 204]. While DC-derived MV do not express FasL, are rich in co-stimulatory and HLA class II molecules can promote T-cell proliferation and have been used for immunization in mice and man [27, 161], tumor-derived MV are immunosuppressive. They express TAA, death ligands and HLA class I molecules but not co-stimulatory epitopes, and they promote apoptosis of activated CD8+ antitumor effector T cells [196]. MV can also interfere with monocyte differentiation to DC, diverting it from a stimulatory to suppressive pathway [173]. As a result, monocytes fail to up-regulate HLA class II molecules, produce TGF-β and acquire the ability to suppress lymphocyte proliferation [173]. In effect, tumor-derived MV can turn monocytes into CD14-negativeHLA-DR^{low} TGF-β+ MSC [173]. The characteristic molecular profile of tumor-derived MV in cancer patients' sera could have a prognostic value. For example, patients with advanced HNC, i.e. those with tumor-involved lymph nodes or systemic metastases, had significantly higher levels of biologically active FasL+ MV as well as T-cell targeted apoptosis than patients with early stage disease [70]. In aggregate, these data suggest that MV represent another mechanism used by tumors to subvert differentiation and antitumor activities of immune cells.

1.4.3 A loss of Tumor Recognition by Immune Cells

As discussed, TAA-specific T lymphocytes are often detectable in the peripheral circulation and at the tumor site in subjects with cancer [55] and elevated antibody titers to TA may also be present [138]. Thus, neither the repertoire deletion nor tolerance to self hypotheses can adequately explain the absence of clinically meaningful antitumor immune responses in these subjects. A possible explanation is provided by observations that tumors can effectively "hide" from immune attack by cells or antibodies. Mechanisms responsible for this type of tumor escape involve downregulation of cell surface molecules that serve as recognition beacons for immune cells. In addition, tumors have also devised strategies to evade immune cells mediating innate immunity.

1.4.3.1 Changes in Expression of HLA Molecules

Tumor progression is associated with changes in HLA class I antigen expression [25, 109], which may range from a total loss or downregulation of all HLA class I allospecificities expressed by one cell to a selective loss or downregulation of a single HLA class I allospecificity and from a loss or downregulation of the gene products of HLA-A, -B or -C loci to a loss of only one haplotype [25, 67]. These changes have been described to occur in most human solid tumors but at a distinct frequency ranging between 16% to 80% for the various types of tumors analyzed *in situ* [109], using mAb recognizing monomorphic HLA determinants. However, technical differences in IHC methods among laboratories and the heterogeneity in the HLA molecule expression levels in different tumors of the same histologic type, suggest that interpretation of HLA class I abnormalities and their frequency in tumors should be conservative, as discussed elsewhere [109]. Studies performed with tumor cell lines have identified a number of distinct molecular mechanisms underlying the abnormal HLA class I antigen phenotypes of malignant cells [42, 184]. These mechanisms may be differentially present or absent in various types of tumors. They include defects in β_2-microglobulin (β_2m) and/or HLA class I heavy chain synthesis, epigenetic alterations involving the HLA class I heavy chain loci, dysfunction of regulatory mechanisms that control HLA class I antigen expression or abnormalities in expression levels of one or more of the APM components [reviewed in 184]. These abnormalities do not represent artifacts of *in vitro* cell culture, since several of them have also been identified in surgically removed tumors. Most investigators have been able to confirm their existence and their emerging prognostic and clinical significance [184]. HLA abnormalities are associated with an unfavorable disease course and decreased survival in several malignancies, and they appear to have an increased frequency in malignant lesions of unresponsive patients treated with T-cell-based immunotherapy [174]. This latter finding implies that the malignant cells which escape immune recognition and expand following adoptive T-cell-based immunotherapy, can do so because they harbor HLA class I defects [67].

1.4.3.2 Defects of APM in Tumor Cells

For immune effector cells to be able to eliminate the tumor, there has to be a recognition signal generated that enables cellular engagement followed by the lytic machinery activation in effector cells. The epitopes on the tumor cell surface that are necessary for recognition by effector T lymphocytes are presented as a trimolecular complex (peptide-β2 microglobulin-HLA class I). This complex is a final result of the APM activity in the cytoplasm [42]. In tumors, defects often exist in either HLA molecule or APM component expression or both, so that these two molecular pathways required for recognition and elimination of malignant cells are altered and dysfunctional. Evidence from ex vivo models of tumor effector T-cell interactions indicate that even when class I HLA molecules and cognate tumor epitopes are expressed on the cell surface, tumor cells such as, e.g. squamous cell carcinomas of the head and neck (SCCHN), may be resistant to lysis [42]. This has been linked to defects in APM components in target cells, resulting in a lack of recognition of the tumor cell by effector T cells [42]. Because of defective or altered antigen processing, the tri-molecular complexes on the tumor cell surface are not presented, insufficient or incorrectly displayed. This is most likely a result of defective peptide transport and/or loading on β2 m-asociated HLA-class I heavy chains. In such situations, recognition by immune cells even when they are present is bound to fail, and the tumor cell escapes. In some tumors, surface expression of HLA molecules may be normal, effector cells able to destroy the tumor are present, but one or more APM components may be dysfunctional, assuring tumor escape. If an existing abnormality in APM component expression is repaired either by transfection of an aberrant component or by exposure of tumor cells to IFN-γ, the sensitivity of tumor targets to CTL is restored [41, 42, 87]. Abnormalities in APM components are rarely structural [185] and more often quantitative (functional), including early proteosomal as well as late APM components and may be related to genetic instability evident in most tumor cells. These defects can now be identified and quantified, because antibodies for APM component expression are available for *in situ* studies [189]. Importantly, reduced or absent APM component and/or HLA molecule expression in tumor cells has been correlated to poor prognosis and shortened survival in patients with several types of cancer [42, 114]. Thus it appears that abnormalities in APM components identified *in vitro* have clinical significance.

1.4.3.3 Loss or Decrease in Expression of TAA

Decreases and/or losses in surface antigens that could be targets for immune cells are a well-known escape strategy used by many tumors [reviewed in 184]. Tumor cells evade the host by being poor targets for effector T cells (CTL). Genetic alterations or environmental factors that regulate tumor growth underlie the variety of mechanisms potentially responsible for the lack or alteration of a protein expression that would otherwise lead to immune recognition. The loss of TAA has been mainly described in melanoma, where expression of differentiation antigens such as gp100, MART-1, TRP-1 and tyrosinase has been found to be decreased or absent in metastatic lesions [e.g. 67], implying that a loss of these epitopes in tumor may be a poor prognostic sign. In SCID mice, expansion of MART-1-loss variant of human melanoma was causally lined to the presence of adoptively transferred MART-1-specific CD8+ T cells [135]. In other solid tumor types, MUC-1 expression was found to be down-regulated in progressive mammary tumors positive for c-Neu in a mouse model [3]. TAA mutations which result in a loss of epitopes recognized by CTL may also occur in tumors. Although a mutated TAA may still be expressed, the mutation site can abolish the generation of immunogenic epitopes that are recognized by cognate CTL, as illustrated by the mutation in the p53 protein, which inhibits proteasome-mediated generation of the HLA-A*0202-restricted, immunogenic $p53_{264-272}$ peptide [58]. Because these defects in immunogenic epitope expression are largely due to genetic instability of tumor cells, they can occur at the mRNA level, affect protein expression or use posttranslational mechanisms such as biochemical alterations in protein glycosylation, activation of metalloproteinases (MMP) and other tissue restructuring enzymes or accelerated ubiquitination leading to a rapid degradation and loss of TAA.

1.4.3.4 Suppression of NK-cell Activity

Natural killer (NK) cells play an important role in immune surveillance against tumors [53, 127, 185]. NK cells are large granular lymphocytes which are capable of killing a broad range of malignant targets by at least two distinct mechanisms (the perforin/granzyme and death ligand pathways) but spare normal cells. NK cells express a variety of activating and inhibitory receptors responsible for regulation of this selective activity targeted at abnormal cells. In accordance with a "missing self" hypothesis, NK cells preferentially kill targets which down-regulate, lose or alter MHC class I or related molecules [85]. Ligands, such as MICA, MICB and/or ULBP1–3, that are expressed on cells with altered or mutated HLA molecules are recognized by activating receptors (e.g. NKG2D) and natural cytotoxicity receptors (NKp30, NKp44 and NKp46) [76]. Inhibitory receptors (KIR, ILT2/LIR1 and CD94/NKG2A) down-regulate NK-cell cytolytic activity, preventing lysis of normal HLA class I expressing cells [76]. To engage a target, NK cells do not require prior sensitization. They constitutively express IL2Rβγ and rapidly respond to IL-2, IL-15 and also to IFN-α and IFN-γ. When activated, NK cells produce an array of cytokines, including IFN-γ and TNF-α which also contribute to tumor cell death [155]. To effectively mediate innate immunity, NK cells interact with DC, by serving as a source of cytokines and contributing apoptotic tumor cells for the up-take and processing by DC [32]. In cancer, the absolute number of NK cells may be reduced and their activity on per cell basis is often impaired [185]. Downregulation of NK-cell function in subjects with advancing disease suggests that tumors can interfere with NK-cell activity. Recent data suggest that TGF-β1 production by the tumor, with elevated levels of this cytokine found in sera of subjects with cancer, results in downregulation of NKG2D expression on

NK cells and, consequently, low NK activity, which obviously favors tumor growth [77]. This molecular mechanism of tumor escape from NK cells is but one example of many known today, including downregulation of NK-cell activity through interference with receptor–ligand interactions, lowering MICA or MICB expression on tumor cells, inhibition of NK–DC interactions in the tumor microenvironment or elimination of activated NK cells via death ligands expressed on tumor cells [126].

1.4.4 Resistance of Tumor Cells to Immune Intervention

In the presence of existing or developing antitumor immune responses, tumors develop resistance to immune mechanisms that otherwise bring about their destruction. As discussed above, many pathways and molecular mechanisms may lead to such immune resistance. However, one of the most interesting aspects of tumor escape is the ability of tumors to utilize the host immune system for promoting their own progression. Referred to as an immune selection hypothesis, this phenomenon allows for emergence of tumor cells and clones of cells that survive an immune attack and prosper because of it. In an analogous way, tumor stem cells appear to be a back-up mechanism called into play when the tumor is being eliminated and repopulate the microenvironment with new tumor cells resistant to further immune intervention.

1.4.4.1 Lack of Susceptibility to Immune Effector Cells

Immune effector cells which recognize tumor targets can eliminate them using the perforin/granzyme pathway, apoptosis or cytokine-mediated mechanisms [20, 97, 180]. There are at least two distinct pathways (intrinsic and extrinsic) by which apoptosis proceeds [120]. In the former, cellular stress, DNA damage or drugs act on the mitochondria, inducing the release of pro-apoptotic factors such as cytochrome c or SMAC/DIABLO. The extrinsic pathway is activated by ligation of death receptors on the cell surface. Subsequent intracytoplasmic events lead to cell death. Although the preferential use of various death-inducing pathways by effector cells mediating immune surveillance has been debated, it is clear that many tumors develop resistance to apoptosis, necrosis or cytokine-mediated death signals [87]. Multiple mechanisms responsible for tumor resistance to apoptosis have been uncovered, including overexpression of FLIP, release of soluble FasL and/or soluble Fas, a loss of expression or function of Fas or TRAIL, the presence of decoy death receptors, expression of surviving as well as the large IAP family of natural apoptosis inhibitors [92, 203, reviewed in 120]. Likewise, tumors are adept at escaping the granule exocytosis pathway mediated by CTL or NK cells via mechanisms involving expression of granzyme B inhibitors (serine proteases) or perforin inhibitors (cathepsin B, cystein protease). Human leukemia blasts are known to be particularly resistant to NK cell-mediated cytotoxicity [180].

1.4.4.2 Immunoselection of Resistant Tumor Cells

The process of tumor cell immune destruction mediated by the host is seldom efficient, given the many mechanisms of tumor escape. Further, as tumor cells sensitive to killing by the granule-mediated, death ligand-mediated, cytokine-dependent or antibody-dependent pathways are eliminated, those that survive gain advantage [20, 108, 175, 180]. In a way, the host facilitates tumor progression by utilizing killer mechanisms for "making room" and selecting the fittest. Immunoselection of resistant tumor cells is best illustrated by experiments performed in mice bearing chemically induced tumors. Methylcholantrene, a carcinogen implicated in cancer development in man, also induces murine tumors with a high incidence of genetic alterations and sensitivity to wild-type (wt)-sequence p53-specific CTL [33]. DNA- and DC-based vaccines targeting wt p53peptides were given to tumor-bearing mice in the protection or therapy setting [30]. The efficacy of these vaccines was found to be severely compromised by vaccine-induced tumor escape: a high incidence of epitope-loss tumors was striking in the vaccinated mice. Tumor escape was attributable to downregulation of HLA class I molecules on tumor cells and to an increased frequency of mutations within or flanking the p53 epitope-coding gene [30]. These results are consistent with the immunoselection hypothesis and suggest that similar phenomena might occur in subjects with cancer receiving tumor peptide-based vaccines. When we examined the frequency of tetramer+ T cells specific for the wt p53 $_{264-272}$ peptide in the circulation of HLA-A2+ subjects with SCCHN, an inverse correlation was observed between the frequency of these T cells and the mutational site/level of p53 expressed in the tumor, which could be explained by *in vivo* immunoselection and outgrowth of more aggressive tumors with mutated p53 [55]. A pre-existing immune response to the tumor was responsible for elimination of tumor cells susceptible to CTL early on in tumor progression, leaving behind memory T cells plus the tumor resistant to these effector cells. While other explanations are also possible, the phenomenon of immunoselection fits with these unexpected and provocative results.

1.4.4.3 Tumor Stem Cells

Recent data suggest that cancers arise from rare self-renewing stem cells that are biologically distinct from the more differentiated tumor cells within the tumor [21]. A fraction of tumor cells (about 1%) might have the ability to regrow the tumor when placed in a suitable environment. Cells with stem-like properties of self renewal, commitment and ability to differentiate into tumor cells can be sorted from human solid tumors based on the expression of certain surface markers and the absence of others [153, 202]. Further, stem cells are more resistant to drugs and irradiation than differentiated tumor cells and presumably are responsible for metastasis [60]. It is not know whether tumor stem cells are also resistant to immune effector cells. However, if tumor stem cells exist, then therapies of cancer, which are currently aimed at decreasing the tumor mass, are unlikely to work. The development of novel therapeutics targeting tumor stem cells will be necessary to overcome this biologic hurdle and effectively prevent tumor escape.

1.4.5 Tumor-derived Suppressive Factors

Early experiments, dating back more than 30 years, provided evidence that tumor-associated factors can alter normal functions of immune cells. In these experiments, supernatants of various tumor cells were found to suppress the proliferation and/or cytotoxicity of normal human PBMC *in vitro* [reviewed in 188]. These observations demonstrated that immunosuppressive factors were produced by human tumors but not by normal tissue cells. Over the years, the number of tumor-associated factors with immunosuppressive activity has grown and their role in tumor escape confirmed [reviewed in 194].

Various categories of inhibitory factors have been identified, including cytokines, enzymes, retroviral-like peptides, over-produced normal metabolites, INOS. This list is not comprehensive, as other factors have been described in the literature but remain incompletely characterized. In some cases, tumor-derived inhibitory factors have been purified, and a few are available as synthetic proteins/peptides, e.g. the retroviral-related p15E-derived peptides [29, 54]. Tumors vary in their ability to produce inhibitory molecules and not all of the identified and described factors are made by all human tumors. Strongly immunosuppressive tumors may simultaneously produce several different factors. Often, sera of cancer patients contain soluble or vesicle-associated molecules [163, 196] which interfere with immune cell differentiation or induce apoptosis in activated T cells. The immunosuppressive effects of small molecules such as PGE_2, histamine or H_2O_2 have been recognized long ago [170]. Enzymes, including IDO and arginases, which suppress T-cell responses [107, 136] and inhibitory cytokines (IL-10, TGF-β or GM-CSF) are all known to be present in the tumor microenvironment [105, 208]. Viral-related products also mediate suppressive effects [29, 90]. Finally, tumor-associated molecules such as gangliosides have been shown to induce immune suppression [34, 149]. The hypoxic microenvironment, which exists in the presence of a tumor, also contributes to immune dysfunction *in situ*, as HIF-1-α and other factors released by stromal cells exert inhibitory effects on lymphocytes [91, 143, 154, 210].

1.5 Conclusions

In this review, evidence is presented for the existence in the tumor microenvironment of multiple mechanisms that interfere with surveillance functions of immune cells. These mechanisms are all biased in favor of the tumor. Given the importance of immune intervention for the control of tumor growth and progression, it is clear that unless these mechanisms are recognized, identified and eliminated, anti-cancer therapies are unlikely to succeed. The multifaceted nature of tumor-associated immunosuppressive activities represents a considerable challenge to basic and clinical investigators. Nevertheless, examples of successful attempts at the elimination of immune suppression with drugs or biologic therapies are being reported and some are associated with clinical responses. Immunotherapy of cancer with cytokines, ex vivo activated immune cells, vaccines and, more recently, genetically modified vaccines, has been under study for almost 30 years. Partial or complete responses, which have lasted for several years in some patients with advanced disease have been documented (10–30%, depending on the type of cancer as well as the center in which therapy has been given). In addition, disease stabilization and upregulation of non-specific as well as specific immune responses have been observed in a substantial but variable proportion of cancer patients [reviewed in 47, 106, 134]. These results are encouraging, because they indicate that immune therapies with biologic agents have the potential to ameliorate disease progression [134]. Taking advantage of this potential, it should be possible in the near future to deliver biologic therapies in the adjuvant setting to patients with early disease with the objectives of avoiding tumor-induced suppression and of sustaining survival and functions of immune effector and helper cells. In a recent review, Vieweg and colleagues discuss therapeutic strategies that are currently emerging for the reversal of tumor-mediated immunosuppression [178]. Although these strategies only concern abrogation of immune suppression by Treg and MSC elimination prior to antitumor vaccinations, they serve as a working model for enhancing therapeutic effects of antitumor immunity. Combinations of therapies targeting the tumor and elimination tumor-induced immunosuppression are needed to restore immune competence and to generate potent and sustained memory T-cell responses in cancer patients. Recent insights into mechanisms of immune suppression in the tumor microenvironment provide new opportunities for future design of more effective therapies for patients with cancer.

References

1. Aalamian M, Pirtskhalaishvili G, Nunez A, Esche C, Shurin GV, Huland E, Huland H, Shurin MR. Human prostate cancer inhibits maturation of monocyte-derived dendritic cells. *Prostate*, 46: 68–75, 2001.
2. Abrahams VM, Straszewski SL, Kamsteeg M, Hanczaruk B, Schwartz PE, Rutherford TJ, Mor G. Epithelial ovarian cancer cells secrete Fas ligand. *Cancer Res.*, 63: 5573–5581, 2003.
3. Adriance MC, Gendler SJ. Downregulation of Muc1 in MMTV-c-Neu tumors. *Oncogene*, 23(3): 697–705, 2004.
4. Aggarwal BB. Nuclear factor-κB: the enemy within. *Cancer Cell*, 6: 203–208, 2004.
5. Albers A, Schaefer C, Kirkwood JM, Gooding W, DeLeo AB, Whiteside TL. Spontaneous apoptosis of epitope-specific CD8+ tetramer+ T lymphocytes in the peripheral circulation of patients with cancer. Submitted, 2007.
6. Albers AE, Kim GG, Ferris RL, Chikamatsu K, DeLeo AB, Whiteside TL. Immune responses to p53 in patients with cancer enrichment in tetramer+ p53 peptide-specific T cells and regulatory CD4+ CD25+ cells at tumor sites. *Cancer Immunol Immunother*, 54: 1072–1081, 2005.

7. Albers AE, Visus C, Tsukishiro T, Ferris RL, Gooding, W, Whiteside TL, DeLeo AB. Alterations in the T-cell receptor variable β gene-restricted profile of CD8+ T lymphocytes in the peripheral circulation of patients with squamous cell carcinoma of the head and neck. *Clin Cancer Res*, 12: 2394–2403, 2006.
8. Alleva C, Berger C, Elgert K. Tumor-induced regulation of suppression macrophage nitric oxide and TNF-α production. Role of tumor-derived IL-10, TGF-β and PGE_2. *J Immunol*, 153:1674–1685, 1994.
9. Allison J, Georgiou HM, Strausser A, Vaux DL. Transgenic expression of CD95 ligand on islet beta cells induces a granulocytic infiltration but does not confer immune privilege upon islet allografts. *PNAS USA*, 94: 3943–3947, 1997.
10. Almand B, Clark JI, Nikitina E, et al. Increased production of immature myeloid cells in cancer patients: a mechanism of immunosuppressionin cancer. *J Immunol*, 166: 678–689, 2001.
11. Almand B, Resser JR, Lindman B, Nadaf S, Clark JI, Kwon ED, Carbone DP, Gabrilovich DI. Clinical significance of defective dendritic cell differentiation in cancer. *Clin Cancer Res*, 6: 1755–1766, 2000.
12. Andreola G, Rivoltini L, Castelli C, Huber V, Perego P, Deho P, Squarcina P, Accornero P, Lozupone F, Lugini L, Stringaro A, Molinari A, Arancia G, Gentile M, Parmiani G, Fais S. Induction of lymphocyte apoptosis by tumor cell secretion of FasL-bearing microvesicles. *J Exp Med*, 195: 1303–1316, 2002.
13. Balkwill F, Mantovani A. Inflammation and cancer: back to Virchow? *Lancet*, 357: 539–545, 2001.
14. Banchereau J, Steinman RM. Dendritic cells and the control of immunity. *Nature*, 392: 245–252, 1998.
15. Baniyash M. Chronic inflammation, immunosuppression and cancer: novel insights and outlook. *Sem Cancer Biol*, 16: 80–88, 2006.
16. Battaglia M, Stabilini A, Migliavacca B, Horejs-Hoeck J, Kaupper T, Roncarolo MG. Rapamycin promotes expansion of functional CD4+ CD25+ Foxp3+ regulatory T cells of both healthy subjects and type 1 diabetic patients. *J Immunol*, 177: 8338–8347, 2006.
17. Bauernhofer T, Kuss I, Henderson B, Baum AS, Whiteside TL. Preferential apoptosis of CD56dim natural killer cell subset in patients with cancer. *Eur J Immunol*, 33: 119–124, 2003.
18. Bergmann C, Strauss L, Zeidler R, Lang S, Whiteside TL. Expansion and characteristics of human T regulatory type 1 cells in co-cultures simulating tumor microenvironment. *Cancer Immunol Immunother*, In Press, 2007.
19. Bergmann C, Strauss L, Zeidler R, Lang S, Whiteside TL. Expansion of human T regulatory type 1 cells in the microenvironment of COX-2 overexpressing head and neck squamous cell carcinoma. Submitted, 2007.
20. Berke G. The CTL's kiss of death. *Cell*, 81: 9–11, 1995.
21. Bjerkvig R, Tysnes BB, Aboody KS, Najbauer J, Terzis AJA. The origin of the cancer stem cell: current controversies and new insights. *Nat Cancer*, 5: 899–904, 2005.
22. Bodmer JL, Holler N, Reynards S, Vinciguerra P, Schneider P, Juo P, Blenis J, Tschopp J. TRAIL receptor-2 signals apoptosis through FADD and caspase-8. *Nat Cell Biol*, 2: 241–243, 2000.
23. Bronte V, Serafini P, Mazzoni A, Segal D, Zanovello P. L-arginine metabolism in myeloid cells controls T lymphocyte functions. *Trends Immunol*, 6: 301–305, 2003.
24. Bukowski RM, Rayman P, Uzzo R, Bloom T, Sandstrom K, Peereboom D, Olencki T, Budd GT, McLain D, Elson P, Novick A, Finke JH. Signal transduction abnormalities in T lymphocytes from patients with advanced renal cell carcinoma: clinical relevance and effects of cytokine therapy. *Clin Cancer Res*, 4: 2337–2347, 1998.
25. Campoli M, Chang G-C, Ferrone S. HLA-class I antigen loss, tumor immune escape and immune selection. *Vaccine*, 20: A40–A45, 2002.
26. Chang CC, Ciubotariu R, Manavalan JS, Yuan J, Colovai AI, Piazza F, Lederman S, Colonna M, Cortesini R, Dalla-Favera R, Suciu-Foca N. Tolerization of dendritic cells by T(S) cells: the crucial role of inhibitory receptors ILT3 and ILT4. *Nat Immunol*, 3: 237–243, 2002.

27. Chaput N, Taieb J, Schartz N, Flament C, Novault S, Andre F, Zitvogel L. The potential of exosomes in immunotherapy of cancer. *Blood Cells Mol Dis*, 35: 111–115, 2005.
28. Chen YL, Chen SH, Wang JY, Yang BC. Fas ligand on tumor cells mediates inactivation of neutrophils. *J Immunol*, 171:1183–1191, 2003.
29. Cianciolo GJ, Copeland TD, Oroszlan S, Snyderman R. Inhibition of lymphocyte proliferation by a synthetic peptide homologous to retroviral envelope proteins. *Science*, 230:453–455, 1985.
30. Ciccinnati VR, Dworacki G, Albers A, Beckebaum S, Tuting T, Daczmarek E, DeLeo AB. Impact of p53-based immunization on primary chemically-induced tumors. *Int J Cancer*, 113: 961–970, 2005.
31. Curiel TJ, Coukos G, Zou L, Alvarez X, Cheng P, Mottram P, Evdemon-Hogan M, Conejo-Garcia JR, Zhang L, Burow M, Zhu Y, Wei S, Kryczek I, Daniel B, Gordon A, Myers L, Lackner A, Disis ML, Knutson KL, Chen L, Zou W. Specific recruitment of regulatory T cells in ovarian carcinoma fosters immune privilege and predicts reduced survival. *Nat Med*,10: 942–949, 2004.
32. Degli-Esposti MA, Smyth MJ. Close encounters of different kinds: dendritic cells and NK cells take centre stage. *Nat Immunol*, 5: 112–124, 2005.
33. DeLeo AB. p53-based immunotherapy of cancer. In: Teicher B (ed). Cancer Drug Discovery and Development, Humana Press, Totowa, NJ, pp. 491–505, 2005.
34. DeMaria R, Lenti L, Malisan F, d'Agostino F, Tomassini B, Zeuner A, Rippo MR, Testi R. Requirement for GD3 ganglioside in CD95-and ceramide-induced apoptosis. *Science*, 277:1652–1655, 1997.
35. Dieckmann D, Plottner H, Berchtold S, Berger T, Schuler G. Ex vivo isolation and characterization of CD4(+)CD25(+) T cells with regulatory properties from human blood. *J Exp Med*,193: 1303–1310, 2001.
36. Dranoff G, Jaffee E, Lazenby A, Golumbek P, Levitsky H, Brose K, Jackson V, Hamada H, Pardoll D, Mulligan RC. Vaccination with irradiated tumor cells engineered to secrete murine granulocyte-macrophage colony-stimulating factor stimulates potent, specific, and long-lasting anti-tumor immunity. *Proc Natl Acad Sci USA*, 90: 3539–3543, 1993.
37. Esche C, Lokshin A, Shurin G, Gastman BR, Rabinovich H, Lotze MT, Shurin MR. Tumors' other immune targets: Dendritic cells. *J Leukoc Biol*, 66: 336–344, 1999.
38. Esche C, Shurin GV, Kirkwood JM, Wang GQ, Rabinowich H, Pirtskhalaishvili G, Shurin MR. TNF-α-promoted expression of Bcl-2 and inhibition of mitochondrial cytochrome c release mediated resistance of mature dendritic cells to melanoma-induced apoptosis. *Clin Cancer Res*, 7: 974s–979s, 2001.
39. Faria AM, Weiner H. Oral tolerance and TGF-beta-producing cells. *Inflamm Allergy Drug Targets*, 5: 179–190, 2006.
40. Ferguson TA, Green DR. Fas ligand and immune privilege: the eyes have it. *Cell Death Differ*, 8: 771–772, 2001.
41. Ferris RL, Hunt JL, Ferrone S. Human leukocyte antigen (HLA) class I defects in head and neck cancer: molecular mechanisms and clinical significance. *Immunol Res*, 33: 113–133, 2005.
42. Ferris RL, Whiteside TL, Ferrone S. Immune escape associated with functional defects in antigen processing machinery in head and neck cancer. *Clin Cancer Res*, 12: 3890–3895, 2006.
43. Finke JH, Zea AH, Stanley J, Longo DL, Mizoguchi H, Tubbs RR, Wiltrout RH, O'Shea JJ, Kudous, Klein E, Bukowski RM, Ochoa A. Loss of T-cell receptor ζ chain and $p56^{lck}$ in T-cell infiltrating human renal cell carcinoma. *Cancer Res*, 53: 5613–5616, 1993.
44. Gabrilovich D, Ishida T, Oyama T, Ran S, Kravtsov V, Nadaf S, Carbone DP. Vascular endothelial growth factor inhibits the development of dendritic cells and dramatically affects the differentiation of multiple hematopoietic lineages *in vivo*. *Blood*, 92: 4150–4166, 1998.
45. Gabrilovich DI. Mechanisms and functional significance of tumor-induced dendritic cell differentiation in cancer. *Nat Rev Immunol*, 4: 941–952, 2004.
46. Gabrilovich DI, Chen HL, Girgis KR, Cunningham T, Meny GM, Nadaf S, Kavanaugh D, Carbone DP. Production of vascular endothelial growth factor by human tumors inhibits the functional maturation of dendritic cells. *Nature Med*, 2: 1096–1103, 1996.

47. Gattinoni L, Powell DJ Jr, Rosenberg SA, Restifo NP. Adoptive immunotherapy for cancer: building on success. *Nat Rev Immunol*, 6: 383–393, 2006.
48. Greten FR, Eckmann L, Greten TF, Park JM, Li ZW, Egan LJ, Kagnoff MF, Karin M. IKKbeta links inflammation and tumorigenesis in a mouse model of colitis-associated cancer. *Cell*, 118: 285–96, 2004.
49. Griffith TS, Brunner T, Fletcher SM, Green DR, Ferguson TA. Fas ligand-induced apoptosis as a mechanism of immune privilege. *Science*, 270: 1189–1192, 1995.
50. Groux H, Bigler M, de Vries JE, Roncarolo MG. Interleukin-10 induces a long-term antigen-specific anergic state in human CD4+ T cells. *J Exp Med*, 184: 19–29, 1996.
51. Haefner B. The transcription factor NF-κB as drug target. *Prog Med Chem*, 43: 137–188, 2005.
52. Hahne M, Rimoldi D, Schroter M, Romero P, Schreier LE, French P, Schneider T, Bornand A, Fontana A, Lienard D, Cerottini J-C, Tschopp J. Melanoma cell expression of Fas (Apo-1/CD95) ligand: Implications for tumor immune escape. *Science*, 274: 1363–1366, 1996.
53. Hamerman JA, Ogasawara K, Lanier LL. NK cells in innate immunity. *Curr Opin Immunol*, 17: 29–35, 2005.
54. Haraguchi S, Good RA, James-Yarish M, Cianciolo GJ, Day NK. Differential modulation of Th1- and Th2-related cytokine mRNA expression by a synthetic peptide homologous to a conserved domain within retroviral envelope protein. *Immunology*, 92: 3611–3615, 1995.
55. Hoffmann TK, Donnenberg AD, Finkelstein SD, Donnenberg VS, Friebe-Hoffmann F, Myers EN, Appella E, DeLeo AB, Whiteside TL. Frequencies of tetramer+ T cells specific for the wild-type sequence $p53_{264-272}$ peptide in the circulations of patients with head and neck cancer. *Cancer Res*, 62: 3521–3529, 2002.
56. Hoffmann TK, Dworacki G, Meidenbauer N, Gooding W, Johnson JT, Whiteside TL. Spontaneous apoptosis of circulating T lymphocytes in patients with head and neck cancer and its clinical importance. *Clin Cancer Res*, 8: 2553–2562, 2002.
57. Hoffmann TK, Muller-Berghaus J, Ferris RL, Johnson JT, Storkus WJ, Whiteside TL. Alterations in the frequency of dendritic cell subsets in the peripheral circulation of patients with squamous cell carcinomas of the head and neck. *Clin Cancer Res*, 8: 1787–1793, 2002.
58. Hoffmann TK, Nakano K, Elder E. Dworacki G, Finkelstein SD, Apella E, Whiteside TL and DeLeo AB. Generation of T cells specific for the wild-type sequence $p53_{264-272}$ peptide in cancer patients – implication for immunoselection of epitope-loss variants. *J Immunol*, 165: 5938–5944, 2000.
59. Houghton AN, Guevara-Patino JA. Immune recognition of self in immunity against cancer. *J Clin Invest*, 114: 468–471, 2004.
60. Huff CA, Matsui WH, Douglas Smith B, Jones RJ. Strategies to eliminate cancer stem cells: Clinical implications. *Eur J Cancer*, In Press, 2006.
61. Iwai Y, Ishida M, Tanaka Y, Okazaki T, Honjo T, Minato N. Involvement of PD-L1 on tumor cells in the escape fromhost immune system and tumor immunotherapy by PD-L1 blockade. *Proc Natl Acad Sci USA*, 99: 12293–12297, 2002.
62. Jaeger E, Bernahrd H, Romero P, Ringhoffer M, Arand M, Karbach J, Ilsemann C, Hagedorn M, Knuth A. Generation of cytotoxic T-cell responses with synthetic peptides *in vivo*: implications for tumor vaccines with melanoma-associated antigens. *Int J Cancer*, 66:162–169, 1996.
63. Jaeger E, Ringhoffer M, Dienes HP, et al. Granulocyte-macrophage colony stimulating factor enhances immune responses to melanoma-associated peptides. *Int J Cancer*, 67: 54–62, 1996.
64. Jonuleit H, Schmitt E, Schuler G, Knop J, Enk AH. Induction of interleukin 10-producing, nonproliferating CD4(+) T cells with regulatory properties by repetitive stimulation with allogeneic immature human dendritic cells. *J Exp Med*, 192: 1213–1222, 2000.
65. Katsenelson NS, Shurin GV, Bykovskaia SN, Shogan J, Shurin MR. Human small cell lung carcinoma and carcinoid tumor regulate dendritic cell maturation and function. *Modern Path*, 14: 40–45, 2001.
66. Keane MM, Ettenberg SA, Lowrey GA, Russell EK, Lipkowitz S. Fas expression and function in normal and malignant breast cell lines. *Cancer Res*, 56: 4791–4798, 1996.

67. Khong HT, Wang QJ, Rosenberg SA. Identifications of multiple antigens recognized by tumor infiltrating lymphocytes form a single patient: tumor escape by antigen loss and loss of MHC expression. *J Immunother*, 27: 184–190, 2004.
68. Kim J-W, Ferris RL, Whiteside TL. Chemokine receptor 7 (CCR7) expression and protection of circulating CD8+ T lymphocytes from apoptosis. *Clin Cancer Res*, 11: 7901–7910, 2005.
69. Kim, J-W, Tsukishiro, T, Johnson JT, Whiteside TL. Expression of pro- and anti-apoptotic proteins in circulating CD8+ T cells of patients with squamous cell carcinoma of the head and neck (SCCHN). *Clin Cancer Res*, 10: 5101–5110, 2004.
70. Kim J-W, Wieckowski E, Taylor DD, Reichert TE, Watkins S, Whiteside TL. FasL+ membraneous vesicles isolated from sera of patients with oral cancer induce apoptosis of activated T lymphocytes. *Clin Cancer Res*, 11: 1010–1020, 2005.
71. Kreuz S, Siegmund D, Rumpf J-J, Samel D, Leverkus M, Janssen O, Hacker G, Dittrich-Breiholz O, Kracht M, Scheurich P, Wayant H. NFκB activation by Fas is mediated through FADD, caspase-8, and RIP and is inhibited by FLIP. *J Cell Biol*, 166: 369–380, 2004.
72. Kuss I, Saito T, Johnson JT, Whiteside TL. Clinical significance of decreased zeta chain expression in peripheral blood lymphocytes of patients with head and neck cancer. *Clin Cancer Res*, 5: 329–334, 1999.
73. Kuss I, Hathaway B, Ferris RL, Gooding W, Whiteside TL. Decreased absolute counts of T lymphocyte subsets and their relation to disease in squamous cell carcinoma of the head and neck. *Clin Cancer Res*, 10: 3755–3762, 2004.
74. Kuss I, Schaefer C, Godfrey TE, Ferris RL, Harris J, Gooding W, Whiteside TL. Recent thymic emigrants and subsets of naïve and memory T cells in the circulation of patients with head and neck cancer. *Clin Immunol*, 116: 27–36, 2005.
75. Lang S, Atarashi Y, Nishioka Y, Stanson J, Meidenbauer N, Whiteside TL. B7.1 on human carcinomas: costimulation of T cells and enhanced tumor-induced cell death. *Cell Immunol*, 201: 132–143, 2000.
76. Lanier LL. NK cell recognition. *Annu Rev Immunol*, 23: 225–274, 2005.
77. Lee J-C, Lee K-M, Kim D-W, Heo SD. Elevated TGF-β1 secretion and down-modulation of NK62D underlies impaired NK cytotoxicity in cancer patients. *J Immunol*, 172: 7335–7340, 2004.
78. Lenardo M, Chan KM, Hornung F, McFarland H, Siegel R, Wang J, Zheng L. Mature T lymphocyte apoptosis– immune regulation in a dynamic and unpredictable antigen environment. *Annu Rev Immunol*, 17: 221–253, 1999.
79. Levings MK, Roncarolo MG. Phenotypic and functional differences between human CD4+ CD25+ and type 1 regulatory T cells. *Curr Top Microbiol Immunol*, 293: 303–326, 2005.
80. Levings MK, Sangregorio R, Galbiati F, Squadrone R, deWaal Malefyt R, Roncarolo MG. IFN-alpha and IL-10 induce differentiation of human type 1 T regulatory cells. *J Immunol*,166: 5530–5539, 2001.
81. Levings MK, Sangregorio R, Sartirana C, Moschin AL, Battaglia M, Orban PC, Roncarolo MG. Human CD25+ CD4+ T suppressor cell clones produce transforming growth factor beta, but not interleukin 10, and are distinct from type 1 T regulatory cells. *J Exp Med*,196: 1335–1346, 2002.
82. Li X, Liu J, Park J-K, Hamilton TA, Rayman P, Klein E, Edinger M, Tubbs RR, Bukowski R, Finke J. T cells from renal cell carcinoma patients exhibit an abnormal pattern of NFκB specific DNA binding activity. *Cancer Res*, 54: 5424–5429, 1994.
83. Li Q, Withoff S, Verma IM. Inflammation-associated cancer: NF-κB is the lynchpin. *Trends Immunol*, 26: 318–325, 2005.
84. Liyanage UK, Moore TT, Joo H-G, Tanaka Y, Herrmann V, Doherty G, Drebin JA, Strasberg SM, Eberlein TJ, Goedegebuure PS, Linehan DC. Prevalence of regulatory T cells is increased in peripheral blood and tumor microenvironment of patients with pancreas or breast adenocarcinoma. *J Immunol*, 169: 2756–2761, 2002.
85. Ljunggren HG, Karre K. In search of the "missing self": MHC molecules and NK recognition. *Immunol Today*, 11: 237–244, 1990.

86. Lopez CB, Rao TD, Feiner H, Shapiro R, Marks JR, and Frey AB. Repression of interleukin-2 mRNA translation in primary human breast carcinoma tumor-infiltrating lymphocytes. *Cell Immunol*, 190:141–155, 1998.
87. Lopez-Albaitero A, Nayak JV, Ogino T, Machandia A, Gooding W, DeLeo AB, Ferrone S, Ferris RL. Role of antigen-processing machinery in the *in vitro* resistance of squamous cell carcinoma of the head and neck cells to recognition by CTL. *J Immunol*, 176: 3402–3409, 2006.
88. Lotze MT, DeMarco RA. Dying dangerously: necrotic cell death and chronic inflammation promote tumor growth. *Discovery Med*, 4: 448–456, 2004.
89. Luo JL, Maeda S, Hsu LC, Yagita H, Karin M. Inhibition of NF-kappa B in cancer cells converts inflammation-induced tumor growth mediated by TNF alpha to TRAIL-mediated tumor regression. *Cancer Cell*, 6: 297–305, 2004.
90. Lybarger L, Wang X, Harris M, Hansen TH. Viral immune evasion molecules attack the ER peptide-loading complex and exploit ER-associated degradation pathways. *Curr Opin Immunol*, 17: 71–78, 2005.
91. Mabjeesh NJ, Amir S. Hypoxia-inducible factor (HIF) in human tumorigenesis. *Histol Histopathol*, 22: 559–572, 2007.
92. Malmberg K-J, Ljunggren H-G. Escape from immune and non-immune tumor surveillance. *Sem Cancer Biol*, 16: 16–31, 2006.
93. Mann B, Gratchev A, Bohm C, Hanski ML, Foss HD, Demel G, Trojanek B, Schmidt-Wolf I, Stein H, Riecken EO, Buhr HJ, Hanski C. FasL is more frequently expressed in liver metastases of colorectal cancer than in matched primary carcinomas. *Br J Cancer*, 79: 1262–1269, 1999.
94. Mantovani A, Bottazzi B, Colotta F, Sozzani S, Ruco L. The origin and function of tumor-associated macrophages. *Immunol Today*, 13: 265–270, 1992.
95. Mantovani A. Tumor-associated macrophages in neoplastic progression: a paradigm for the *in vivo* function of chemokines. *Lab Invest*, 71: 5–16, 1994.
96. Marchand M, Weynants P, Rankin E, Arienti F, Belli F, Parmiani G, Cascinelli N, Bourlond A, Vanwijck R, Humblet Y, et al. Tumor regression responses in melanoma patients treated with a peptide encoded by gene MAGE-3. *Int J Cancer*, 63: 883–885, 1995.
97. Martinez-Lorenzo MJ, Alva MA, Gamen S, Kim KJ, Chuntharapai A, Pineiro A, Naval J, Anel A. Involvement of APO2 ligand/TRAIL in activation-induced death of Jurkat and human peripheral blood T cells. *Eur J Immunol*, 28: 2714–2725, 1998.
98. Matsuda M, Petersson M, Lenkei R, Raupin J-L, Magnusson I, Mellstedt H, Anderson P, Kiessling R. Alterations in the signal-transducing molecules of T cells and NK cells in colorectal tumor-infiltrating gut mucosal and peripheral lymphocytes: correlation with the stage of the disease. *Int J Cancer*, 61: 765–772, 1995.
99. Mayordomo JI, Loftus DJ, Sakamoto H, Lotze MT, Storkus WJ, Appella E, DeLeo AB. Therapy of murine tumors with dendritic cells pulsed with p53 wild type and mutant sequence peptides. *J Exp Med*, 183: 1357–1365, 1996.
100. McCawley LJ, Martisian LM. Tumor progression: defining the soil round the tumor seed. *Curr Biol*, 11: R25–R27, 2001.
101. Medema JP, de Jong J, van Hall T, Melief CJ, Offringa R. Immune escape of tumors *in vivo* by expression of cellular FLICE-inhibitory protein. *J Exp Med*, 190: 1033–1038, 1999.
102. Micheau O, Tschopp J. Induction of TNF receptor I-mediated apoptosis via two sequential signaling complexes. *Cell*, 114: 181–190, 2003.
103. Miwa K, Asano M, Horai R, Lwakura Y, Nagata S, Suda T. Caspase 1-independent IL-1 beta release and inflammation induced by the apoptosis inducer Fas ligand. *Nat Med*, 4: 1287–1292, 1998.
104. Mizoguchi H, O'Shea JJ, Longo DL, Loeffler CM, McVicar DW, Ochoa A. Alterations in signal transduction molecules in T lymphocytes from tumor bearing mice. *Science*, 258: 1795–1798, 1992.
105. Mocellin S, Ohnmacht GA, Wang E, Marincola FM. Kinetics of cytokine expression in melanoma metastases classifies immune responsiveness. *Int J Cancer*, 93: 236–242, 2001.

106. Morisaki T, Matsumoto K, Onishi H, Kuroki H, Baba E, Tasaki A, Kubo M, Nakamura M, Inaba S, Yamaguchi K, Tanaka M, Katano M. Dendritic cell-based combined immunotherapy with autologous tumor-pulsed dendritic cell vaccine and activated T cells for cancer patients: rationale, current progress, and perspectives. *Hum Cell*, 16: 175–182, 2003.
107. Munn DH. Indoleamine 2.3 dioxygenase, tumor-induced tolerance and counter-regulation. *Curr Opin Immunol*, 18: 220–225, 2006.
108. Nagata S, Goldstein P. The Fas death factor. *Science*, 267: 1449–1456, 1995.
109. Natali PG, Bigotti A, Nicotra MR, Viora M, Manfredi D, Ferrone S. Distribution of human class I (HLA-A,-B,-C) histocompatibility antigens in normal and malignant tissues of non-lymphoid origin. *Cancer Res*, 44: 4679–4687, 1984.
110. Ng WF, Duggan PJ, Ponchel F, Matarese G, Lombardi G, Edwards AD, Isaacs JD, Lechler RI. Human CD4(+)CD25(+) cells: a naturally occurring population of regulatory T cells. *Blood*, 98: 2736–2744, 2001.
111. Niehans GA, Brunner T, Frizelle SP, Liston JC, Salerno CT, Kanpp DJ, Green DR, Kratzke RA. Human lung carcinomas express Fas ligand. *Cancer Res*, 57: 1007–1012, 1997.
112. O'Connell J, Bennett MW, O'Sullivan GC, Collins JK, Shanahan F. The Fas counterattack: a molecular mechanism of tumor immune privilege. *Mol Med*, 3: 294–300, 1997.
113. O'Connell J, O'Sullivan GC, Collins JK, Shanahan F. The Fas counterattack: Fas-mediated T cell killing by colon cancer cells expressing Fas ligand. *J Exp Med*, 184: 1075–1082, 1996.
114. Ogino T, Bandoh N, Hayashi T, Miyokawa N, Harabuchi Y, Ferrone S. Association of tapasin and HLA class I antigen down-regulation in primary maxillary sinus squamous cell carcinoma lesions with reduced survival of patients. *Clin Cancer Res*, 9: 4043–4051, 2003.
115. Okazaki T, Honjo T. The PD-1-PD-L pathway in immunological tolerance. *Trends Immunol*, 27: 195–201, 2006.
116. Pak AS, Wright MA, Matthews JP, Collins SL, Petruzzelli GJ, Young MR. Mechanisms of immune suppression in patients with head and neck cancer: presence of CD34+ cells which suppress immune functions within cancers that secrete granulocyte-macrophage colony-stimulating factor. *Clin Cancer Res*, 1: 95–103, 1995.
117. Parmiani G, Castelli C, Dalerba P, Mortarini R, Rivoltini L, Marincola RM, Anichini A. Cancer immunotherapy with peptide-based vaccines: what have we achieved: where are we going? *J Nat Cancer Inst*, 94: 805–818, 2002.
118. Parry RV, Chemnitz JM, Frauwieth KA, Lanfranco AR, Braunstein I, Kobayashi SV, Linsley PS, Thompson CB, Riley JL. CTLA-4 and PD-1 receptors inhibit T-cell activation by distinct mechanisms. *Mol Cell Biol*, 25: 9543–9553, 2005.
119. Perkins ND. NF-κB: tumor promoter or suppressor? *Trends Cell Biol*, 14: 64–69, 2004.
120. Peter ME, Krammer PH. The CD95(APO-1/Fas) DISC and beyond. *Cell Death Differ*, 10: 26–35, 2003.
121. Piali L, Fichtel A, Terpe H-J, Imhof BA, Gisler RH. Endothelial vascular adhesion molecule 1 expression is suppressed by melanoma and carcinoma. *J Exp Med*, 181: 811–816, 1995.
122. Pikarsky E, Porat RM, Stein I, Abramovitch R, Amit S, Kasem S, Gutkovich-Pyest E, Urieli-Shoval S, Galun E, Ben-Neriah Y. NF-KappaB functions as tumour promoter in inflammation-associated cancer. *Nature*, 431: 461–466, 2004.
123. Pirtskhalaishvili G, Gambotto A, Esche C, Yurkovetsky ZR, Lotze MT, Shurin MR. IL-12 and Bcl-xl gene transfection of murine dendritic cells protects them from prostate cancer-induced apoptosis and improves their antitumor activity. *J Urol*, 163: 105, 2000.
124. Pirtskhalaishvili G, Shurin GV, Esche C, Salup RR, Lotze MT, Shurin MR. Cytokine-mediated protection of human dendritic cells from prostate cancer-induced apoptosis is regulated by the Bcl-2 family of proteins. *Br J Cancer*, 83: 506–513, 2000.
125. Pittet MJ, Valmori D, Dunbar PR, Speiser DE, Lienard D, Lejeune F, Fleischhauer K, Cerundolo V, Cerottine JC, Romero P. High frequencies of naïve Melan-A/MART-1-specific CD8+ T cells in a large proportion of human histocompatibility leukocyte antigen HLA-A2 individuals. *J Exp Med*, 190: 705–715, 1999.

126. Poggi A, Massaro AM, Negrini S, Contini P, Zocchi MR. Tumor-induced apoptosis of human IL-2 activated NK cells: role of natural cytotoxicity receptors. *J Immunol*, 174: 2653–60, 2005.
127. Rabinowich H, Vitolo D, Altarac S, Herberman RB, Whiteside TL. Role of cytokines in the adoptive immunotherapy of an experimental model of human head and neck cancer by human IL-2-activated natural killer cells. *J Immunol*, 149: 340–349, 1992.
128. Rabinowich H, Suminami Y, Reichert TE, Crowley-Nowick P, Bell M, Edwards R, Whiteside TL. Expression of cytokine genes or proteins and signaling molecules in lymphocytes associated with human ovarian carcinoma. *Int J Cancer*, 68: 276–284, 1996.
129. Rabinowich H, Reichert TE, Kashii Y, Bell MC, Whiteside TL. Lymphocyte apoptosis induced by Fas ligand-expressing ovarian carcinoma cells: implications for altered expression of TcR in tumor-associated lymphocytes. *J Clin Invest*, 101: 2579–2588, 1998.
130. Reichert TE, Rabinowich H, Johnson JT, Whiteside TL. Human immune cells in the tumor microenvironment: mechanisms responsible for signaling and functional defects. *J Immunother*, 21: 295–306, 1998.
131. Reichert TE, Strauss L, Wagner EM, Gooding W, Whiteside TL. Signaling abnormalities and reduced proliferation of circulating and tumor-infiltrating lymphocytes in patients with oral carcinoma. *Clin Cancer Res*, 8: 3137–3145, 2002.
132. Reichert TE, Day R, Wagner E, Whiteside TL. Absent or low expression of the ζ chain in T cells at the tumor site correlates with poor survival in patients with oral carcinoma. *Cancer Res*, 58: 5344–5347, 1998.
133. Renkvist N, Castelli C, Robbins PF, Parmiani G. A listing of human tumor antigens recognized by T cells. *Cancer Immunol Immunother*, 50: 51–59, 2001.
134. Rivoltini L, Canese P, Huber V, Iero M, Pilla L, Valenti R, Fais S, Lozupone F, Casati C, Castelli C, Parmiani G. Escape strategies and reasons for failure in the interaction between tumor cells and immune system: how can we tilt the balance towards immune-mediated cancer control. *Exp Opin Biol Ther*, 5: 463–476, 2005.
135. Rivoltini L, Carrabba M, Huber V, Castelli C, Novellino L, Dalerba P, Mortarini R, Arancia G, Anichini A, Fais S, Parmiani G. Immunity to cancer: attack and escape in T lymphocyte-tumor cell interaction. *Immunol Rev*, 188: 97–113, 2002.
136. Rodriguez PC, Ochoa AC. T cell dysfunction in cancer: Role of myeloid cells and tumor cells regulating amino acit availability and oxidative stress. *Sem Cancer Biol*, 16: 66–72, 2006.
137. Rosenberg SA, Yang JC, Restifo NP. Cancer immunotherapy: moving beyond current vaccines. *Nat Med*, 10: 909–915, 2004.
138. Sahin U, Turecio P, Pfzeundschuch M. Serological identification of human tumor antigens. *Curr Opin Immunol*, 9: 709–716, 1997.
139. Saito T, Dworacki G, Gooding W, Lotze M, Whiteside TL. Mononuclear cells undergo spontaneous *ex vivo* apoptosis in the peripheral blood of patients with metastatic melanoma. *Clin Cancer Res*, 6: 1351–1364, 2000.
140. Saito T, Kuss I, Dworacki G, Gooding W, Johnson JT, Whiteside TL. Spontaneous *ex vivo* apoptosis of peripheral blood mononuclear cells in patients with head and neck cancer. *Clin Cancer Res*, 5: 1263–1273, 1999.
141. Sakaguchi S, Sakaguchi N, Shimizu J, Yamazaki S, Sakihama T, Itoh M, Kuniyasu Y, Nomura T, Toda M, Takahashi T. Immunologic tolerance maintained by CD25+ CD4+ regulatory T cells: their common role in controlling autoimmunity, tumor immunity, and transplantation tolerance. *Immunol Rev*, 182: 18–32, 2001.
142. Schaefer C, Kim GG, Albers A, Hoermann K, Myers EN, Whiteside TL. Characteristics of CD4+ CD25+ regulatory T cells in the peripheral circulation of patients with head and neck cancer. *Br J Cancer*, 92: 913–920, 2005.
143. Semenza GL. HIF-1 and tumor progression: pathophysiology and therapeutics. *Trends Mol Med*, 8: S62–S67, 2002.
144. Serafini P, Borello J, Bronte V. Myeloid suppressor cells in cancer: recruitment, phenotype, properties and mechanisms of immune suppression. *Sem Cancer Biol*, 16: 53–65, 2006.

145. Serafini P, DeSanto C, Marigo I, Cingarlini S, Dolcetti L, Gallina G, Zanovello P, Bronte V. Derangement of immune responses by myeloid suppressor cells. *Cancer Immunol Immunother*, 53: 64–72, 2004.
146. Shevach EM. Fatal attraction: tumors beckon regulatory T cells. *Nat Med*,10: 900–901, 2004.
147. Shevach EM. CD4+ CD25+ suppressor T cells: more questions than answers. *Nat Rev Immunol*, 2: 389–400, 2002.
148. Shevach EM. Regulatory T cells in autoimmmunity. *Annu Rev Immunol*, 18: 423–449, 2000.
149. Shurin GV, Shurin MR, Lotze MT, Barksdale EM. Gangliosides mediate neuroblastoma-induced inhibition of dendritic cell generation. *Cancer Res*, 61: 363–369, 2001.
150. Shurin MR, Esche C, Lokshin A, Lotze MT. Tumors induce apoptosis of dendritic cells *in vitro. J Immunother*, 20: 403, 1997.
151. Shurin MR, Gabrilovich DI. Regulation of dendritic cell system by tumor. *Cancer Res Ther Control*, 11: 65–78, 2001.
152. Siegel PM, Massague J. Cytostatic and apoptotic actions of TGF-β in homeostasis and cancer. *Nat Rev Cancer*, 3: 807–821, 2003.
153. Singh SK, Clarke ID, Hide T, Dirks PB. Cancer stem cells in nervous system tumors. *Oncogene*, 23: 7267–7273, 2004.
154. Sitkovsky M, Lukashev D. Regulation of immune cells by local-tissue oxygen tension: HIF1 alpha and adenosine receptors. *Nat Rev Immunol*, 5: 712–721, 2005.
155. Smyth MJ, Godfrey DI, Trapani JA. A fresh look at tumor immunosurveillance and immunotherapy. *Nat Immunol*, 6: 17–21, 2001.
156. Steinman RM. Dendritic cells and immune-based therapies. *Exp Hematol*, 24: 859–862, 1996.
157. Strand S, Vollmer P, Van den Abeelen L, Gottfried D, Alla V, Heid H, Kuball J, Theobald M, Galle PR, Strand D. Cleavage of CD95 by matrix metaboproteinase-7 induces apoptosis resistance in tumour cells. *Oncogene*, 23: 3732–3736, 2004.
158. Strauss L, Bergmann C, Szczepanski M, Gooding W, Johnson TJ, Whiteside TL. A unique subset of CD4+ $CD25^{high}$Foxp3+ T cells secreting IL-10 and TGF-β1 mediates suppression in the tumor microenvironment. *Clin Cancer Res*, In Press, 2007.
159. Strauss L, Whiteside TL, Knights A, Bergmann C, Knuth A, Zippelius A. Selective expansion of naturally occurring human CD4+ CD25+ FOXP3+ regulatory T cells with rapamycin *in vitro. J Immunol*, 178: 320–329, 2006.
160. Suda T, Okazaki T, Naito Y, Yokota T, Arai N, Ozaki S, Nakao K, Nagata S. Expression of the Fas Ligand in cells of T-cell lineage. *J Immunol*, 154: 3806–3813, 1995.
161. Taieb J, Chaput N, Zitvogel L. Dendritic cell-derived exosomes as cell-free peptide-based vaccines. *Crit Rev Immunol*, 25: 215–223, 2005.
162. Takeda K, Smyth MJ, Cretney E, Hayakawa Y, Yamaguchi N, Yagita H, Okumura K. Involvement of tumor necrosis-related apoptosis-inducing ligand in NK-cell mediated and IFN-gamma-dependent suppression of subcutaneous tumor growth. *Cell Immunol*, 214: 194–200, 2001.
163. Taylor DD, Gercel-Taylor C, Lyons KS, Stanson J, Whiteside TL. T-cell apoptosis and suppression of T-cell receptor/CD3-ζ by Fas Ligand-containing membrane vesicles shed from ovarian tumors. *Clin Cancer Res*, 9: 5113–5119, 2003.
164. Taylor PA, Lees CJ, Blazar BR. The infusion of *ex vivo* activated and expanded CD4(+)CD25(+) immune regulatory cells inhibits graft-versus-host disease lethality. *Blood*, 99: 3493–3499, 2002.
165. Taylor PA, Lees CJ, Fournier S, Allison JP, Sharpe AH, Blazar BR. B7 expression on T cells down-regulates immune responses through CTLA-4 ligation via T-T interactions. *J Immunol*, 172: 34–39, 2004.
166. Taylor PA, Noelle RJ, Blazar BR. CD4(+)CD25(+) immune regulatory cells are required for induction of tolerance to alloantigen via costimulatory blockade. *J Exp Med*, 193: 1311–1318, 2001.

167. Thompson RH, Gillett MD, Cheville JC, Lohse CM, Dong H, Webster WS, Krejci KG, Lobo JR, Sengupta S, Chen L, Zincke H, Blute ML, Strome SE, Leibovich BC, Kwon ED. Costimulatory B7-H1 in renal cell carcinoma patients: Indicator of tumor aggressiveness and potential therapeutic target. *Proc Natl Acad Sci USA*, 101: 17174–17179, 2004.
168. Tourkova IL, Shurin GV, Chatta GS, Periz L, Whiteside TL, Ferrone S, Shurin MR. Restoration by IL-15 of MHC class I antigen processing machinery in human dendritic cells inhibited by tumor-derived gangliosides. *J Immunol*, 175: 3045–3052, 2005.
169. Tsukishiro T, Donnenberg AD, Whiteside TL. Rapid turnover of the CD8+ CD28- T-cell subset of effector cells in the circulation of patients with head and neck cancer. *Cancer Immunol Immunother*, 52: 599–607, 2003.
170. Uotila P. The role of cyclic AMP and oxygen intermediates in the inhibititon of cellular immunity in cancer. *Cancer Immunol Immunother*, 43: 1–9, 1996.
171. Uzzo R, Rayman P, Kolenko V, Clark PE, Cathcart MK, Bloom T, Novick AC, Bukowski RM, Hamilton T, Finke JH. Renal cell carcinoma-derived gangliosides suppress nuclear factor-κB activation in T cells. *J Clin Invest*, 104: 769–776, 1999.
172. Uzzo RG, Clark PE, Rayman P, Bloom T, Rybicki, Novick A, Bukowski R, Finke J. Evidence that tumor inhibits NFκB activation in T lymphocytes of patients with renal cell carcinoma. *JNCI*, 91: 718–721, 1999.
173. Valenti R, Huber V, Filipazzi P, Pilla L, Sovena G, Villa A, Corbelli A, Fais S, Parmiani G, Rivoltini L. Human tumor-release microvesicles promote the differentiation of myeloid cells with transforming growth factor-beta-mediated suppressive activity on T lymphocytes. *Cancer Res*, 66: 9290–9298, 2006.
174. Valmori D, Dutoit V, Schnuriger V, Quiquerez AL, Pittet MJ, Guillaume P, Rubio-Godoy V, Walker PR, Rimoldi D, Lienard D, Cerottini JC, Romero P, Dietrich PY. Vaccination with a Melan-A peptide selects an oligoclonal T cell population with increased functional avidity and tumor reactivity. *J Immunol*, 168: 4231–4240, 2002.
175. Van den Broek ME, Kagi D, Zinkernagel RM, Hengartner H. Perforin dependence of natural killer cell-mediated tumor control *in vivo*. *Eur J Immunol*, 25: 3514–3525, 1995.
176. Van der Merve PA, Davis SJ. Molecular interactions mediating T cell antigen recognition. *Annu Rev Immunol*, 21: 659–684, 2003.
177. Van Parijs L, Ibrahimov A, Abbas AK. The roles of costimulation and Fas in T cell apoptosis and peripheral tolerance. *Immunity*, 93: 951–955, 1996.
178. Vieweg J, Su Z, Dahm P, Kusmartsev S. Reversal of tumor-mediated immunosuppression. *Clin Cancer Res*, 13: 727s–732s, 2007.
179. Von Herrath MG, Harrison LC. Regulatory Lymphocytes: Antigen-induced regulatory T cells in autoimmunity. *Nat Rev Immunol*, 3: 223–232, 2003.
180. Vujanovic NL. Role of TNF family ligands in antitumor activity of natural killer cells. *Int Rev Immunol*, 20: 415–437, 2001.
181. Wajant H. CD95/FasL and TRAIL in tumour surveillance and cancer therapy. In: Dalgleish, A.G., Haefner, B. (eds). *The Link Between Inflammation and Cancer*. Springer, New York, pp. 141–166, 2006.
182. Wakkach A, Cottrez F, Groux H. Differentiation of regulatory T cells 1 is induced by CD2 costimulation. *J Immunol*, 167: 3107–3113, 2001.
183. Wang HY, Lee DA, Peng G. Tumor-specific human CD4+ regulatory T cells and their ligands: implications for immunotherapy. *Immunity*, 20: 107–118, 2004.
184. Whiteside TL, Campoli M, Ferrone S. Tumor induced immune suppression and immune escape: mechanisms and possible solutions. In: Nagorsen E, Marincola F (eds). *Analyzing T Cell Responses*. Springer, pp. 43–82, 2005.
185. Whiteside TL, Herberman RB. The role of natural killer cells in immune surveillance of cancer. *Curr Opin Immunol*, 7: 704–710, 1995.
186. Whiteside TL, Odoux C. Dendritic cell biology and cancer therapy. *Cancer Immunol Immunother*, 53: 240–248, 2004.
187. Whiteside TL, Parmiani G. Tumor-infiltrating lymphocytes: their phenotype, function and clinical use. *Cancer Immunol Immunother*, 39: 15–21, 1994.

188. Whiteside TL, Rabinowich H. The role of Fas/FasL in immunosuppression induced by human tumors. *Cancer Immunol Immunother*, 46: 175–184, 1998.
189. Whiteside TL, Stanson J, Shurin MR, Ferrone S. Antigen processing machinery (APM) in human dendritic cells: up-regulation by maturation and down-regulation by tumor cells. *J Immunol*, 173: 1526–1534, 2004.
190. Whiteside TL. Apoptosis of immune cells in the tumor microenvironment and peripheral circulation of patients with cancer: implications for immunotherapy. *Vaccine*, 20: A46–A51, 2002.
191. Whiteside TL. Immune responses to malignancies. *J Allergy Clin Immunol*, 111: S677–S686, 2003.
192. Whiteside TL. Signaling defects in T lymphocytes of patients with malignancy. Symposium-in-writing. *Cancer Immunol Immunother*, 48: 346–352, 1999.
193. Whiteside TL. The role of immune cells in the tumor microenvironment. *Cancer Treat Res*, 130: 103–124, 2006.
194. Whiteside TL. Tumor-induced death of immune cells: its mechanisms and consequences. *Sem Cancer Biol*, 12: 43–50, 2002.
195. Whiteside TL. Down-regulation of ζ chain expression in T cells: A biomarker of prognosis in cancer? *Cancer Immunol Immunother*, 53: 865–876, 2004.
196. Whiteside TL. Exosomes in cancer and their role in tumor escape. In: Kasid Notario U, Haimovitz-Friedman Bar-Eli A (eds). *Recent Developments in Cancer Research.* , Research Signpost, In Press, 2007.
197. Whiteside TL. Tumor-infiltrating lymphocytes as antitumor effector cells. *Biotherapy* 5: 47–61, 1992.
198. Whiteside TL. *Tumor-Infiltrating Lymphocytes in Human Malignancies*. R.G. Landes, Austin, TX, 1993.
199. Whiteside TL. The role of immune cells in the tumor microenvironment. In: Dalgleish AG, Haefner B (eds). *The Link Between Inflammation and Cancer: Wounds That Do Not Heal.* Springer, pp. 103–124, 2006.
200. Whiteside TL. Immune suppression in cancer: effects on immune cells, mechanisms and future therapeutic intervention. *Sem Cancer Biol*, 16: 3–15, 2006.
201. Whiteside TL. The role of death receptor ligands in shaping tumor microenvironment. *Immunol Invest*, 36: 25–46, 2007.
202. Wicha MS, Liu S, Dontu G. Cancer stem cells: an old idea–a paradigm shift. *Cancer Res*, 66: 1883–1890, 2006.
203. Wieckowski E, Atarashi Y, Stanson J, Sato T-A, Whiteside TL. FAP-1-mediated activation of NF-κB induces resistance of head and neck cancer to Fas-induced apoptosis. *J Cell Biochem*, 100: 16–28, 2007.
204. Wieckowski E, Whiteside TL. Human tumor-derived vs. dendritic cell-driven exosomes have distinct biologic roles and molecular profiles. *Immunol Res*, 36: 247–254, 2006.
205. Woo EY, Chu CS, Goletz TJ, Schlienger K, Yeh H, Coukos G, Rubin SC, Kaiser LR, June CH. Regulatory CD4(+)CD25(+) T cells in tumors from patients with early- stage non-small cell lung cancer and late-stage ovarian cancer. *Cancer Res*, 61: 4766–4772, 2001.
206. Woo EY, Yeh H, Chu CS, Schlienger K, Carroll RG, Riley JL, Kaiser LR, June CH. Regulatory T cells from lung cancer patients directly inhibit autologous T cell proliferation. *J Immunol*, 168: 4272–4276, 2002.
207. Wood KJ, Sakaguchi S. Regulatory lymphocytes: regulatory T cells in transplantation tolerance. *Nat Rev Immunol*, 3: 199–210, 2003.
208. Xu WQ, Jiang XC, Zheng L, Yu YY, Tang JM. Expression of TGF-beta1, TGF-betaRII and Smad4 in colorectal carcinoma. *Exp Mol Pathol*, In Press, 2007.
209. Younes M, Schwartz MR, Ertan A, Finnie D, Younes A. Fas ligand expression in esophageal carcinomas and their lymph node metastases. *Cancer*, 88: 524–528, 2000.
210. Yu P, Rowleys DA, Fu Y-X, Schreiber H. The role of stroma in immune recognition and destruction of well-established solid tumors. *Curr Opin Immunol*, 18: 226–231, 2006.

211. Zea AH, Cutri BD, Longo DL, Alvord WG, Strobl SL, Mizoguchi H, Creekmore SP, O'Shea JJ, Powers GC, Urba WJ, Ochoa AC. Alterations in T cell receptor and signal transduction molecules in melanoma patients. *Clin Cancer Res*, 1: 1327–1335, 1995.
212. Zippelius A, Pittet MJ, Batard P, Rufer N, de Smedt M, Guillaume P, Ellefsen K, Valmore D, Lienard D, Plum J, MacDonald, HR, Speiser DE, Cerottini J-C, Romero P. Thymic selection generates a large T cell pool recognizing a self-peptide in humans. *J Exp Med*, 195: 485–494, 2002.
213. Zitvogel L, Mayordomo JI, Tjandrawan T, DeLeo AB, Clarke ML, Lotze MT, Storkus WJ. Therapy of murine tumors with tumor peptide-pulsed dendritic cells: dependence on T cells, B7 costimulation, and T helper cell 1-associated cytokines. *J Exp Med*, 183: 87–97, 1996.

Chapter 2
Histocompatibility Antigens, Tumor Microenvironment and Escape Mechanisms Utilized by Tumor Cells

Soldano Ferrone and Theresa L. Whiteside

Abstract The disappointing clinical responses to T cell-based immunotherapy have stimulated interest in the characterization of the escape mechanisms ultilized by tumor cells to avoid immune recognition and destruction. In this paper we review changes in the expression of antigen processing machinery components, classical and non-classical HLA class I antigens and NK cell activating ligands by cytokines present in the microenvironment. In addition the potential effects on immune cells of classical and non-classical HLA class I antigens and NK cell activating ligands released in the tumor microenvironment are described. The implications of these findings for the design of immunotherapeutic strategies for the treatment of malignant diseases are discussed.

Keywords Antigen processing machinery, classical HLA class I antigens, cytokines, immunoescape, immunotherapy, NK-cell activating ligands, non-classical HLA class I antigens

2.1 Introduction

In recent years several developments have converged to stimulate the enthusiastic application of immunotherapy for the treatment of malignant diseases, with special emphasis on T cell-based immunotherapy. They include (i) the successful application of T cell-based immunotherapy to control tumor growth in animal model systems, (ii) the availability of well-defined clinically relevant tumor antigens to be used as immunogens, (iii) the development of effective immunization strategies, and (iv) the limited, if any, efficacy of most of the conventional therapies to control tumor growth. As a result a large number of patients have been enrolled in clinical trials at many centers. Contrary to the expectations, however, the results have been rather disappointing. The percentage of clinical responses has generally been rather low. Furthermore in most,

Hillman Cancer Center, University of Pittsburgh, Cancer Institute, Pittsburgh, PA 15213-1863, USA

E. Yefenof (ed.), *Innate and Adaptive Immunity in the Tumor Microenvironment.*

if not all the clinical trials no correlation has been found between tumor antigen-specific immune responses and clinical responses. In fact, the lack of clinical response in patients who had developed a tumor antigen-specific immune response following active specific immunotherapy has been the rule, rather than the exception. Along the same line in some patients metastases have recurred or disease has progressed in spite of the induction or persistence of a tumor antigen-specific immune response. There is general agreement among tumor immunologists that these disappointing clinical results are caused, at least in part, by the escape mechanisms tumor cells utilize to avoid recognition and destruction by the host's immune system. The multiple escape mechanisms which have been identified and characterized can be grouped as follows: (i) defects in the induction of a tumor antigen-specific immune response, (ii) defects in infiltration of malignant lesions by immune cells, (iii) suppression of immune effector cells, (iv) defective recognition of tumor cells and (v) tumor cell resistance to cell death [34].

We [4] have focused our studies on the characterization of abnormalities in the expression and function of HLA antigens by tumor cells, of the molecular mechanisms underlying these abnormalities and on the assessment of their clinical significance. We have found that defects in the expression of HLA antigens are frequently associated with malignant transformation of cells, are caused by multiple distinct molecular abnormalities, and in some cases, are associated with the clinical course of the disease. However one question which, to the best of our knowledge, has not been addressed is related to the potential changes, which may occur in the expression and function of HLA antigens in malignant lesions because of cytokines present in the tumor microenvironment and the potential effects of classical and non-classical HLA class I antigens and NK-cell activating ligands released by tumor cells on immune cells present in this tumor microenvironment. Therefore, in the present paper we will discuss these two topics, since they may contribute to our understanding of the potential mechanisms underlying a frequent clinical scenario, i.e. the lack of control of tumor growth and disease progression in the face of appropriate HLA class I antigen and tumor antigen expression by malignant cells and the presence of functional tumor antigen-specific CTL in tumor bearing hosts [9]. For our discussion we will utilize the available information about (i) cytokines present in the tumor microenvironment and their ability to modulate the expression and function of HLA class I antigens and NK-cell activating ligands as well as (ii) the effects of these molecules on immune cells. This discussion is preceded by a short summary of the information related to the expression of classical and non-classical HLA class I antigens and of NK-cell activating ligands in malignant lesions and their functional and clinical relevance.

2.2 Classical and Non-Classical HLA Class I Antigens and NK-Cell Activating Ligands in Malignant Lesions

In man, like in other species, major histocompatibility class I antigens, which include the classical and non-classical HLA class I antigens, and NK-cell activating ligands play a crucial role in the interactions of malignant cells with the host's immune system.

This observation has stimulated interest in the characterization of the expression of these molecules by malignant cells. The expression of classical HLA class I antigens by tumor cells has been most extensively investigated, since the potential role of classical HLA class I antigens in the pathogenesis and clinical course of malignant disease has been identified much earlier than that of non-classical HLA class I antigens or NK-cell activating ligands. Furthermore, reagents to characterize classical HLA class I antigen expression by malignant cells have become available many years before those required to characterize the expression of the other two types of molecules. Convincing evidence derived from the analysis of cell lines in long term culture and from immunohistochemical staining of surgically removed malignant lesions with monoclonal antibodies recognizing framework, locus specific and polymorphic determinants of classical HLA class I antigens indicates that their expression is frequently defective in malignant cells. The defects range from a total loss or downregulation of all HLA class I molecules encoded in one cell to the selective loss or downregulation of one HLA class I allele, and from the selective loss or downregulation of the gene products of one locus to a loss or downregulation of one HLA class I haplotype [23]. Selective or total HLA class I antigen loss is generally caused by mutations in genes encoding molecules crucial for HLA class I antigen expression. It cannot be corrected by cytokines. On the other hand, selective or total HLA class I antigen downregulation is usually caused by functional defects of the antigen processing machinery (APM) which can be corrected by cytokines. The APM plays a crucial role in the generation of peptides from proteins, in their loading on β2microglobulin (β2 m) associated HLA class I heavy chains and transport of this trimolecular complex to the cell membrane, as discussed below.

Abnormalities in HLA class I antigen expression appear to have clinical significance, since they not only increase in their frequency with progression from preneoplastic lesions to metastases, but they also display a statistically significant association with disease free interval and with patients' survival in at least some malignant diseases [4, 23, 34]. It has been suggested that these associations reflect the detrimental effects of classical HLA class I antigen abnormalities on the interactions of tumor cells with HLA class I antigen restricted, tumor antigen-specific cytotoxic T lymphocytes (CTL) and with NK cells.

More recently, the non-classical HLA class I antigens, HLA-G, have been unexpectedly reported to be expressed both on cell lines and in surgically removed malignant lesions [48]. Although results in the literature conflict, there is a general agreement that malignant transformation of cells may be associated with the appearance of HLA-G. The frequency of this phenomenon markedly varies among different tumor types, as it ranges from a maximum of about 40% in cutaneous lymphoma, clear cell renal carcinoma and ovarian carcinoma to a lack of detection in uveal melanoma and laryngeal carcinoma. It is also noteworthy that the frequency of HLA-G expression is lower in cell lines than in surgically removed malignant lesions. Although one cannot exclude that this difference is due to technical reasons, such as differences in the sensitivity of the assays used or mis-scoring of HLA-G-bearing macrophages as tumor cells, we favour the possibility that this difference reflects biological variability. Specifically, we believe that malignant cells cultured *in vitro*

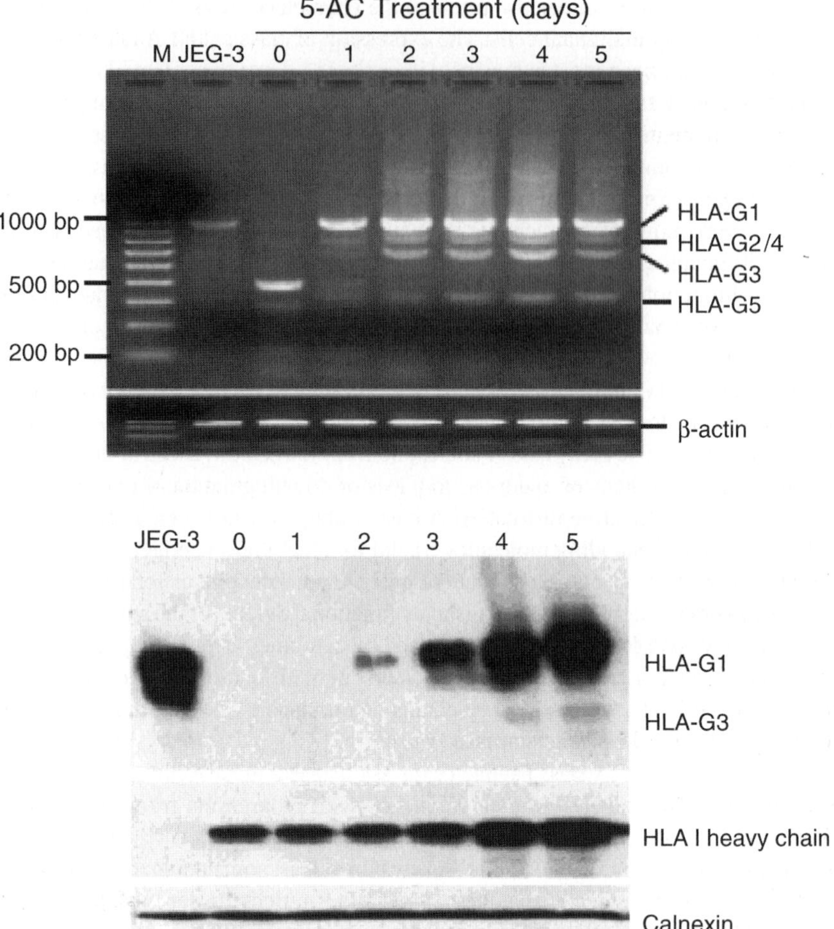

Fig. 2.1 Kinetics of HLA-G mRNA and protein induction in OCM-1A uveal melanoma cells by 5' azacytidine OCM-1A uveal melanoma cells were incubated for up to 5 days at 37 °C with 5' azacytidine. Every 24 h an aliquot of cells was harvested and tested for expression of HLA-G at the mRNA and protein level

may lose HLA-G expression because of methylation of its promoter, as we have observed in the uveal melanoma cell line OCM1-A (Fig. 2.1).

Although the serum level of soluble HLA-G is associated with survival in glioblastoma [59] and with disease stage and tumor load in melanoma [56], to the best of our knowledge, there is no convincing evidence that HLA-G expression in malignant lesions is associated with the clinical course of the disease. However, it is generally assumed that HLA-G expression in malignant lesions is likely to have a negative impact on the clinical course of the disease, since HLA-G antigens may provide tumor cells with escape mechanisms from immune recognition and destruction. As recently reviewed [41], these mechanisms include (i) inhibition of cytotoxic activity of CTL and NK cells, (ii) inhibition of CD4$^+$ T cell proliferation and cytokine release, (iii) inhibition of cell cycle progression in human alloreactive T cells, (iv) generation of a new type of regulatory T cells, either CD4$^+$ or CD8$^+$, through transfer of membrane HLA-G from antigen presenting cells to activated T cells ("trogocytosis"), and (v) induction of a "Th2" cytokine profile at the tumor site through stimulation of IL-3, IL-4 and IL-10 secretion.

The stress-inducible MHC class I chain-related surface glycoproteins MICA and MICB and the UL16-binding proteins ULBP1, ULBP2, ULBP3 and ULBP4 are ligands of the C-type lectin-like receptor NKG2D. The latter potently activates NK cells, even overcoming inhibitory signals by MHC class I molecules. MICA, and most likely MICB, have a restricted distribution in normal tissues, as their expression is induced by inflammatory stress only in gastric and small and large intestinal epithelium [19]. Information about the expression of these molecules by malignant cells is derived from the analysis of only a limited number of surgically removed malignant lesions. Nevertheless, the results of these studies, which have been corroborated by those derived from the analysis of cell lines in long term culture of different embryological origin [39], indicate that MICA and MICB have a much broader distribution in malignant tumors, as they are expressed by many types of epithelial tumors and by tumors of neuroectodermal origin [20, 46]. This expression pattern reflects the induction of these molecules by the DNA damage pathway in response to genotoxic insults, which represents a critical step during the malignant transformation of cells [14]. The expression of ULBP has been analyzed only on cell lines in long term culture [39], except for a limited number of glioblastoma, neuroblastoma and leukemic samples [11, 46, 51]. These molecules are in general expressed with a lower frequency than MIC molecules. Furthermore, the expression of ULBP and MIC molecules is not coordinated with that of classical and non-classical HLA class I antigens by malignant cells.

The lack of correlation between the clinical significance of membrane bound MICA and MICB and that of their soluble counterparts in serum parallels that mentioned above for HLA-G. Specifically, although the level of soluble MICA and/or MICB in serum is correlated with cancer stage and metastasis [24, 25] and has a prognostic significance in patients with prostate carcinoma or multiple myeloma [47, 60], there is no evidence that MICA expression in malignant lesions has clinical significance, except for an association between the MICA loss and disease progression in melanoma [4]. Nevertheless, it is expected that MIC and ULBP induction in tumors has a beneficial effect on the course of the disease, since these molecules are considered to be a type of "induced-self" tumor antigens that serve as targets for host immune surveillance. This possibility is supported by the role Rae-1, the mouse homolog of MIC, plays in NKG2D-bearing $\gamma\delta$ T cells' control of skin cancer in mice [18].

2.3 Antigen Presentation by HLA Class I Antigens

Recognition of tumor cells, like that of other types of targets, by CTL is mediated by a complex resulting from the loading of HLA class I molecules with antigenic peptides. This complex is generated by the HLA class I APM through four steps: (i) protein cleavage into peptide, (ii) transport of the resultant peptides into the endoplasmic reticulum (ER), (iii) assembly of the peptides with HLA class I heavy chains and β2m and (iv) transport of the HLA class I antigen-peptide complexes to the cell membrane (Fig. 2.2). Each of these steps can be modulated by cytokines. We have recently described these steps in detail and we refer the interested reader to our paper also as a source of references [2]. Here, we will briefly present the most important information.

Peptide ligands for association with HLA class I molecules are generated from proteins by proteasomes which are comprised of a barrel shaped 20 S core particle with caps on the ends to regulate the entry of substrate proteins. The 20 S core proteasome consists of four heptameric rings stacked atop of one another, with the inner two rings containing the three proteolytic subunits MB1, delta and zeta which are responsible for chymotryptic, tryptic and post-glutamyl cleavage. Proteasome composition is modulated by IFN-γ which induces the replacement of MB1, delta and zeta subunits with Low Molecular Mass Polypeptides (LMP) 7, LMP2 and LMP10, respectively, and also induces expression of the proteasome activator subunits, PA28α and PA28α. The latter replace the constitutive 19 S caps. This change of subunits results in the formation of an "immunoproteasome" which generates a different spectrum of peptides than the constitutive proteasome. As a result, the epitopes expressed by the peptides generated by proteasomes and immunoproteasomes from a given tumor antigen may be different.

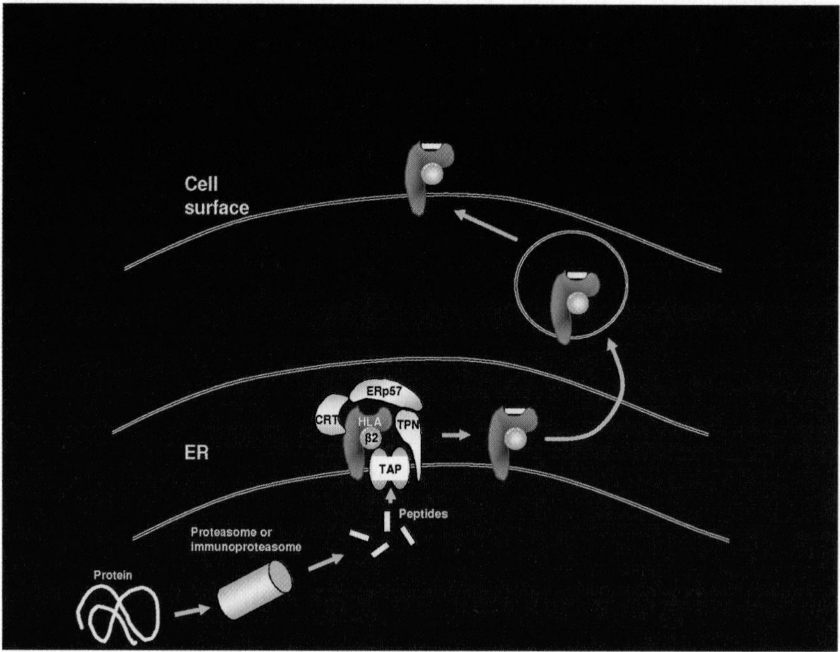

Fig. 2.2 Schematic representation of the steps which lead to the synthesis and expression of HLA class I antigen-peptide complexes

After degradation by proteasomes or immunoproteasomes into relatively large peptides (12–20 amino acids in length), the N-terminus of most peptides is further trimmed by aminopeptidases. Peptides of the appropriate size are transported into the ER lumen by an IFN-γ inducible member of the ATP-Binding Cassette (ABC)-family of transporters termed the Transporter Associated with Antigen Processing (TAP) which is composed of the two subunits TAP1 and TAP2. In the absence of TAP, HLA class I antigens are not loaded with peptides. As a result, they are poorly exported from the ER and have a low surface expression. This appears to be the rule for all of the allospecificities that have been tested with the exception of HLA-A2. The latter alloantigen can bind peptides derived from signal sequences whose presence in the ER is independent of TAP function. Therefore, TAP deficient cells exhibit a limited supply and repertoire of peptides that can be presented to CD8+ T cells. On the other hand, when TAP is downregulated rather than lost, the level of HLA class I antigen expression may not be affected as dramatically, but the repertoire of presented peptides may still be reduced leading to ineffective recognition of target cells by CTLs. This is clearly shown by the resistance to CTL killing of head and neck squamous cell carcinoma (SCCHN) cell lines in spite of HLA class I antigen expression. The lack of recognition of SCCHN cells by CTL is associated with downregulation of some APM components, such as LMP2, TAP1, TAP2 and tapasin in tumor cells. Furthermore, recognition of SCCHN cells by CTL can be restored by transfecting tumor cells with TAP1 cDNA [33].

After transport into the ER by TAP, peptides may be further trimmed by the heterodimeric ER aminopeptidase ERAP1/ERAP2 before they bind to HLA class I antigens. The loading of peptides onto HLA class I antigens is facilitated by the chaperones calreticulin, ERp57 and tapasin which are members of the peptide-loading complex. The lack of their expression causes a reduction in the repertoire of peptides and/or HLA class I antigen expression at the cell surface, leading to impaired CTL recognition. Once assembled with peptides, HLA class I molecules dissociate from TAP and travel to the cell membrane and present peptides to CTL.

2.4 Modulation of APM Components by Cytokines in the Tumor Microenvironment

The tumor microenvironment is shaped by interactions of tumor cells, stroma and infiltrating immune cells. It contains many types of cytokines produced by malignant and normal cells. The susceptibility of some APM components, including HLA class I heavy chains and $\beta 2$ m, HLA-G and MICA to modulation by cytokines and epigenetic mechanisms raises the question of how their expression by tumor cells may change in the tumor microenvironment and how these changes affect the interactions of tumor cells with immune cells. Neither question has been addressed with experimental investigations in animal models or in a clinical setting. Therefore, we will address them utilizing information derived from other systems.

2.4.1 IL-10

IL-10 is known to be present in the tumor microenvironment (Fig. 2.3), because it may be secreted by tumor cells and by tumor-infiltrating mononuclear cells. IL-10 expression in melanoma *in situ* and lymphoma cells has been well established [1, 5, 17]; it is likely that other tumors also produce IL-10. Tumor-infiltrating leukocytes are also a rich source of IL-10. Specifically, plasmacytoid DC, which accumulate in tumor draining lymph nodes produce IL-10 as do myeloid suppressor cells, which accumulate in the tumor microenvironment. The presence of IL-10 in the tumor milieu is associated with the recruitment and expansion of regulatory T cells (Treg). We and others have shown that Treg *in situ* and after isolation from the tumor are able to secrete IL-10 [52, 53]. This feature correlates with strong suppressor function of Treg in human TIL [52]. Neutralization of IL-10 production by CD4+CD25+Foxp3+ T cells with IL-10-specific antibodies inhibits their suppressor function, indicating that Treg-derived IL-10 is a key factor in immune-mediated suppression at the tumor site [52, 53].

Evidence for down-regulatory effects of IL-10 on APM component expression in tumor cells comes from *in vitro* experiments with mouse lymphoma and mastocytoma

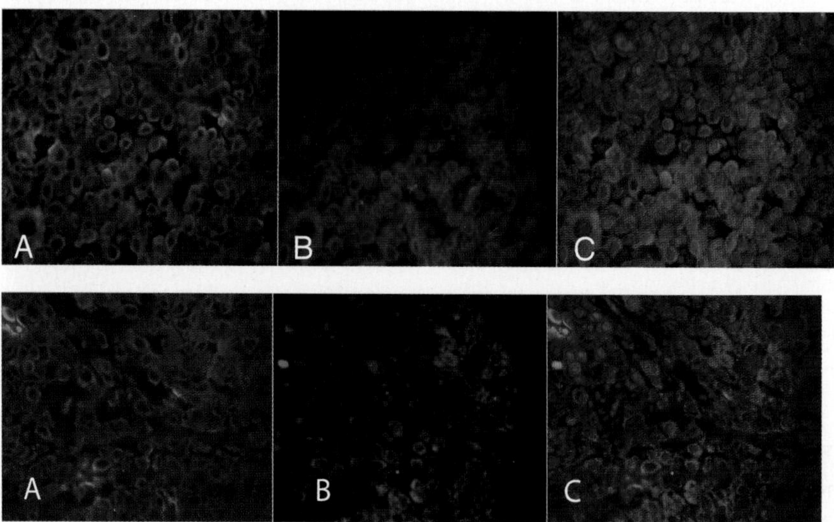

Fig. 2.3 Multicolor immunofluorescence for expression of IL-10 and IL-10R in a representative surgically removed SCCHN lesion. Cryostat sections of the tumor tissue were stained with IL-10-(top panels) and IL-10R-(bottom panels) specific labeled antibodies. Tumor cells are cytokeratin$^+$ (left top and bottom panels); IL-10-(top middle panel) and IL-10R-(bottom middle panel) are expressed in the tumor; the overlay shows cytokeratin$^+$ tumor cells stained with IL-10-(right top panel) or with IL-10R-(right bottom panel) specific labeled antibodies. Tumor cell nuclei are stained with DAPI (right top and bottom panels). X400. Courtesy of Dr. M. Szczepanski

cell lines. Transfection of these cells with a mouse IL-10 cDNA reduced the expression level of TAP1 and TAP2 by 30–60% and 25–35%, respectively, and of MHC class I antigens by up to 90% [49]. The causal role of IL-10 in these changes is suggested by the correlation of the degree of HLA class I antigen downregulation with levels of IL-10 secreted by the cells as well as by HLA class I antigen upregulation induced by transfection of cells with IL-10 anti-sense RNA. However, the drastic HLA class I antigen downregulation at the cell surface in spite of the limited reduction in TAP subunit expression in these cells likely reflects an additional effect of IL-10 on HLA class I heavy chain, β2m, or tapasin transcription, because in other experiments, even very low levels of TAP have been shown to support high HLA class I antigen expression [13].

In agreement with the results obtained with the mouse cell lines tested, a dose-dependent downregulation of TAP1, TAP2 and HLA class I antigens was found in a human melanoma cell line incubated with the synthetic peptide homologous to the C-terminus of human IL-10 [30]. On the other hand, treatment of primary human B lymphocytes with human IL-10 led to TAP1 downregulation, but caused no detectable change in TAP2 expression [63]. These divergent results may reflect differences in the regulation of TAP2 in normal and malignant cells. Furthermore IL-10, which downregulates multiple genes, did not cause detectable changes in LMP7 expression in human B cells. Whether IL-10 modulates the level of other APM components and whether the findings in malignant cells and primary B lymphocytes are paralleled by similar results in dendritic cells residing in the tumor microenvironment remains to be determined.

The changes in the TAP1 levels induced by IL-10 have functional implications, since the ATP-dependent transport of peptides to the ER is reduced. As a result, immature HLA class I molecules accumulate in the ER and are poorly expressed on the cell membrane. Furthermore, TAP downregulation induced by IL-10 in human and mouse cell lines is associated with their reduced recognition by MHC class I antigen restricted, tumor antigen-specific CTL. At the same time, these cells show increased susceptibility to NK cell-mediated lysis [30, 36, 40, 49]. Although the latter finding may be explained by the downregulation of classical HLA class I antigens which inhibit NK cell-mediated cytotoxicity, one might speculate that induction or upregulation of NK-cell activating ligands by IL-10 could also contribute to enhanced NK cell-mediated killing. The latter possibility as well as the effect on HLA-G expression should be explored, since IL-10 has been shown to modulate the expression of several immunologically relevant molecules.

2.4.2 *Transforming Growth Factor-β (TGF-β)*

Like IL-10, TGF-β is a cytokine that: (a) accumulates in the tumor microenvironment; (b) inhibits T cell activation, proliferation and differentiation; and (c) negatively regulates HLA class I antigen expression [31]. TGF-β is produced by tumor cells, normal stromal cells and infiltrating mononuclear cells in the tumor microenvironment. Like IL-10, it is secreted by Treg *in situ* and is, in part, responsible for Treg-mediated immune suppression *in situ* [52, 53]. Tumors express receptors for TGF-β. Negative regulation of HLA class I expression by TGF-β is suggested by the enhanced MHC class I antigen expression in TGF-&bdotbeta; null mice [15] and in tumor cells transfected with TGF-&bdotbeta; anti-sense RNA [55]. Whether the effects of TGF-β on HLA class I antigen expression *in vitro* are mediated, in part, through downregulation of APM components is not known. This possibility should be investigated given the reduced antigen presentation of dendritic cells incubated with TGF-β [16]. Whatever the mechanism, TGF-β upregulation in the tumor microenvironment is likely to have a negative impact on HLA class I antigen expression on tumor cells and most likely on the recognition of tumor cells by CTLs. This effect in conjunction with the tumor growth-enhancing activity of TGF-β may promote tumor progression.

TGF-β has also been shown to downregulate NK-cell activating ligands on tumor cells. The effects are selective, since TGF-β downregulates MICA, ULBP2 and ULBP4 expression, but have no detectable effects on MICB, ULBP1 and ULBP3 expression [8]. Interference with the synthesis of TGF-β by small interfering RNA technology has been found to upregulate MICA expression on glioma cells. The changes induced by TGF-β appear to have functional significance. MICA downregulation in conjunction with reduced expression of its receptor NKG2D, which is also induced by TGF-β on $CD8^+$ T cells and NK cells, is associated with inhibition of NK cell-mediated lysis of target cells and T cell costimulation. Lysis was restored following TGF-β gene silencing [12].

2.4.3 *Interferon-γ (IFN-γ)*

IFN-γ is another cytokine which is known to modulate classical and non-classical HLA class I antigens and to be produced in the course of an immune response, as it is secreted by activated T cells. However, T cells in the tumor microenvironment, are unlikely to produce much IFN-γ because Th1-type responses are suppressed. Paucity of Th1-type cytokines (IL-2, IFN-γ) in tumors has been observed [45]. *In vitro* experiments indicate that IFN-γ can induce, or at least enhance, the expression of immunoproteasome subunits, TAP subunits, tapasin, HLA class I heavy chain and β2m. The presence of IFN-γ in the tumor microenvironment is expected to enhance tumor antigen-derived peptide presentation by HLA class I molecules to CTLs, provided that the IFN-γ signaling pathway is not mutated, as has been shown in some tumor cell lines [7, 22]. In addition, upregulation of immunoproteasomes may have differential effects on the presentation of distinct antigens and on the recognition of tumor cells by CTL. For example, the presentation of tyrosinase-related protein 2 $(TRP2)_{360-368}$ epitope is enhanced upon cleavage of TRP2 by immunoproteasomes, while the expression of Melanoma Antigen Recognized by T $cells_{26-35}$ epitope, like that of the LCMV epitope, is lost upon cleavage by immunoproteasomes [3, 37, 54]. These differences may have marked effects on the interactions of tumor cells with CTL. Likewise, the downregulation of immunoproteasome subunits may alter the recognition of some tumor-derived epitopes by CTLs in a negative or positive manner.

2.5 Potential Effects of Released Classical and Non-Classical HLA Class I Antigens and NK-Cell Activating Ligands in the Tumor Microenvironment

Classical and non-classical HLA class I antigens as well as NK-cell activating ligands are expressed not only on cell membranes, but also in plasma [4, 41–43]. Classical HLA class I antigens are expressed as soluble molecules or bound to membranous nanoparticles or microvesicles [28] (Fig. 2.4). Whether HLA-G and NK-cell activating ligands are also bound to microvesicles is not known. Distinct mechanisms underlie their release from cells. They include secretion, proteolytic cleavage and metalloprotease-mediated cleavage of membrane bound molecules and alternative splicing. As a result different isoforms of classical and non-classical HLA class I antigens are present in plasma [41–43]. In view of the frequent loss of β2 m by malignant cells, it is noteworthy that both classical and non-classical HLA class I antigens have been detected in plasma not only as HLA class I heavy chain-β2 m-peptide complexes, but also as β2 m-free heavy chains. The latter maintain most, if not all the functional activities of β2 m-associated HLA class I heavy chains [41–43].

The functional role of soluble classical and non-classical HLA class I antigens as well as NK-cell activating ligands is not known. It is likely that they play a role in immunological phenomena, since they are immunologically active, as indicated by their reactivity with antibodies and/or with T cell receptors as well as by their immunogenicity in autologous, allogeneic and xenogeneic combinations [2, 27]. Furthermore, as already mentioned, the serum levels of classical and non-classical HLA class I antigens and of NK-cell activating ligands change in some pathological conditions; in some diseases these changes are associated with the clinical

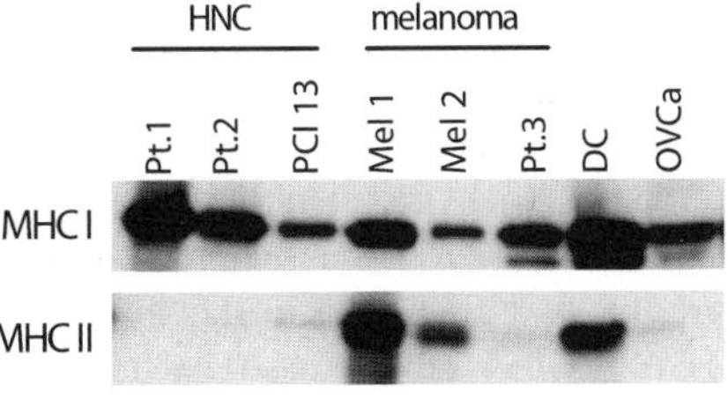

Fig. 2.4 HLA class I and class II antigen expression on microvesicles isolated from sera of (pt) patients with head and neck cancer (HNC) or melanoma (mel) or from supernatants of the SCCHN cell line PCI-13, of the melanoma cell lines mel 1 and mel 2, of the ovarian carcinoma cell line OVCa and of human dendritic cells (DC). Western blot analysis was performed with the HLA class I heavy chain-specific mAb HC-10 and with the HLA class II ßchain-specific mAb LGII-612.14

course of the disease [2, 4, 24, 25, 41]. However, to the best of our knowledge, in no case a cause–effect relationship between these two phenomena has been described.

Convincing evidence indicates that classical HLA class I antigens can induce apoptosis of activated CD8 T cells *in vitro* utilizing Fas–Fas ligand interactions [44, 63]. A similar result has been obtained for the non-classical HLA class I antigens HLA-G by Fournel et al. [10] and by us [6], although this finding has not been confirmed by Hunt [26] who has used recombinant HLA-G in her experiments. There is conflicting information about the mechanisms underlying the induction of apoptosis of activated T cells. The series of reactions leading to apoptosis of activated T cells is triggered by the interaction of HLA class I antigens with T cell receptors (TCR), according to Zavazava and Kronke [62], but with CD8 according to our own data [6, 44]. Furthermore there are conflicting data about the potency of classical and non-classical HLA class I antigens in inducing apoptosis of activated T cells. According to Fournel et al. [10] non-classical HLA class I antigens are more potent than classical HLA class I antigens, while the reverse is true according to our own data [6]. At any rate HLA class I antigens released from tumor cells in their microenvironment provide a mechanism for the presence of apoptotic T cells in the tumor microenvironment [58].

NK-cell activating ligands are also released from tumor cells by the activity of metalloproteases [21, 50, 57]. This causes not only downregulation of their expression on tumor cell membrane, but also promotes a reversible downregulation of NKG2D on NK and CD8 T cells because of its internalization [21, 50]. These changes have a negative impact on the immunosurveillance mediated by cytotoxic cells. Specifically the reduced expression of NK-cell activating ligands decreases the immunogenicity of tumor cells which is influenced by the density of NK-cell activating ligands on cell membrane. Furthermore the reduced surface NKGD2 expression impairs the cytotoxic activity of tumor antigen-specific CTL and of NK cells [21]. The causal role played by soluble NK-cell activating ligands in these phenomena is indicated by the results recently published by Dranoff and his collaborators [27]. The development of MICA-specific antibodies in patients treated with anti-CTLA-4 monoclonal antibodies or immunized with autologous melanoma cells engineered to secrete granulocyte-macrophage colony-stimulating factor was associated with a decrease in the levels of soluble MICA in serum and an increase in the expression of NKG2D on NK cells and $CD8^+$ T cells and their cytotoxicity.

2.6 Conclusions

The data we have discussed suggest that changes which are likely to take place in the tumor microenvironment may have a negative impact on the recognition of tumor cells by host's immune system and on the ability of cytotoxic cells to control tumor growth. Accumulation in the microenvironment of cytokines, such as IL-10 and TGF-β, which are produced by tumor cells an/or stromal cells as well as infiltrating leukocytes, may downregulate the expression and/or function of some APM components. As a result the generation of HLA class I antigen-tumor antigen-derived peptide complexes may be defective and tumor cells may not be recognized by HLA class I antigen-restricted, tumor antigen-specific CTL in spite of HLA class I antigen expression. These phenotypic changes of tumor cells in the tumor microenvironment provide a potential mechanism for the lack of correlation between *in vitro* susceptibility of tumor cells to lysis mediated by HLA class I antigen-restricted, tumor antigen-specific CTL and control of tumor growth *in vivo*. Because of the lack of exposure to cytokines which are present in the tumor microenvironment, tumor cells cultured *in vitro* restore the expression and/or function of the downregulated and/or dysfunctional APM component(s) and reacquire their ability to synthesize and express HLA class I antigen-tumor antigen derived peptide complexes on their membranes. Furthermore some cytokines, such as TGF-β, may downregulate the expression of NK-cell activating ligands on tumor cells, thus reducing their immunogenicity and their recognition by NKGD2 bearing cytotoxic cells. Lastly, some cytokines may synergize with other changes in the microenvironment to induce expression of HLA-G by tumor cells. These non-classical HLA class I antigens may interfere with the interactions of tumor cells with cytotoxic cells.

The potential changes which may be induced in tumor cells by the microenvironment have implications for the design of approaches to characterize their phenotype and of immunotherapeutic strategies for the treatment of malignant diseases. Specifically, analysis of the antigenic profile of tumor cells should rely on probes specific for the HLA class I antigen-tumor antigen derived peptide complex being targeted, since analysis of HLA class I antigen expression may generate misleading information. Furthermore we should reevaluate the emphasis on the development of T cell-based immunotherapeutic strategies which target stromal cells in the microenvironment to counteract the multiple immunoescape mechanisms utilized by tumor cells caused by their genetic instability [29, 32, 38, 61]. The results we have reviewed are compatible with the possibility that the cytokines present in the tumor microenvironment may affect not only tumor cells, but also normal stromal and dendritic cells present in the microenvironment. If so, targeting stromal cells may suffer from limitations similar to those that have been experienced with T cell-based immunotherapy targeting tumor cells. Therefore, strategies to counteract the negative effects of cytokines in the microenvironment may have to rely on intralesional injection of IFN-γ to restore APM function, and/or on the use of monoclonal antibodies or anti-sense constructs directed towards cytokines, such as IL-10 and/or TGF-β, which modulate the expression of HLA antigens, APM components and NK-cell activating ligands.

The data we have reviewed indicate that classical and non-classical HLA class I antigens as well as NK-cell activating ligands released by tumor cells in the microenvironment may interfere with the function of cytotoxic cells utilizing multiple mechanisms. As a result, cytotoxic cells are not likely to be able to control tumor growth in spite of the expression of HLA class I antigen-tumor antigen derived peptide complexes and/or of NK-cell activating ligands by tumor cells.

In this case, however, the data published by Dranoff and his collaborators [27] and corroborated by Marten et al.'s [35] recent results, suggest that administration or induction of antibodies recognizing the molecules involved, if not responsible for tumor progression may represent an effective therapeutic approach.

In conclusion, the changes which may take place in tumor cells in the microenvironment as well as the potential effects of histocompatibility antigens released by tumor cells on immune cells provide mechanisms for the lack of clinical responses in spite of a tumor antigen-specific immune response. In addition, the information we have discussed emphasizes the need to characterize tumor cells within the microenvironment in order to optimize immunotherapeutic strategies for the treatment of malignant disease.

Acknowledgements This study has been supported by PHS grants RO1CA67108, RO1CA110249 and PO1CA109688 awarded by the National Cancer Institute, DHHS.

References

1. Asadullah, K., Sterry, W., Volk, H.D. *Interleukin-10 therapy – review of a new approach.* Pharmacol. Rev. 2003; 55:241–269.
2. Bangia, N., Ferrone, S. *Antigen presentation machinery (APM) modulation and soluble HLA molecules in the tumor microenvironment: Do they provide tumor cells with escape mechanisms from recognition by cytotoxic T lymphocytes?* Immunol. Invest. 2006; 35:485–503.
3. Basler, M., Youhnovski, N., Van Den, B. M., Przybylski, M., Groettrup, M. *Immunoproteasomes down-regulate presentation of a subdominant T cell epitope from lymphocytic choriomeningitis virus.* J. Immunol. 2004; 173:3925–3934.
4. Chang, C. C., Campoli, M., Ferrone, S. *Classical and nonclassical HLA class Iantigen and NK cell-activating ligand changes in malignant cells: current challenges and future directions.* Adv. Cancer Res. 2005: 93189–93234.
5. Conrad, C. T., Ernst, N. R., Dummer, W,, Bröcker, E. B., Becker, J. C. *Differential expression of transforming growth factor beta 1 and interleukin 10 in progressing and regressing areas of primary melanoma.* J. Exp. Clin. Cancer Res. 1999; 18:225–232.
6. Contini, P., Ghio, M., Poggi, A., Filaci, G., Indiveri, F., Ferrone, S., Puppo, F. *Soluble HLA-A,-B,-C and -G molecules induce apoptosis in T and NK CD8+ cells and inhibit cytotoxic T cell activity through CD8 ligation.* Eur. J. Immunol. 2003; 33:125–134.
7. Dovhey, S. E., Ghosh, N. S., Wright, K. L. *Loss of interferon-gamma inducibility of TAP1 and LMP2 in a renal cell carcinoma cell line.* Cancer Res. 2002; 60:5789–5796.
8. Eisele, G., Wischhusen, J., Mittelbronn, M., Meyermann, R., Waldhauer, I., Steinle, A., Weller, M., Friese, M. A. *TGF-beta and metalloproteinases differentially suppress NKG2D ligand surface expression on malignant glioma cells.* Brain. 2006; 129:2416–2425.
9. Ferris, R., Whiteside, T. L., Ferrone, S. Immune escape associated with functional defects in antigen processing machinery in head and neck cancer. Clin. Cancer Res. 2006; 12:3890–3895.
10. Fournel, S., Aguerre-Girr, M., Huc, X., Lenfant, F., Alam, A., Toubert, A., Bensussan, A., Le Bouteiller, P. *Cutting edge: soluble HLA-G1 triggers CD95/CD95 ligand-mediated apoptosis in activated CD8+ cells by interacting with CD8.* J. Immunol. 164:6100–6104.
11. Friese, M. A., Platten, M., Lutz, S. Z., Naumann, U., Aulwurm, S., Bischof, F., Bühring, H. J., Dichgans, J., Rammensee, H. G., Steinle, A., Weller, M. *MICA/NKG2D-mediated immunogene therapy of experimental gliomas.* Cancer Res. 2003; 63:8996–9006.

12. Friese, M. A., Wischhusen, J., Wick, W., Weiler, M., Eisele, G., Steinle, A., Weller, M. *RNA interference targeting transforming growth factor-beta enhances NKG2D-mediated antiglioma immune response, inhibits glioma cell migration and invasiveness, and abrogates tumorigenicity in vivo.* Cancer Res. 2004; 64:7596–7603.
13. Garbi, N., Tanaka, S., van den, B. M., Momburg, F., Hammerling, G. J. *Accessory molecules in the assembly of major histocompatibility complex class I/peptide complexes: how essential are they for CD8(+) T-cell immune responses*? Immunol. Rev. 2005; 20:777–788.
14. Gasser, S., Orsulic, S., Brown, E. J., Raulet, D. H. *The DNA damage pathway regulates innate immune system ligands of the NKG2D receptor.* Nature 2005; 25(436):1186–1190.
15. Geiser, A. G., Letterio, J. J., Kulkarni, A. B., Karlsson, S., Roberts, A. B., Sporn, M. B. *Transforming growth factor beta 1 (TGF-beta 1) controls expression of major histocompatibility genes in the postnatal mouse: aberrant histocompatibility antigen expression in the pathogenesis of the TGF-beta 1 null mouse phenotype.* Proc. Natl. Acad. Sci. USA 1993; 90:9944–9948.
16. Geissmann, F., Revy, P., Regnault, A., Lepelletier, Y., Dy, M., Brousse, N., Amigorena, S., Hermine, O., Durandy, A. *TGF-beta 1 prevents the noncognate maturation of human dendritic Langerhans cells.* J. Immunol. 1999; 162:4567–4575.
17. Gerlini, G., Tun-Kyi, A., Dudli, C., Burg, G., Pimpinelli, N., Nestle, F. O. *Metastatic melanoma secreted IL-10 down-regulates CD1 molecules on dendritic cells in metastatic tumor lesions.* Am. J. Pathol. 2004; 165:1853–1863.
18. Girardi, M., Oppenheim, D. E., Steele, C. R., Lewis, J. M., Glusac, E., Filler, R., Hobby, P., Sutton, B., Tigelaar, R. E., Hayday, A. C. *Regulation of cutaneous malignancy by gammadelta T cells.* Science 2001; 294:605–609.
19. Groh, V., Bahram, S., Bauer, S., Herman, A., Beauchamp, M., Spies, T. *Cell stress-regulated human major histocompatibility complex class I gene expressed in gastrointestinal epithelium.* Proc Natl. Acad. Sci. USA 1996; 93:12445–12450.
20. Groh, V., Rhinehart, R., Secrist, H., Bauer, S., Grabstein, K. H., Spies, T. *Broad tumor-associated expression and recognition by tumor-derived gamma delta T cells of MICA and MICB.* Proc. Natl. Acad. Sci. USA 1999; 96:6879–6884.
21. Groh, V., Wu, J., Yee, C., Spies, T. *Tumour-derived soluble MIC ligands impair expression of NKG2D and T-cell activation.* Nature 2002; 419:734–738.
22. Hayashi, T., Kobayashi, Y., Kohsaka, S., Sano, K. *The mutation in the ATP binding region of JAK1, identified in human uterine leiomyosarcomas, results in defective interferon-gamma inducibility of TAP1 and LMP2.* Oncogene 2006; 25(29):4016–4026.
23. Hicklin, D. J., Marincola, F. M., Ferrone, S. *HLA class I antigen downregulation in human cancers: T-cell immunotherapy revives an old story.* Mol. Med. Today 1999; 5:178–186.
24. Holdenrieder, S., Stieber, P., Peterfi, A., Nagel, D., Steinle, A., Salih, H. R. *Soluble MICA in malignant diseases.* Int. J. Cancer 2006; 118:684–687.
25. Holdenrieder, S., Stieber, P., Peterfi, A., Nagel, D., Steinle, A., Salih, H. R. *Soluble MICB in malignant diseases: analysis of diagnostic significance and correlation with soluble MICA.* Cancer Immunol. Immunother. 2006; 55:1584–1589.
26. Hunt, J. S. *Stranger in a strange land.* Immunol. Rev. 2006; 213:36–47.
27. Jinushi, M., Hodi, F. S., Dranoff, G. *Therapy-induced antibodies to MHC class I chain-related protein A antagonize immune suppression and stimulate antitumor cytotoxicity.* Proc Natl Acad Sci USA 2006; 103:9190–9195.
28. Kim, J.-W., Wieckowski, E., Taylor, D. D., Reichert, T. E., Watkins, S., Whiteside, T. L. *FasL+ membraneous vesicles isolated from sera of patients with oral cancer induce apoptosis of activated T lymphocytes.* Clin. Cancer Res. 2005; 11:1010–1020.
29. Ko, E., Wei, L., Peng, L., Wang, X., Ferrone, S. *Anti-tumor effects of anti-angiogenic immunity elicited in mice by dendritic-endothelial cell hybrids and 4–1BB-specific mAb.* Cancer Res. in press 2007.
30. Kurte, M., Lopez, M., Aguirre, A., Escobar, A., Aguillon, J. C., Charo, J., Larsen, C. G.,Kiessling, R., Salazar-Onfray, F. *A synthetic peptide homologous to functional domain of human IL-10 down-regulates expression of MHC class I and Transporter associated with Antigen Processing 1/2 in human melanoma cells.* J. Immunol. 2004; 73:1731–1737.

31. Li, M. O., Wan, Y. Y., Sanjabi, S., Robertson, A. K., Flavell, R. A. *Transforming growth factor-beta regulation of immune responses*. Annu. Rev. Immunol. 2006; 24:99–146.
32. Li, Y., Wang, M. N., Li, H., King, K. D., Bassi, R., Sun, H., Santiago, A., Hooper, A. T., Bohlen, P., Hicklin, D. J. *Active immunization against the vascular endothelial growth factor receptor flk1 inhibits tumor angiogenesis and metastasis*. J. Exp. Med. 2002; 195:1575–1584.
33. Lopez-Albaitero, A., Nayak, J. V., Ogino, T., Machandia, A., Gooding, W., DeLeo, A. B., Ferrone, S., Ferris, R. L. *Role of antigen-processing machinery in the in vitro resistance of squamous cell carcinoma of the head and neck cells to recognition by CTL*. J. Immunol. 2006; 176:3402–3409.
34. Marincola, F. M., Jaffee, E. M., Hicklin, D. J., Ferrone, S. *Escape of human solid tumors from T- cell recognition: molecular mechanisms and functional significance*. Adv. Immunol. 2000; 74:181–273.
35. Märten, A., von Lilienfeld-Toal, M., Büchler, M. W., Schmidt, J. *Soluble MIC is elevated in the serum of patients with pancreatic carcinoma diminishing gammadelta T cell cytotoxicity*. Int. J. Cancer 2006; 119:2359–2365.
36. Matsuda, M., Salazar, F., Petersson, M., Masucci, G., Hansson, J., Pisa, P., Zhang, Q. J., Masucci, M. G., Kiessling, R. *Interleukin 10 pretreatment protects target cells from tumor- and allo-specific cytotoxic T cells and downregulates HLA class I expression*. J. Exp. Med. 1994; 180:2371–2376.
37. Morel, S., Levy, F., Burlet-Schiltz, O., Brasseur, F., Probst-Kepper, M., Peitrequin, A. L., Monsarrat, B., Van Velthoven, R., Cerottini, J. C., Boon, T., Gairin, J. E., Van den Eynde, B. J. *Processing of some antigens by the standard proteasome but not by the immunoproteasome results in poor presentation by dendritic cells*. Immunity 2000; 12:107–117.
38. Niethammer, A. G., Xiang, R., Becker, J. C., Wodrich, H., Pertl, U., Karsten, G., Eliceiri., B. P., Reisfeld, R. A. *A DNA vaccine against VEGF receptor 2 prevents effective angiogenesis and inhibits tumor growth*. Nat. Med. 2002; 8:1369–1375.
39. Pende, D., Rivera, P., Marcenaro, S., Chang, C. C., Biassoni, R., Conte, R., Kubin, M., Cosman, D., Ferrone, S., Moretta, L., Moretta, A. *Major histocompatibility complex class I-related chain A and UL16-binding protein expression on tumor cell lines of different histotypes: analysis of tumor susceptibility to NKG2D-dependent natural killer cell cytotoxicity*. Cancer Res. 2002; 62:6178–6186.
40. Petersson, M., Charo, J., Salazar-Onfray, F., Noffz, G., Mohaupt, M., Qin, Z., Klein, G., Blankenstein, T., Kiessling, R. *Constitutive IL-10 production accounts for the high NK sensitivity, low MHC class I expression, and poor transporter associated with antigen processing (TAP)-1/2 function in the prototype NK target YAC-1*. J. Immunol. 1998; 161:2099–2105.
41. Pistoia, V., Morandi, F., Wang, X., Ferrone, S. *Soluble HLA-G: are they clinically relevant?* Semin. Cancer Biol. 2007.
42. Puppo, F., Indiveri, F., Scudeletti, M., Ferrone, S. *Soluble HLA antigens: new roles and uses*. Immunol. Today 1997; 18:154–155.
43. Puppo, F., Scudeletti, M., Indiveri, F., Ferrone, S. *Serum HLA class I antigens: markers and modulators of an immune response?* Immunol. Today 1995; 16:124–127.
44. Puppo, F., Contini, P., Ghio, M., Brenci, S., Scudeletti, M., Filaci, G., Ferrone, S., Indiveri, F. *Soluble human MHC class I molecules induce soluble Fas ligand secretion and trigger apoptosis in activated CD8(+) Fas (CD95)(+) T lymphocytes*. Int. Immunol. 2000; 12:195–203.
45. Rabinowich, H., Suminami, Y., Reichert, T. E., Crowley-Nowick, P., Bell, M., Edwards, R., Whiteside, T. L. *Expression of cytokine genes or proteins and signaling molecules in lymphocytes associated with human ovarian carcinoma*. Intl. J. Cancer 1996; 68:276–284.
46. Raffaghello, L., Prigione, I., Airoldi, I., Camoriano, M., Levreri, I., Gambini, C., Pende, D., Steinle, A., Ferrone, S., Pistoia, V. *Downregulation and/or release of NKG2D ligands as immune evasion strategy of human neuroblastoma*. Neoplasia 2004; 6:558–568.
47. Rebmann, V., Schütt, P., Brandhorst, D., Opalka, B., Moritz, T., Nowrousian, M. R., Grosse-Wilde, H. *Soluble MICA as an independent prognostic factor for the overall survival and progression-free survival of multiple myeloma patients*. Clin. Immunol. 2007; 123:114–120.

48. Rouas-Freiss, N., Moreau, P., Ferrone, S. and Carosella, E. D. *HLA-G proteins in cancer: do they provide tumor cells with an escape mechanism?* Cancer Res. 2005; 66:10139–10144.
49. Salazar-Onfray, F., Charo, J., Petersson, M., Freland, S., Noffz, G., Qin, Z., Blankenstein, T., Ljunggren, H. G., Kiessling, R. *Down-regulation of the expression and function of the transporter associated with antigen processing in murine tumor cell lines expressing IL-10.* J. Immunol. 1997; 159:3195–3202.
50. Salih, H. R., Antropius, H., Gieseke, F., Lutz, S. Z., Kanz, L., Rammensee, H. G., Steinle, A. *Functional expression and release of ligands for the activating immunoreceptor NKG2D in leukemia.* Blood 2003; 102:1389–1396.
51. Salih, H. R., Rammensee, H. G., Steinle, A. *Cutting edge: down-regulation of MICA on human tumors by proteolytic shedding.* J. Immunol. 2002; 169:4098–4102.
52. Strauss, L., Bergmann, C., Szczepanski, M., Gooding, W., Johnson. J. T., Whiteside, T. L. *A unique subset of CD4+CD25^{high}Foxp3+ T cells secreting IL-10 and TGF-β_1 mediates suppression in the tumor microenvironment.* Clin. Cancer Res. 2007.
53. Strauss, L., Whiteside, T. L., Knights, A., Bergmann, C., Knuth, A., Zippelius, A. *Selective survival of naturally occurring human CD4+CD25+Foxp3+ regulatory T cells cultured with rapamycin.* J. Immunol. 2007; 178:320–329.
54. Sun, Y., Sijts, A. J., Song, M., Janek, K., Nussbaum, A. K., Kral, S., Schirle, M., Stevanovic, S., Paschen, A., Schild, H., Kloetzel, P. M., Schadendorf, D. *Expression of the proteasome activator PA28 rescues the presentation of a cytotoxic T lymphocyte epitope on melanoma cells.* Cancer Res. 2002; 62:2875–2882.
55. Tzai, T. S., Shiau, A. L., Liu, L. L., Wu, C. L. *Immunization with TGF-beta antisense oligonucleotide-modified autologous tumor vaccine enhances the antitumor immunity of MBT-2 tumor-bearing mice through upregulation of MHC class I and Fas expressions.* Anticancer Res. 2000; 20:1557–1562.
56. Ugurel, S., Rebmann, V., Ferrone, S., Tilgen, W., Grosse-Wilde, H., Reinhold, U. *Soluble human leukocyte antigen-G serum level is elevated in melanoma patients and is further increased by interferon-alpha immunotherapy.* Cancer 2001; 92:369–376.
57. Waldhauer I, Steinle A. Proteolytic release of soluble UL16-binding protein 2 from tumor cells. Cancer Res. 2006; 66:2520–2526.
58. Whiteside, T. L. *Apoptosis of immune cells in the tumor microenvironment and peripheral circulation of patients with cancer: implications for immunotherapy.* Vaccine 2002; 20 Suppl. 4: A46–A51.
59. Wiendl, H., Mitsdoerffer, M., Hofmeister, V., Wischhusen, J., Bornemann, A., Meyermann, R., Weiss, E. H., Melms, A., Weller, M. *A functional role of HLA-G expression in human gliomas: an alternative strategy of immune escape.* J. Immunol. 2002; 168:4772–4780.
60. Wu, J. D., Higgins, L. M., Steinle, A., Cosman, D., Haugk, K., Plymate, S. R. *Prevalent expression of the immunostimulatory MHC class I chain-related molecule is counteracted by shedding in prostate cancer.* J. Clin. Invest. 2004; 114:560–568.
61. Yu, P., Rowley, D. A., Fu, Y. X., Schreiber H. *The role of stroma in immune recognition and destruction of well-established solid tumors.* Curr. Opin. Immunol. 2006; 18:226–231.
62. Zavazava, N., Kronke, M. *Soluble HLA class I molecules induce apoptosis in alloreactive cytotoxic T lymphocytes.* Nat. Med. 1996; 2:1005–1010.
63. Zeidler, R., Eissner, G., Meissner, P., Uebel, S., Tampe, R., Lazis, S., Hammerschmidt, W. *Downregulation of TAP1 in B lymphocytes by cellular and Epstein-Bar rvirus-encoded interleukin-10.* Blood 1997; 90:2390–2397.

Chapter 3
Local Tumor Growth and Spontaneous Systemic T Cell Responses in Cancer Patients: A Paradox and Puzzle

Philipp Beckhove and Volker Schirrmacher

Abstract We describe and discuss the paradox situation that in many cancer patients functional antitumor memory T cells can be detected in their bone marrow which coexist with a growing tumor in the periphery. This phenomenon, known as "concomitant immunity" suggests that the tumor and its microenvironment prevent systemic antitumor immunity to become effective. Strategies of intervention at the tumor microenvironment are being discussed.

Keywords Memory T cells, bone marrow, tumor dormancy, intervention, danger signals, tumor infiltration

3.1 Introduction

Until the early 1990s, it remained largely unclear how the immune system could recognize antigens on autologous cells and specific tumor cell recognition by T cells was doubted by many researchers. In 1991, the first description of antigens specifically expressed on melanoma cells formed the basis for a new era of molecularly defined tumor immunology [1]. In the following years, numerous antigens, either overexpressed or specifically expressed on tumor cells, have been characterized and respective HLA-restricted epitopes that are capable of triggering CD8 and CD4 T cell responses have been identified [2]. With the introduction of dendritic cell culture and demonstration of their T cell stimulatory capacity by Steinmann et al., therapeutic approaches have been developed during the last years that aim at inducing tumor-antigen (TAA) specific T cell responses for the treatment of tumor patients [3]. These approaches are based on the assumption that the immune system of tumor patients is mainly unresponsive for antigens on autologous tumors.

Division of Cellular Immunology, German Cancer Research Center, D-69120 Heidelberg, Germany

E. Yefenof (ed.), *Innate and Adaptive Immunity in the Tumor Microenvironment.*

The induction of protective T cell responses depends on fully activated antigen presenting cells, e.g. dendritic cells (DCs). During infectious diseases DCs are activated in inflamed tissues by so called "danger signals". Danger signals are provided by conserved molecular signatures of pathogens such as double stranded RNA molecules or hypomethylated DNA motifs and recognized by respective toll like receptors on cells of the innate immune system, including DCs, which enables them to recognize the presence and general class of the infectious pathogen [4].

Tumor tissue, as autologous "self" is widely assumed to lack danger signals. This assumption was substantiated by findings that significant proportions of natural TAA-specific CD8 T cells as isolated by the HLA-tetramer technology from the peripheral blood of the patients were non-functional [5].

3.2 Spontaneous T Cell Immunity and the Tumor Microenvironment

3.2.1 Spontaneous Systemic T Cell Responses in Cancer Patients

During the last few years, numerous studies have emerged that clearly demonstrate the spontaneous induction of functional, *TAA-specific T cell responses* in many tumor patients. We observed in the bone marrow of mice bearing disseminated lymphoma cells TAA-reactive CD8 T cells that conferred protection against (i) expansion of tumor cells *in situ* as well as (ii) against the formation of distant metastases [6, 7]. Based on this observation, we analyzed the possibility that the presence of tumor antigen in the bone marrow might also trigger the generation of functional, tumor-specific T cells in tumor patients. We therefore evaluated by HLA-tetramer analyses as well as by short term IFN-γ Elispot assays the presence of such cells in the peripheral blood and bone marrow of breast cancer patients [8]. We detected high frequencies of tumor-specific CD8 T cells in the blood and bone marrow of most patients. However, only BM-resident T cells exerted functional capacities such as IFN-γ secretion or cytotoxic activity. Altogether, T cells reactive against the entirety of TAAs in autologous tumor cells (as presented by autologous tumor cell lysate-pulsed dendritic cells) could be detected in app. 40% of primary operated *breast cancer patients*. These T cells existed in frequencies of 1:200 to 1:10,000 of total T cells and included cytotoxic $CD8^+$ T cells as well as $CD4^+$ T helper cells. Similar frequencies of tumor-reactive T cells were detected in patients with different tumors. Using the IFN-γ Elispot assay, we investigated reactivity of T cells from bone marrow and peripheral blood to melanoma lysate-pulsed autologous DCs in *melanoma patients* [9]. We detected tumor-reactive T cells in altogether 25% of the patients. In the bone marrow, melanoma-reactive T cells were present in 18% while in peripheral blood such cells were present in only 10% of the patients.

Pancreatic cancer is characterized by aggressive growth and treatment resistance. A dominance of TH2 cytokines in patients' sera, as reported previously, suggests systemic tumor-induced immunosuppression, potentially inhibiting the induction of tumor-reactive T cells. Despite these potentially powerful immunosuppressive factors we detected high numbers of tumor-reactive T cells in 100% of bone marrow samples and in 50% of the blood samples of pancreatic cancer patients [10]. These comprised CD8 and CD4 T cells which secreted the TH1 cytokine IFN-γ rather than TH2 cytokines and exerted cytotoxic capacity upon stimulation with tumor antigens [10, 11]. Freshly isolated T cells from cancer patients recognized autologous tumor cells and the pancreatic cancer-associated tumor antigen MUC1. Thus, tumor antigen-specific T cell responses occur regularly during pancreatic cancer disease and lead to enrichment of tumor cell-reactive memory T cells in the bone marrow.

Besides carcinomas, *hematological malignancies* can also be recognized by the immune system and lead to the induction of spontaneous T cell responses. We analyzed bone marrow and peripheral blood of HLA-A2-positive multiple myeloma (MM) patients for the presence and functional activity of CD8 T cells specific for the MUC1-derived peptide LLLLTVLTV [12]. 40% of the patients with MM contained elevated frequencies of MUC1-specific CD8 T cells in freshly isolated samples from peripheral blood (PB) or bone marrow (BM) compared with corresponding samples from healthy donors. Similar to our findings from breast carcinoma, pancreatic carcinoma and malignant melanoma, BM-residing T cells possessed a higher functional capacity upon specific reactivation than PB-derived T cells with regard to IFN-γ secretion, perforin production, and cytotoxicity.

It can therefore be concluded that virtually all tumor types induce spontaneous and functional T cell responses. These result in the accumulation of tumor-reactive T cells especially in lymphoid organs such as the BM.

Detailed phenotypic analyses revealed that the majority (approximately 70%) of tumor-specific BM T cells belonged to the population of *memory T cells*, including app. 40% of central- and 60% of effector memory T cells [13]. The presence of effector memory T cells is particularly interesting, since this subset is generated by appropriate reactivation of long-lived bone marrow-resident central memory T cells that subsequently differentiate into the rather short-lived latter subset and are capable of mediating protective effector functions. Thus, the dominance of effector memory T cells in the bone marrow of cancer patients suggests that presentation of tumor antigens to T cells is a rather frequent event in the BM parenchyme.

In order to characterize the *memory T cell repertoire* of tumor patients in more detail, we analyzed the bone marrow of primary operated breast cancer patients and healthy female donors for the presence and frequencies of spontaneously induced effector/memory T lymphocytes with peptide-HLA-A2-restricted reactivity against 10 different breast TAAs and a variety of normal breast tissue-associated antigens by short-term IFN-γ Elispot analysis [14]. 67% of these patients recognized TAAs with a mean frequency of 1:7,000. TAA reactive T cells recognized simultaneously an average of half of the tested TAAs. The T-cell repertoire was highly polyvalent and exhibited pronounced interindividual differences in the pattern of TAAs recognized by each patient. Strong differences of reactivity were noticed between TAAs, ranging from 100% recognition of prostate-specific antigen (pp. 141–149) to only 25% recognition of MUC1 (pp. 12–20) or Her-2/neu (pp. 369–377). In comparison with TAAs, reactivity to normal breast tissue-associated antigens was lower with respect (i) to the proportions of responding patients (30%) and (ii) to the frequency of T cells recognizing the normal tissue antigens (only 1:12,000 T cells). Interestingly, healthy individuals also contained TAA-reactive T cells but this repertoire was more restricted and the frequencies were in the same range as T cells reacting to normal breast tissue-associated antigens. Thus, the natural repertoire of tumor-reactive memory T cells in the BM of cancer patients is polyvalent and highly individual and compared to healthy individuals clearly shifted towards tumor recognition. Taken together, the bone marrow appears as a lymphatic organ particularly involved in the induction and/or maintenance of natural T cell immunity against tumor antigens.

3.2.2 The Bone Marrow as a Source of Tumor-reactive T Cells

The bone marrow was known for a long time as a primary lymphoid organ, but its potential to serve as a secondary immune organ has been detected only a few years ago. Figure 3.1 shows a scheme of *T cell priming in the bone marrow* parenchyma against blood-borne antigens. The bone marrow stroma expresses constitutively multiple adhesion molecules (ICAM-1, VCAM-1, MadCAM-1, P-selectin) that are relevant for the homing and infiltration of BM by blood derived T lymphocytes, and also the co-stimulatory molecule CD80, that is relevant for T cell activation [15]. We and others found out that naïve and also memory T cells selectively migrated to BM. T cell homing to BM involved the integrins LFA-1 alpha and alpha4 which interact with the above constitutively expressed cell adhesion molecules (CAMs [15]). We could demonstrate that naive, antigen-specific T cells home to bone marrow, where they can be primed in follicle-like structures in BM parenchyma involving clusters of DCs and CD8 and CD4 T cells and differentiate into memory T cells [8, 15]. These structures form on the basis of specialized microdomains of bone marrow endothelium expressing respective adhesion molecules that attract T cells as well as antigen-pulsed dendritic cells and allow their selective entry and subsequent cognate interactions [16].

Bone marrow resident CD11c+ DCs are highly efficient in taking up exogenous blood-borne antigen and processing it via major histocompatibility complex (MHC) class I and class II pathways. T-cell activation correlates with *DC-T cell clustering* in bone marrow stroma. Primary CD4+ and CD8+ T-cell responses generated in BM occur in the absence of secondary lymphoid organs. The responses are not tolerogenic and result in generation of cytotoxic T cells, protective antitumor immunity and immunological memory. Together, these findings suggest that DC-T cell interactions in BM play an important role in immune responses to blood borne antigen and in the establishment of systemic immunity and long-term memory.

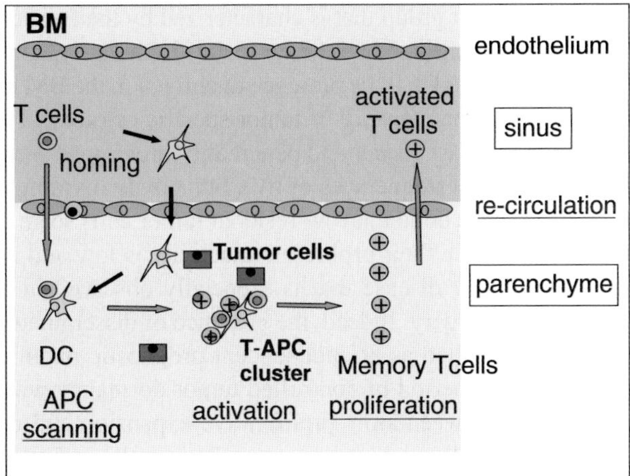

Fig. 3.1 Induction of T cell responses against blood borne or local tumor antigens in the bone marrow. BM; bone marrow, DC; dendritic cell, APC; antigen presenting cell

We further evaluated the *requirements for the maintenance* of tumor-reactive memory T cells in the ESb lymphoma mouse model [17]. The transfer of CD8 T cells reactive to a model tumor antigen (LacZ/ß-Gal) prevented the outgrowth of Gal-expressing syngeneic tumor cells (ESbL-Gal) in athymic nu/nu mice and resulted in long term persistence of high numbers of transferred CD8+ memory T cells in the BM. In contrast, in the absence of the model tumor antigen, no such memory generation and – persistence was detectable. Long-term immune memory and tumor protection could be maintained over four successive transfers between tumor-inoculated recipients, which involved periodic antigenic restimulation *in vivo* prior to reisolating the cells for adoptive transfer. These findings suggest that TAA from residual dormant tumor cells are implicated in maintaining high frequencies of long-term surviving Gal-specific memory CD8(+) T cells. Using a cell line (ESbL-Gal-BM) that was established from dormant tumor cells isolated from the bone marrow, it could be demonstrated that the tumor cells had up-regulated the expression of MHC class I molecules and down-regulated the expression of several adhesion molecules during the *in vivo* passage. Our results suggest (i) that the BM microenvironment has special features that are of importance for the maintenance of tumor dormancy and immunological T-cell memory, and (ii) that a low level of persisting TAA favors the maintenance of TAA-specific memory T cells over irrelevant memory T cells [17].

Dormant tumor cells in the BM or significant amounts of soluble TAAs in the serum seem to be also required for the induction and maintenance of tumor-reactive T cells in tumor patients, since, according to a recent observation, melanoma-reactive T cell could be detected almost exclusively in the BM of late stage (stage IV) melanoma patients – a patient group that is characterized by the frequent occurrence of disseminated melanoma cells and high tumor load [9]. Similarly, a recent study on myeloma patients demonstrated that the presence of antigen in the BM microenvironment was important for the maintenance of tumor-specific cytotoxic T cells [18].

Many tumors, such as breast cancer and pancreatic carcinomas are characterized by the establishment of small frequencies of live, but slowly dividing *disseminated tumor cells in the BM*. These bone marrow-resident tumor cells seem to be particularly treatment resistant, since their proliferative activity is low and may therefore provide a basis for recurrent disease that is especially observed in breast cancer many years after primary surgery. Indeed, the presence of disseminated tumor cells in the BM has been established as an independent prognostic parameter in breast cancer [19]. This implies a period of controlled tumor dormancy which is characterized by a net balance between slow proliferative expansion and elimination of tumor cells. Such control of disseminated tumor cells as well as of minimal residual disease might be mediated by functional tumor-reactive CD4 and CD8 memory T cells. Indeed, in a lymphoma mouse model of long term bone marrow tumor dormancy, persistence of dormant tumor in the bone marrow correlated with the duration of antitumor immunity [7]. Depletion of CD8 T cells in this model was immediately associated with outgrowth of distant tumor metastases and subsequent death of animals. These experiments show that BM and lymph nodes are privileged sites where potentially lethal tumor cells are controlled in a dormant state by the immune system. Metastasis may be a consequence of the breakdown of this immune control [6, 7, 20].

In order to evaluate the *therapeutic potential* of the natural repertoire of tumor specific T cells in the bone marrow of tumor patients, we transferred these cells after appropriate ex vivo reactivation with TAA – pulsed autologous DCs into NOD/Scid mice xenotransplanted with small pieces of breast- or pancreatic carcinomas or normal skin transplants from the same patients. Separated BM-derived CD45RA(-) memory but not CD45RA(+) naive T cells infiltrated autologous tumor but not skin tissues after the transfer [8, 13]. These tumor-infiltrating cells had a central or effector memory phenotype and produced perforin. Many of them expressed the P-selectin glycoprotein ligand 1 and were found around P-selectin(+) tumor endothelium. Tumor infiltration included cluster formation in tumor tissue by memory T cells with cotransferred DCs [21]. A single transfer of restimulated BM memory T cells was associated with the induction of tumor cell apoptosis and significant tumor reduction. T cells from peripheral blood showed much lower antitumor reactivity. Thus, reactivated memory T cells selectively home to human tumors and reject them on the basis of specific recognition of TAAs antigens on tumor cells and on tumor-resident DCs, suggesting that in case of their reactivation *in situ* the tumor-specific T cell repertoire in the BM might have a protective capacity.

Up to now, clinical studies that correlate the presence of tumor antigen-reactive T cells in the bone marrow of cancer patients with the course and duration of their tumor disease have not been published. A case report demonstrated that pre-existing immunity to tyrosine-related protein (TRP)-2 and NY-ESO-1 melanoma antigen in PBMC of a melanoma patient was associated with a dramatic clinical and immunological response to vaccination with respective TAAs [22]. An attempt to correlate tumor-reactive T cells in the blood of colorectal carcinoma patients with their clinical outcome did not show any significant influence of specific T cell immunity in the peripheral blood [23]. Another study revealed a potential duality of CTL-activity in bone metastases [24]. While tumor-specific CTLs could be recruited to bone metastases, they exerted osteoclastic bone resorption by release of IL-7 and IL-8. Such bone degradation resulted in the release of TGFß1 from the bone matrix which suppressed T cell function and respective T cell mediated tumor cell lysis while maintaining T cell driven bone resorption. Thus, the well established role of bisphosphonates, the state of the art treatment in metastatic bone disease for improving the overall patient survival, may be mediated by their inhibition of local TGFß-release.

3.2.3 Systemic T Cell Immunity and T Cell Entry into Tumor Tissue

The potential of naturally induced tumor-specific effector and memory T cells to influence progression of established tumors depends on their *capacity to infiltrate tumor tissues* and to mediate tumor cell rejection *in situ*. The difficulty of this has been convincingly demonstrated in a mouse model of spontaneous insulinoma [25]. In this model, high numbers of strongly activated CTL carrying transgenic TCR with high affinity to a model foreign tumor antigen did not accumulate in the growing insulinomas and did not mediate tumor rejection. Only after activation of the tumor endothelium through inflammatory stimuli, these T cells infiltrated efficiently into the tumors and mediated tumor rejection. Interestingly, gene expression analyses revealed that strong overexpression of the transcription factor RGS5 by tumor endothelium associated pericytes might be a key regulator of the observed inhibition of T cell entry in this model [26] – a first hint that tumor endothelium not only resembles that of normal, non-inflamed tissues but that the tumor microenvironment influences the vasculature to prevent effector T cell transmigration. Another molecule, Galectin -1 is expressed by endothelial cells in tumor tissue under hypoxic conditions and mediates suppression of T cell tumor infiltration in HNSCC patients [27].

Despite these observations, numerous studies have demonstrated that tumor specific T cells can be isolated from the tumor tissues of various cancers. Among 123 tumor-infiltrating T cell *(TIL) cultures* from melanoma biopsies, 57% recognized HLA-A2 restricted epitopes of the highly immunogenic melanoma-associated antigens MART-1 and gp100, while much less cultures reacted to antigens presented in the context of other HLA-alleles [28]. T cell responses were frequently dominated by a single HLA allele and besides Melan-A, no dominant antigens could be characterized among a broad variety of TAAs that were simultaneously tested [29]. Multiple HLA-A2-restricted melanoma differentiation antigens were recognized by TILs from melanoma patients [30]. Interestingly, the frequencies of Mart 1-reactive HLA-A2-restricted CD8 TIL as detectable by HLA-tetramer analysis in approximately half of the freshly excised melanomas were between 1:50 and 1:200, which is similar to the frequencies of such cells in the peripheral blood [31].

Although tumor-reactive T cells may not be selectively enriched in tumor tissues, (which might as well be due to a rapid loss *in situ* of such cells after their encounter of cognate antigen), it may be that cognate recognition of TAAs induces their activation and subsequent effector function. In a mouse model, antigen specific CD4 and CD8 T cells eliminated not only antigen-expressing tumor cells by direct tumor cell lysis, but also antigen loss variants indirectly through destruction of the tumor supportive stromal environment. This stroma destruction was mediated by recognition of TAAs crosspresented by non malignant, bone marrow-derived stroma cells [32].

By CDR3-region analysis, PB T cell clones capable of lysing autologous ovarian carcinoma cells were detected in ovarian carcinomas [33]. In a study on *colorectal carcinomas*, we did not detect differences in absolute numbers of CD8 T cells between carcinoma tissue and corresponding colonic mucosa of the same patients [34]. However, TIL showed increased levels of activation as measured by CD69 expression as well as increased proportions of cells with cytotoxic degranulation *in situ* as measured by membrane expression of CD107a. Proportions of such TIL recognized ex vivo the colon carcinoma associated antigen MUC1 and responded with IFN-γ secretion. Interestingly, the presence of functional CD4 and CD8 memory T cells with specificity for MUC1 in the bone marrow or peripheral blood of the patients was associated with increased numbers of activated or cytotoxically active CD8 TIL in colorectal carcinomas. This suggests that systemic tumor-specific immunity is a prerequisite for the presence and functional activity of tumor specific intratumoral T cells. These findings may explain in part the meanwhile well established prognostic significance of increased numbers of activated CD8 memory T cells infiltrating colorectal carcinomas [35, 36].

A recent study demonstrated in patients with *multiple myeloma*, demonstrated that activated bone marrow infiltrating T lymphocytes effectively targeted autologous human plasma cells and their clonogenic precursors as shown by profound inhibition in a clonogenic assay, suggesting their potential therapeutic role *in situ* [19]. In *follicular lymphoma* (FL), low numbers of CD8 TIL are a significant negative prognostic factor of overall survival [37]. Two different types of FL could be identified with regard to their composition of stroma cell infiltrates: FL1 type lymphomas contained a reactive microenvironment consisting mainly of CD8 and CD4 T cells and macrophages and were significantly associated with favorable prognosis. FL2 type lymphomas contained mainly CD57+ NK cells and were associated with a high incidence of adverse clinicobiological manifestations such as "B" symptoms. In breast canrcinomas increased T cell infiltration was correlated with lymph node negative tumor stage [38], while intraepithelial CD8 TIL and a high CD8+/regulatory T cell ratio are associated with a favorable prognosis in ovarian cancer [39]. TIL counts in testicular seminoma and large B cell lymphoma also possessed a favorable prognostic significance [40]. Interestingly, the presence and numbers of melanoma-specific TIL but not of circulating melanoma specific T cells predicted survival in advanced stage melanoma patients [41]. Taken together, these data clearly point to a *prognostic role* of spontaneously induced tumor-reactive T cells in human tumors. The numbers of such cells in the tumor tissue might be gradually regulated through influences of the tumor microenvironment on the efficiency of T cell infiltration and accumulation within the tumor tissue. Besides that, the general capacity of a patient's immune system to induce functional T cell responses and to accumulate such cells in lymphoid organs such as the BM appears to be a fundamental question, since such cells are only detectable in proportions of patients.

3.2.4 The Tumor Microenvironment and the Induction and Function of Tumor Specific T Cells

It has become apparent that development of cancer is not simply a result of genetic alterations within the tumor but is associated with complex changes in host stromal, endothelial and inflammatory cells [42] The development of an invasive cancer involves increasing release of intracellular *(endogeneous) danger signals* from necrotic cells which is associated with the formation of a disordered tumor microenvironment. Such environment is characterized by promotion of angiogenesis and stromal proliferation and influences local immune responses.

It has become evident that cells dying by nonapoptotoc pathways (principally Necrosis) release substances that elicit host responses. Among these are the nuclear protein HMGB1, the S100 family of molecules, purinergic metabolites ATP, AMP adenosine and uric acid and heat shock proteins. Upon their release from the cytosol of necrotic cells they activate respective receptors on immune cells and result in significant immune responses [42]. S100 family members, secreted by macrophages at sites of inflammation are strongly enriched in a variety of tumors. They activate endothelial cells and phagocytes and serve as chemoattractants for tumor-infiltrating leukocytes. Purine metabolites, such as nucleosides and nucleotides interact respectively with specialized P1 and P2 receptors of many immune effector cells. At low concentrations, these molecules enhance recruitment, maturation and emigration of antigen presenting cells via P1 receptors on immature DCs. However, on mature DCs, they are bound by P2 receptors. In this context, simultaneous binding of TLR9 by CpG diminishes the secretion of pro-inflammatory cytokines such as IL-12, IFN-alpha or IL-6 [42]. Uric acid augments the development of antigen specific T cells and maturates DCs for enhanced APC function. Heat shock proteins have been found to drive immune function and immune reactivity. These chaperones bind a broad variety of peptides, including those from TAAs, within the cytosol. Upon their release, they can be recognized by DCs, which leads to their efficient uptake, enhancement of presentation of antigenic peptides and DC maturation.

Lately, we have demonstrated surprisingly high frequencies of CTL in breast cancer patients recognizing HLA-restricted peptides from the matrix-degrading enzyme *heparanase* [43]. Increased expression and secretion of heparanase (Hpa) by tumor cells promotes tumor invasion through extracellular matrices, tissue destruction, angiogenesis, and metastasis. Importantly, heparan-sulfates, the degradation products of heparanase are released at sites of inflammation, wound healing and tissue repair and can act as an endogenous activator of antigen presenting cells [43]. In tumor patients, they may represent a major "danger signal" that induces inflammation and T cell activation at sites of tumor metastasis.

The *interplay between endogenous and exogenous danger signals* determines an immune response during a pathogenic situation. Apoptotic cells can efficiently inhibit responses of immune cells against exogenous danger signals such as LPS or CpG-induced secretion of the pro-inflammatory cytokines IL-12 or TNF-alpha by promoting the secretion of TGFß1 and IL-10. In contrast, the presence of endogenous danger signals released from necrotic cells in the vicinity of exogenous danger signals leads to more vigorous adaptive immune responses [42].

Although several studies report the implication of pro-inflammatory cytokines and chemokines as potentiators of carcinogenesis there is a growing body of evidence demonstrating the power of immunological protection from tumor growth [44]. Production of biologically active molecules by tumor cells includes hormones, prostaglandins, bioamines, NO, lactic acid, neuropeptides, gangliosides, cytokines, chemokines and growth factors [44]. By secretion of distinct patterns of these molecules tumors may be able to regulate the recruitment and activation of antigen presenting cells and thereby modulate the strength and type of a systemic antitumor immune response.

Up to date, little is known about the *role of DCs* in the natural immunosurveillance of cancer, although they are believed to play an essential role in the induction of tumor-specific T cell responses. Tumor-infiltrating DCs most likely play an important role in antitumor T cell immunity. First, increased numbers of tumor infiltrating DCs are associated with improved patient outcome with a variety of human tumors. For instance, in HNSCC, low numbers of tumor-infiltrating DC were a better predictor for poor prognosis than lymph node involvement [44]. Similarly, numbers of $CD83^+$ DCs in liver metastases of colorectal carcinoma positively correlated with improved prognosis [44]. The loss of chemokine CXCL14 expression in human head and neck tumors was associated with decreased DC recruitment and deficient induction of antitumor immune responses [44]. MIP3alpha and GM-CSF were shown to recruit immature DCs to tumor sites and to increase natural antitumor T cell responses in mouse models [44]. Another cytokine which is released in tumor tissues by TIL and also by activated macrophages is TNF-α. TNF-alpha is required for the expression of C-C and CXC chemokines and subsequent recruitment of antigen presenting cells in a mouse model [45]. A prognostic relevance of intratumoral DCs can also be deduced from various observations of efficient tumor rejection after intratumoral DC injection [46]. The infiltration of tumors by DCs is of great importance in initiating the primary antitumor immune response and was confirmed as an independent prognostic parameter for survival in various cancer types.

One reason for the limited efficacy of the immune system to cope with tumors could be due to local suppression of DC, resulting in inhibition of antitumor T cell responses. This assumption is corroborated by frequent observations of reduced stimulatory capacities of circulating DCs from cancer patients. These exhibited reduced levels of HLA-II, CD11c, CD83, CD86 and reduced T cell stimulation capacity.

IL-10 and TGFβ1 are among the best characterized tumor-derived cytokines with immunosuppressive function. IL-10 is produced by many tumor cells and involved in regulating tumor cell proliferation, protection from immune recognition and immunosuppression [45]. IL-10 may inhibit CTL induction by downregulation of HLA class I and-II molecules and of ICAM-1 expression on DCs. The former may be due to an IL-10-mediated downregulation of TAP proteins. Moreover, IL-10 activated macrophages posses increased capacity to engulf apoptotic tumor cells, an anti-inflammatory process that suppresses T cell responses against respective antigens [42]. Recently, it was shown that IL-10 regulates the induction of an indoleamine-2,3-dioxygenase (IDO)-secreting DC subset. Interestingly, in combination with IL-10, IFN-γ also increases IDO secretion. Thus, the presence of IL-10 can turn a proinflammatory signal into an immunosuppressive one [42]. Furthermore, IL-10 has been shown to promote the induction of T regulatory cells which may be also involved in the inhibition of DC-function through the release of TGFß1. High concentrations of TGFß1 are also frequently found in cancer patients and are associated with disease progression and poor responses to immunotherapy. TGFß plays

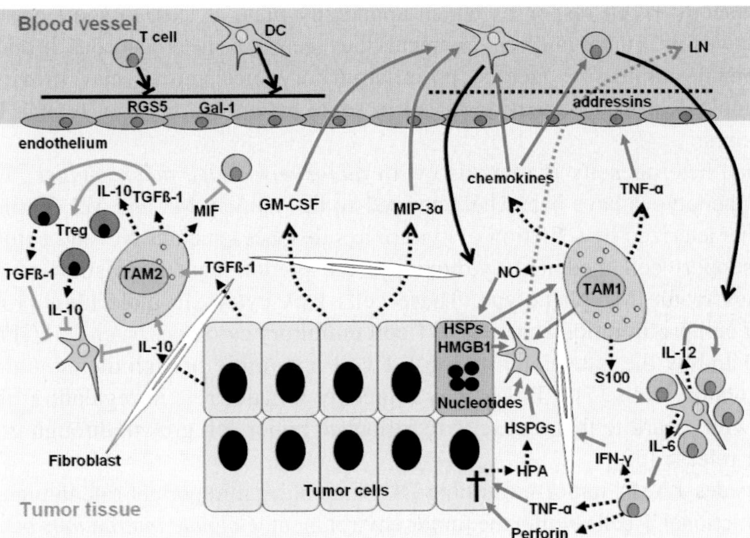

Fig. 3.2 Regulation of antitumor immune responses in the tumor microenvironment. DC; dendritic cell, LN; lymph node, HPA; heparanase, HSPG; heparane sulfate proteoglycan, TAM; tumor associated macrophage, Treg; regulatory T cell, HSPs; heat shock proteins, black arrows; migration into tumor tissue, grey, interrupted arrow; emigration from tumor tissue for T cell priming in lymphoid organs, black, interrupted arrows; release of substance, red arrows; activation of pro-inflammatory pathways, blue arrows; activation of inhibitory factors and pathways, blue lines; inhibition

an important role in regulating the activity of T cells and DCs in the tumor environment. These aspects of the regulation of antitumor immune responses in the tumor microenvironment are illustrated in Fig. 3.2.

The tumor microenvironment regulates antitumor immunity not only at the level of induction of systemic T cell responses but also regulates the infiltration and function *in situ* of tumor specific effector T cells. Numerous studies have demonstrated that tumor-infiltrating T cells exert *reduced functional activity* after their re-isolation from tumor tissue. They fail to be activated by TCR+ anti-CD28 stimulation under conditions that fully activate peripheral blood T cells [46]. TIL from colorectal carcinoma and from melanoma are anergic, express low TCR, perforin and Fas-L [47] and are deficient in perforin-mediated cytolytic activity due to defective microtubule organizing center mobilization and lytic granule exocytosis [48]. This is due to a combination of regulatory mechanisms in the tumor microenvironment and the lack of essential proinflammatory cytokines. T cell anergy due to insufficient B7 costimulation, extrinsic suppression by regulatory cell populations, inhibition by ligands such as programmed death ligand 1, metabolic deregulation by enzymes such as IDO, the action of soluble inhibitory factors such as TGFß1 and IL-10 and tumor derived macrophage migration inhibitory factor (MIF) have been implicated in the generation of this suppressive environment [49, 50].

It has become apparent that in proportions of tumor patients the tumor microenvironment is not immunosuppressive and seems to support the induction of spontaneous T cell responses which apparently play, at least to some extend, a protective role. Tumor microenvironments thus seem to be heterogeneous. In addition to immunosuppressive factors, proinflammatory mechanisms may provide an immunological balance that is shifted in some patients in favor of type 1 T cell responses.

Such heterogeneity is seen also with *tumor-associated macrophages (TAM)*. Two phenotypes have been characterized so far, namely M1 and M2. While M1 are characterized by secretion of cytotoxic substances such as NO and cytokines that support cell mediated cytotoxicity, the predominant population of TAM belongs to the M2 phenotype. These cells lack cytotoxic molecules, promote tumor cell proliferation and secrete T cell inhibitory cytokines, such as TGFß and IL-10 and by the latter inhibit type 1 T cell immunity through the induction of T regulatory cells [51]. TAMs play moreover a major role in regulating fibroblasts with regard to their capacity to influence tumor cell growth through growth factor release [52].

Besides TAMs, tumor-infiltrating DCs may play an important role in maintaining functional T cells within the tumor environment. *Cognate interactions* between antigen-pulsed DCs and tumor-reactive T cells can inhibit apoptosis of T cells. On BM DCs such cognate interactions cause upregulation in the expression of CD83, MHC class II, CD40 and CD86 molecules and secretion of IL-12 and IFN-alpha. Adoptive transfer of breast cancer-reactive memory T cells together with APCs into human breast cancer-bearing NOD/SCID mice led to the generation of intratumoral clusters between transferred DCs, CD4 and CD8 T cells. This was associated with regression of the tumor and prolonged survival of the animals. When such animals had been treated by transfer of reactivated BM T cells without BM-DCs no tumor regression or prolonged survival was observed [21]. Thus, intratumoral, antigen-laden DCs may support the survival and function of tumor-infiltrating T cells. T lymphocytes are defective in cystine uptake and thus require exogenous thiols for activation and function. Upon cognate contact with T cells, antigen-presenting DCs release cysteine into the extracellular space and thereby contribute to a local microenvironment that facilitates immune responses. Il-12 derived from intratumoral DCs may play an essential role for maintaining TIL function, since anergy and non-responsiveness of TIL could be reverted by culture in Il-12 [53, 54] and also by local IL-12 application. Such treatment induced spontaneous rejection of xenotransplanted human tumor pieces in NOD/Scid mice by co-transplanted intratumoral T cells and provoked in congenic tumor mouse models the establishment of protective systemic antitumor immunity. Besides IL-12, IL-6 secreted by TIL may be involved in the generation of a proinflammatory tumor microenvironment by inhibition of tumor-derived TGFß1 and restoration of LAK activity *in situ* [55].

Taken together, tumors are regularly recognized by the immune system and endogenous danger signals released by necrotic tumor cells can promote the induction and accumulation of T cell responses in the bone marrow. The bone marrow parenchyma particularly mediates and supports protective T cell responses against blood borne TAAs and disseminated tumor cells. While type 1 T cell responses seem to occur in virtually all tumor entities, they are generally restricted to proportions of patients. A pronounced heterogeneity in the molecular composition of the tumor microenvironment may determine the net result of either presence or absence of type 1 T cell responses in individual patients. The tumor microenvironment obviously regulates antitumor immunity at several levels: (i) induction of systemic T cell responses, (ii) T cell infiltration and (iii) T cell function. The individual level of pro- and anti-inflammatory factors in the tumor tissue may provide the key to the type of immune response occurring.

3.3 Intervention in Tumor Microenvironment

3.3.1 Counteracting Immune Deregulation

Recent data suggest that the proper trafficking of effector T cells into the tumor microenvironment may not always occur [56, 57]. Furthermore, T cells that effectively home to tumor metastases are often found to be dysfunctional ("effector-phase tolerance"), pointing towards immunosuppressive mechanisms in the tumor microenvironment [58]. Counteracting local immunosuppression might facilitate the effector phase of antitumor immune responses in concert with vaccination or adoptive T cell transfer.

The tumor microenvironment may be altered by modulating tumor cell biology through interference in *signaling pathways*. For example, STAT-3 inhibitors have been shown to augment the expression of chemokines and increase chemotaxis of immune effector cells [59].

The inhibition of immunosuppressive factors (e.g. TGF-β) either by their direct neutralization [60] or by interfering with their transcription (e.g. STAT-3 blockade [61]) was shown in animal models to augment the effect of anti-cancer vaccination.

Intratumoral introduction of *chemokines* through the use of viruses or viral vectors could also serve the purpose of promoting effector cell recruitment. Newcastle Disease Virus (NDV) was shown to induce in tumor cells the expression of IP-10 and RANTES [61], chemokines which recruit NK cells, monocytes and T cells and it also provides a T cell costimulatory signal [62–64]. Introduction of NDV into the tumor or into a tumor vaccine [65] might thus be beneficial. Similarly, introduction of the TNF-superfamily member LIGHT at the tumor site caused a massive T cell infiltration and eradication of established tumors [66]. LIGHT has the dual properties of organizing secondary lymphoid-like structures (including chemokine induction) while simultaneously providing a T cell costimulatory signal [67]. Chemokines may also be of advantage in the tumor microenvironment to recruit antigen-presenting cells such as DCs [68]. In glioblastoma, DC vaccination was able to augment systemic and intra-cranial T cell responses and it was shown that the magnitude of T cell infiltration inversely correlated with the local concentration of TGF-β [69].

Overproduction of *TGF-β* by tumor cells has been shown to be sufficient to drive the generation of Tregs [70]. When tumor-reactive CD8 T cells from mice were isolated, expanded ex vivo and rendered insensitive to TGF-β by introduction of a dominant negative TGF-β type 2 receptor vector and transferred to tumor-bearing mice there was a massive infiltration of CD8 T cells associated with secretion of relevant cytokines, decreased tumor-proliferation, reduced angiogenesis and increased tumor apoptosis [71]. RNA interference (RNAi) techniques could be used as a means of intra-cellular gene "knock-down". Its application for immunotherapy has already given interesting results [72].

Experiments in mouse tumor models have suggested an important role for $CD4^{+}CD25^{+}$ T regulatory cells (*Tregs*) [73] in limiting antitumor immune responses *in vivo* [74]. One strategy to counteract such limitation aims at the depletion of Tregs [74]. This was tested using an IL-2/diphteria toxin fusion protein called ONTAG which was developed for the treatment of CD25 T cell lymphomas [75, 76]. A second approach is the use of anti-CD25 monoclonal antibody coupled magnetic beads which can be used to deplete Tregs from T cell products to be utilized for adoptive transfer protocols. Another strategy may consist of interference with Treg regulatory function or with intra-tumoral migration of Tregs by inhibiting the engagement of chemokine receptors such as CCR4 [77].

TILs are often affected by the tumor micro-environment so that they are incapable of performing cytolytic effector function or even have become anergic. The status of *anergy* may be irreversible or reversible [78, 79]. Strategies that aim at restoration of T cell function include the provision of costimulatory ligands such as CD80 and CD86 [80]. These may be induced upon introduction of LIGHT [66] or by the introduction of viral vectors like recombinant vaccinia virus rV-TRICOM which locally introduces not only CD80 but also the adhesion molecules ICAM-1 and LFA-3 [81]. Such a vector, applied to metastatic melanoma patients induced objective clinical responses in about 30% of patients [82]. Reversion of anergy can also be tried by co-application of anti-CD3 and anti-CD28 signals. These may be attached to a tumor vaccine via bispecific fusion molecules [83]. A tumor vaccine containing anti-CD3 and anti-CD28 bispecific antibodies triggered strong and durable antitumor activity in human lymphocytes [84]. Similarly, super-peptides [85] or bivalent peptides [86] may be capable of delivering a strong signal 1 and additional stimuli via CD4-CD8 interaction.

Another important concept relates to the activation of *antigen-presenting cells* (APCs) in the tumor microenvironment. This can be achieved by Toll-like receptor (TLR) stimulation. TLRs can be activated via molecularly defined ligands such as *imiquimod* or CpG [87]. They can also be activated via bacterial or viral agents which deliver double-stranded (ds) RNA or DNA [88, 89]. Such double-stranded foreign polynucleotides can be recognized by APCs either via cytosolic receptors such as PKR or via endosomal TLRs. Cytosolic RIG-1, in addition, mediates antiviral responses to single-stranded RNA bearing 5′-phosphates [90]. Bacterial and viral components such as superantigens may also be suitable to cause non-specific T cell activation in the tumor microenvironment.

Type I interferons play an important role in host defense and are integrated into both innate and adaptive immunity. This family of cytokines restricts viral spread and is positively linked to the activation and expansion of lymphocytes that are important for control of intracellular infections [91]. IFN-α can have an important influence on DC activation. The central role of DCs in the initiation of immune responses requires that these cells are able to determine the degree of danger in their microenvironment. It was shown that IFN-α is required for maximal secretion of IL-12 and TNF-α by DCs [92]. The secretion and autocrine activity of type I IFN after TLR stimulation enables DCs to orchestrate a hierarchical maturation response with regard to changes in surface phenotype and secretion of cytokines. The activation of nuclear factor κB and p38 pathways in DC can occur either in an additive fashion when DCs are exposed to dual stimulation or can be activated in discrete phases over time. The differential activation of these pathways provides a mechanism for DCs to integrate the activation of multiple stimuli and thus amplify responses. For CD8 T cells encountering MHC-peptide antigens in the context of viral infection, T cell receptor and costimulatory receptor signaling cooperates with type I IFNs to drive their clonal expansion and differentiation.

Intervention with the tumor micro-environment may also be targeted to *tumor stroma* including tumor endothelia. It has been reported that complete tumor rejection by CD8 T cells in murine systems depends on MHC-matched stromal cells infiltrating the tumor [93]. T cell mediated killing of stromal cells alone was successful at long-term control of tumor growth. It is likely that the mechanism of this effect depends on cross-presentation of tumor-derived antigens by stromal cell subpopulations. The threshold for stromal cell killing may be lower than that for tumor cell killing. Initiation of stromal cell destruction by T cells may trigger a sequence of events that leads to better tumor antigen-cross presentation and ultimately better direct tumor cell recognition and lysis. Recently it was shown that local irradiation or a chemotherapeutic drug can cause loading of tumor stroma with cancer antigen within 2 days. This was the optimal time for adoptive T cell transfer to cause eradication of established cancer [94].

Anti-angiogenic compounds such as anti-VEGF monoclonal antibody could theoretically stress tumor cells because of diminished perfusion and thus render T cell mediated tumor cell killing more effective [95]. Inhibition of other stromal support cells including fibroblasts may have a similar effect. PT-100 is an inhibitor of fibroblast activating protein produced by tumor cells [96] that has shown early evidence of clinical efficacy in cancer patients.

3.3.2 Counteracting Metabolic Deregulation

To counteract metabolic deregulation at tumor sites, several therapeutic approaches are being explored. It is likely that tumors subvert natural host metabolic mechanisms (such as *IDO or* arginase [97]) to create tolerance to itself. Therefore, inhibitors of IDO such as 1-methyltryptophan are being purified for study in cancer patients [98]. Inhibition of arginase in the tumor microenvironment produced by mature myeloid cells [99] may improve T cell receptor expression and antigen-specific T cell responses.

Overexpression by malignant cells of *cyclooxigenase* (COX) can lead to high intratumoral levels of prostaglandine-E2. This immunosuppressive prostanoid can block the generation of mature DC starting from intra-tumoral immature DC thus undermining an appropriate TAA presentation to T cells in secondary lymphoid organs. Moreover, COX overexpression might be responsible for malignant cell resistance to apoptotic stimuli and compromise the effectiveness of apoptosis based immune-attack systems. Inhibition of COX activity is thus expected to increase the efficacy of anti-cancer immunotherapy. Another way to counteract immunosuppressive factors is by inhibiting nitric oxide synthase (NOS) [100] and the production of reactive oxygen-species (ROS) which may be produced by tumor-infiltrating macrophages [101]. To counteract the hypoxia [102] of the tumor-microenvironment of established tumors the method of oxygenization via ozone therapy [103] has been established.

3.3.3 Intervention by Physical Means: Radiation, Hyperthermia, Electrochemical Therapy

Ionizing radiation therapy (RT) is an important local modality for the treatment of cancer. RT is largely based on the ability to kill cancer cells by direct cytotoxic effects. A large body of evidence is accumulating on the ability of RT to modify the tumor microenvironment and generate inflammation. This might have far reaching consequences regarding the response of a patient to treatment, especially if radiation induced tumor cell kill were to translate into the generation of effective antitumor immunity. Data from pre-clinical studies provide the proof of principle that different immunotherapeutic strategies can be combined with RT to enhance antitumor effects. An adenovector expressing TNF-α under the control of an irradiation-inducible promoter was developed. A recently published phase I study in patients with solid tumors demonstrated safety and a greater response in lesions treated with a viral vector and RT compared with RT alone [104]. Another clinical trial was designed to examine whether vaccination with Pox-virus encoding prostate specific antigen could be combined with standard external beam RT in patients with prostate cancer. The trial suggests that this combination can generate an antigen cascade with development of T cells directed against other TAAs then those present in the vaccine [105], a phenomenon recently proposed to play a crucial role in determining the therapeutic efficacy of immunotherapy [106, 107].

Hyperthermia application to the site of tumors is another physical means of altering the tumor microenvironment. An elevated local temperature of 39–44 °C can affect tumor cell survival [108] and it can make this environment more susceptible to infiltration by viruses, NK cells, DCs or T-lymphocytes. The attraction of oncolytic viruses in such pre-treated tumor after local or loco-regional virus inoculation may cause NK cell attraction and DC activation due to virus-derived danger signals, interferons, cytokines and chemokines. A rise in temperature as in fever was recently shown to trigger enhanced lymph node recruitment of lymphocytes by augmenting high endothelial vesicle expression of the homing molecules ICAM-1 and CCL21 [109].

Several authors have recently reported encouraging results from *electro-chemical treatment* (EChT) in malignant tumors [110]. The electric field in a tumor microenvironment causes a flux of interstitial water, electro-osmosis, from the anode toward the cathode, since the water-molecules act like a dipole. Consequently, the tissue surrounding the anode dehydrates while edema is obtained around the cathode. Some results suggest that secondary cell destruction is caused by necrosis with cathodic EChT and apoptosis or necrosis with anodic EChT. In China, more than 15,000 patients with various malignant tumors have been treated with this procedure over the last 15 years [111]. The application of electric pulses and currents to a tumor microenvironment can help to increase the effectiveness of uptake of cytotoxic drugs and also that of immune activating agents such as DNA plasmids, TLR activators, etc.

3.3.4 Intra-Tumoral Delivery of Viral Vectors or Slow-release Systems

Viral vectors have been very useful for intra-tumoral application of GM-CSF or CCL16 chemokine in combination with anti-IL-10 and CpG. Introduction of adenoviral vector expressing CD40 L into a murine bladder cancer caused the downregulation of IL-10 and TGF-α and a 200-fold upregulation of IL-12 combined with small tumor regression [112]. Direct intra-tumoral injection of viral vectors may only serve as proof of concept. For clinical application it seems appropriate to develop agents that can be delivered systemically to target tumor metastases. Such agents could be recombinant viruses with substituted ligands that target tumor expressed receptors or it could be antibody constructs that selectively recognize tumor markers and which are complexed with specific chemokines.

Another concept of changing the tumor micro-environment is via introduction of *slow release* cytokine depots. This has been shown with IL-2 gels composed of poly-N-acetylglucosamine-polymers. In malignant mesothelioma this caused the triggering of inflammation and of infiltrating CD8 T cells [113]. Tumor endothelia can be influenced via the secretion of interferon-γ and this has been shown to cause the largest number of gene expression changes in endothelia [114].

3.3.5 Locoregional Interference Via Port Systems

The intervention within the tumor microenvironment by physical, chemical or biological means is relatively easy when the tumor is exposed at the outside of the body, for instance in the skin or subcutaneously. It is more difficult to reach within the body in the case of metastases. In such cases the application of ports, either intravenously or intra-arterially, may be appropriate for systemic or locoregional delivery of chemical or biological substances. This is exemplified in the case of brain tumors where a cytokine immuno-gene therapy has been applied by means of intra-cranial cannula through which tumor cells were inoculated which were modified with either the gene coding for IL-2, IL-4 or TGF-β 2 antisense [115].

3.4 Future Directions

Prospective clinical studies should be carried out to determine if any of the candidate negative regulators can be targeted and is predictive of clinical outcome in response to immune based therapies. Gene expression profiling of metastatic tumors pre- and post-therapy may be an unbiased approach to identifying factors that might predict clinical response. One such study revealed multiple immunologically relevant genes suggesting features of interferon responsiveness and cytolytic potential [116]. With regard to apoptotic or necrotic cell death, a recent provocative hypothesis [42] suggests that most of the derangements that we associate with progression of cancer and the associated immunologic consequences can be ascribed to the consequences of disordered tumor cell death rather than cell growth. Many of the soluble factors released from either apoptotic or necrotic cells are now identifiable in the serum. In the context of this hypothesis such markers might guide the way for future intervention in tumor the microenvironment.

References

1. P. van der Bruggen, C. Traversari, P.Chomez, et al. A gene encoding an antigen recognized by cytolytic T lymphocytes on a human melanoma. *Science* **254**, 1643–1647, (1991).
2. D. Nagorsen, C. Scheibenbogen, F.M. Marincola, et al. Natural T cell immunity against cancer. *Clin Cancer Res* **9**, 4296–4303 (2003).
3. R.M. Steinmann., H. Hemmi. Dendritic cells: translating innate to adaptive immunity. *Curr Top Microbiol Immunol* **311**, 17–58 (2006).
4. H. Wagner. Endogenous TLR ligands and autoimmunity. *Adv Immunol* **91**,159–173 (2006).
5. P.D. Lee, C. Yee, P.A. Savage, et al. Charcterization of circulating T cells specific for tumor-associated antigens in melanoma patients. *Nature Med* **5**, 677–685 (1999).
6. K. Khazaie, S. Prifti, P. Beckhove, A. Griesbach, S. Russell, M. Collins, V. Schirrmacher. Persistence of dormant tumor-cells in the bone marrow of tumor-cell-vaccinated mice correlates with long term immunological protection. *Proc Natl Acad Sci* **91**, 7430–7434 (1994).
7. M. Müller, F. Gounari, S. Prifti, H.J. Hacker, V. Schirrmacher, K. Khazaie. EblacZ tumor dormancy in bone marrow and lymph nodes: active control at proliferating tumor cells by $CD8^+$ immune T cells. *Cancer Res* **58**, 5439–5446 (1998).
8. M. Feuerer, P. Beckhove, L. Bai, et al. Therapy of human tumors in NOD/SCID mice with patient-derived reactivated memory T cells from bone marrow. *Nat Med* **4**, 452–458 (2001).
9. J. Müller-Berghaus, K. Ehlert, S. Ugurel, V. Umansky, V. Schirrmacher, P. Beckhove, D. Schadendorf. Melanoma-reactive T cells in the bone marrow of melanoma patients: association with disease stage. *Cancer Res* **66**(12), 5997–6001 (2006).
10. F. Schmitz-Winnenthal, C. Volk, K. Z'graggen, L. Galindo, D. Nummer, Y. Ziouta, M. Bucur, J. Weitz, V. Schirrmacher, M.W. Büchler, P. Beckhove. High frequencies of functional tumor-reactive T cells in bone marrow and blood marrow and blood of pancreatic cancer patients. *Cancer Res* **65**(21), 10079–10087 (2005).
11. F.H. Schmitz-Winnenthal, L.V. Galindo Escobedo, P. Beckhove, V. Schirrmacher, M. Bucur, Y. Ziouta, C. Volk, B. Schmied, M. Koch, D. Antolovic, J. Weitz, M.W. Büchler, K. Z'graggen. Specific immune recognition of pancreatic carcinoma by patient-derived CD4 and CD8 T cells and its improvement by interferon gamma. *Int J Oncol* **28**(6), 1419–1428 (2006).
12. C. Choi, M. Witzens, M. Bucur, M. Feuerer, N. Sommerfeldt, A. Trojan, A. Ho, V. Schirrmacher, H. Goldschmidt, P. Beckhove. Enrichment of functional CD8 memory T cells specific for MUC1 in bone marrow of multiple myeloma patients. *Blood* **105**(5), 2132–2134 (2005).
13. P. Beckhove, M. Feuerer, M. Dolenc, F. Schuetz, C. Choi, N. Sommerfeldt, J. Schwendemann, K. Ehlert, P. Altevogt, G. Bastert, V. Schirrmacher, V. Umansky. Specifically activated memory T cell subsets from cancer patients recognize and reject xenotransplanted autolgous tumors. *J Clin Invest* **114**, 67–76 (2004).
14. N. Sommerfeldt, F. Schütz, C. Sohn, J. Förster, V. Schirrmacher, P. Beckhove. The shaping of a polyvalent and highly individual T cell repertoire in the bone marrow of breast cancer patients.*Cancer Res* **66**(16), 8258–2365 (2006).
15. M. Feuerer, P. Beckhove, N. Garbi, Y. Mahnke, A. Limmer, M. Hommel, G.J., Hämmerling, B. Kyewski, A. Hamann, V. Umansky, V. Schirrmacher. Bone marrow as a priming site for T cell responses to blood-borne antigen. *Nat Med* **9**, 1151–1157 (2003).
16. L.L. Cavanagh, R. Bonasio, I.B. Mazo, et al. Activation of bone marrow-resident memory T cells by circulating, antigen-bearing dendritic cells. *Nat Immunol* **10**, 1029–1037 (2005).
17. Y.D.Mahnke, J. Schwendemann, P. Beckhove, V. Schirrmacher. Maintenance of long-term tumour-specific T-cell memory by residual dormant tumor cells. *Immunology* **115**, 325–336 (2005).
18. K. Noonan, W. Matsui, P. Serafini, et al. Activated marrow infiltrating lymphocytes effectively target plasma cells and their clonogenic precursors. *Cancer Res* **65**, 2026–2034 (2005).

19. S. Braun, F.D. Vogl, B. Naume, et al. A pooled analysis of bone marrow micrometastasis in breast cancer. *N Engl J Med* **353**, 793–802 (2005).
20. V. Schirrmacher. T-cell immunity in the induction and maintenance of a tumor dormant state. *Sem Cancer Biol* **11**, 285 (2001).
21. L. Bai, P. Beckhove, M. Feuerer, V. Umansky, C. Choi, E.F. Solomayer, I.J. Diel, V. Schirrmacher, Cognate interactions between memory T cells and tumor antigen presenting dendritic cells from bone marrow of breast cancer patients: bi-directional cell stimulation survival and anti-tumor activity in vivo. *Int J Cancer* **103**, 73–83 (2003).
22. G. Angelini, S. Gardella, M. Ardy, et al. Antigen-presenting dendritic cells provide the reducing extracellular microenvironment required for T lymphocyte activation. *PNAS* **99**, 1491–1496 (2002).
23. D. Nagorsen, C. Scheibenbogen, A. Letsch, et al. T cell responses against tumor associated antigens and prognosis in colorectal cancer patients. *J Transl Med* **3**, 3 (2005).
24. P.G. Fournier, J.M. Chirgwin, T.A. Guise. New insights into the vicious circle of bone metastases. *Curr Opin Rheumatol* **18**, 396–404 (2006).
25. R. Ganss, E. Ryschisch, E. Klar, et al. Combination of T-cell therapy and trigger of inflammation induces remodeling of the vasculature and tumor eradication. *Cancer Res* **62**, 1462–1470 (2002).
26. M. M. Berger, G. Bergers, B. Arnold, et al. Regulator of G-protein signaling-5 induction in pericytes coincides with active vessel remodeling during neovascularization. *B Blood* **105**, 1094–1101 (2005).
27. Q. Q. Le, H. Cao, D. Nelson, et al. Galectin -1:A link between tumor hypoxia and tumor immune privilege. *J Clin Oncol* **23**, 8932–8941 (2005).
28. Y. Kawakami, N. Dang, X. Wang, et al. Recognition of shared melanoma antigens in association with major HLA-A alleles by tumour-infiltrating lymphocytes from 123 patients with melanoma. *J Immunother* **23**, 17–27 (2000).
29. H. H. Benlalam, N. Labarriere, B. Linard, et al. Comprehensive analysis of the frequency of melanoma-associated antigen (MAA) by CD8 melanoma infiltrating lymphocytes (TIL): implications for immunotherapy. *Eur J Immunol* **31**, 2007–2015 (2001).
30. P.P.F. Robbins, M. El-Gamil, Y.F. Li, et al. Multiple HLA-II-restricted melanocyte differentiation antigens are recognized by tumour-infiltrating lymphocytes from a patient with melanoma. *J Immunol* **169**, 6036–6047 (2002).
31. S. Seiter, V. Monsurro, M.B. Nielsen, et al. Frequency of MART-1/MelanA and gp100/PMel17-specific T cells in tumnor metastases and cultured tumour-infiltrating lymphocytes. *J Immunother* **25**, 252–263 (2002).
32. M.M.T. Spiotto, H. Schreiber. Rapid destruction of the tumor microenvironment by CTLs recognizing cancer specific antigens cross-presented by stromal cells. *Cancer Immun* **5**, 8 (2005).
33. K.K. Hayashi, K. Yonamine, K. Masuko-Hongo, et al. Clonal expansion of T cells that are specific for autologous ovarian tumour among tumour-infiltrating T cells in humans. *Gynecol Oncol* **74**, 86–92 (1999).
34. M. Koch, P. Beckhove, J. op den Winkel, et al. Tumor infiltrating T-lymphocytes in colorectal cancer: tumor-selective activation and cytotoxic activity in situ. *Ann Surg* **244**, 986–992 (2006).
35. J.J. Galon, A. Costes, F. Sanchez-Cabo, et al. Type, density, and location of immune cells within human colorectal tumors predict clinical outcome. *Science* **313**, 1960–1964 (2006).
36. F.F. Pages, A. Berger, M. Camus, et al. Effector memory T cells, early metastasis and survival in colorectal cancer. *N Eng J Med* **353**, 2654–2666 (2005).
37. T.T. Alvaro, M. Lejeune, M-T. Salvado, et al. Immunohistochemical patterns of reactive microenvironment are associated with clinicobiologic behavior in follicular lymphoma patients. *JCO* **24**, 5350–5357 (2006).
38. X.X. Yan, R. Orentas, B. Johnson. Tumor-derived macrophage migration inhibitory factor (MIF) inhibits T lymphocyte activation. *Cytokine* **33**, 188–198 (2006).

39. E. Sato, S.H. Olson, J. Ahn, et al. Intraepithelial CD8 TIL and a high CD8+ /regulatory T cell ratio are associated with a favourable prognosis in ovarian cancer. *PNAS* **102**, 18538–18543 (2005).
40. Y.Y. Xu, S.H. Kroft, R.W. McKenna, et al. Prognostic significance of tumour infiltrating T lymphocytes and T cell subsets in de novo diffuse large B cell lymphoma: a multiparameter flow cytometry study. *Br J Haematol* **112**, 945–949 (2001).
41. J.B. Haanen, A. Baars, R. Gomez, et al. Melanoma specific TIL but not circulating melanoma specific TIL predict survival in resected advanced-stage melanoma patients. *Cancer Immunol Immunother* **55**, 451–458 (2006).
42. J. Herbert III Zeh, T. Michale Lotze. Addicted to death: Invasive cancer and the immune response to unscheduled cell death. *J Immunotherapy* **28**(1), 1–9 (2005).
43. N. Sommerfeldt, P. Beckhove, Y. Ge, et al. Heparanase: a new metastasis-associated antigen recognized in breast cancer patients by spontaneously induced memory T lymphocytes. *Cancer Res* **66**, 7716–7723 (2006).
44. M.R. Shurin, G.V. Shurin, A. Lokshin, et al. Intratumoral cytokines, chemokines/growth factors and tumour infiltrating dendritic cells: friends or enemies? *Cancer Metastasis Rev* **25**, 333–356 (2006).
45. D.R. Roach, A.G.D. Bean, C. Demangel, et al. TNF regulates chemokine induction essential for cell recruitment, granuloma formation and clearance of mycobacterial infection. *J Immunol* **168**, 4620–4627 (2002).
46. L. Broderick, R.B. Bankert. Memory T cells in human tumor and chronic inflammatory microenvironments: sleeping beauties re-awakened by a cytokine kiss. *Immunol Invest* **35**, 419–436 (2006).
47. F. De Paola, R. Ridolfi, A. Riccobon, et al. Restored T cell activation mechanisms in human tumour infiltrating lymphocytes and colorectal carcinomas after exposure to interleukin-2. *Br J Cancer* **88**, 320–326 (2003).
48. S.S. Radoja, M. Saio, D. Schaer, et al. CD8+ tumour infiltrating T cells are deficient in perforin-mediated cytolytic activity due to defective microtubule-organizing center mobilization and lytic granule exocytosis. *J Immunol* **167**, 5042–5051 (2001).
49. T.T.F. Gajewski, Y. Meng, C. Blank, et al. Immune resistance orchestrated by the tumor microenvironment. *Immunol Rev* **213**, 131–145 (2006).
50. X.X. Yan, R.J. Orentas, B.D. Johnson. Tumor-derived migration inhibitory factor (MIF) inhibits T lymphocyte activation. *Cytokine* **33**, 188–198 (2006).
51. A.A.G. Jarnicki, J. Lysaght, S. Todryk, et al. Suppression of antitumor immunity by IL-10 and TGFß producing T cell infiltrating the growing tumour: influence of tumour microenvironment on the induction of CD4+ and CD8+ regulatory T cells. *J Immunol* **177**, 896–904 (2006).
52. Y. Luo, H. Zhou, J. Krueger, et al. Targeting tumor-associated macrophages as a novel strategy against breast cacner. *JCI* **116**, 2132–2141(2006).
53. L. Broderick, S. Yokota, J. Reineke, et al. Human CD4 effector memory T cells persisting in the microenvironment of lung cancer xenografts are activated by local delivery of IL-12 to proliferate, produce IFN-y and eradicate tumor cells. *J Immunol* **174**, 898–906 (2005).
54. L.O. Broderick, S.P. Brooks, H. Takita, et al. IL-12 reverses anergy to T cell receptor triggering in human lung tumour-associated memory T cells. *Clin Immunol* **118**, 159–169 (2006).
55. Y.W. Hsiao, K.W. Liao, S.W. Hung, et al. Tumour infiltrating lymphocytes secretion of IL-6 antagonizes tumour-derived TGFß1 and restores the lymphokine-activated killing activity. *J Immunol* **172**, 1508–1514 (2004).
56. F.M. Marincola, E. Wang, M. Herlyn, B. Seliger, S. Ferrone. Tumors as elusive targets of T-cell-based active immunotherapy. *Trends Immunol* **24**, 335–342 (2003).
57. V. Monsurro, E. Wang, M.C. Panelli, D. Nagorsen, P. Jin, Z. Katia, et al. Active-specific immunization against melanoma: is the problem at the receiving end? *Semin Biol* **13**, 473–480 (2003).
58. A.B. Frey, Monu Ngozi. Effector-phase tolerance: another mechanism of how cancer escapes antitumor immune response. *J Leukoc Biol* **79**, 652–662 (2006).

59. L. Burdelaya, et al. Stat3 activity in melanoma cells affects migration of immune effector cells and nitric oxide-mediated antitumor effects. *J Immunol* **174**, 3925–3931 (2005).
60. J.J. Kobie, R.S. Wu, R.A. Kurt, et al. Transforming growth factor beta inhibits the antigen-presenting functions and antitumor activity of dendritic cell vaccines. *Cancer Res* **63**, 1860–1864 (2003).
61. B. Washburn, V. Schirrmacher. Human tumor cell infection by Newcastle Disease Virus leads up to upregulation of HLA and cell adhesion molecules and to induction of interferons, chemokines and finally apoptosis. *Int J Oncol* **21**, 85–93 (2002).
62. C. Ertel, N.S. Millar, P.T. Emmerson, V. Schirmracher, P. von Hoegen. Viral hemagglutinin augments peptide specific cytotoxic T-cell responses. *Eur J Immunol* **23**, 2592 (1993).
63. C. Haas, C. Ertel, R. Gerhards, V. Schirrmacher. Introduction of adhesive and costimulatory immune functions into tumor cells by inflection with Newcastle disease virus. *Int J Oncol* **13**, 1105 (1998).
64. C.C. Termeer, V. Schirrmacher, E.B. Bröcker, J.C. Becker. Newcastle disease virus infection induces B7-1/B7-2 independent T-cell costimulatory activity in human melanoma cells. *Cancer Gene Ther* **7**, 316 (2000).
65. V. Schirrmacher. Clinical trials of antitumor vaccination with an autologous tumor cell vaccine modified by virus infection: improvement of patient survival based on improved anti-tumor immune memory. *Cancer Immunol. Immunother* **54**(6) (Apr.), 587–598 (2005).
66. P. Yu, Y. Lee, W. Liu, et al. Priming of naïve T cells inside tumors leads to eradication of established tumors. *Nat Immunol* **5**, 141–149 (2004).
67. B. Aggarwal. Signalling pathways of the TNF superfamily: a double edged sword. *Nat Rev Immunol* **3**, 745–756 (2003).
68. G. Dranoff. GM-CSF-secreting melanoma vaccines. *Oncogene* **22**, 3188–3192 (2003).
69. L.M. Liau, R.M. Prins, S.M. Kiertscher, S.K. Odesa, T.J. Kremen, A.J. Giovannone, J.W. Lin, D.J. Chute, P.S. Mischel, T.F. Cloughesy, M.D. Roth. Dendritic cell vaccination in glioblastoma patients induces systemic and intracranial T-cell responses modulated by the local central nervous system tumor microenvironment. *Clin Cancer Res* **11**(15), 5515–5525 (2005).
70. Y. Peng, Y. Laouar, M.O. Li, E.A. Green, R.A. Flavell. TGF-beta regulates in vivo expansion of Foxp3-expressing CD4+CD25+ regulatory T cells responsible for protection against diabetes. *Proc Natl Acad Sci USA* **101**, 4572–4577 (2004).
71. L. Gorelik, R.A Flavell. Immune-mediated eradication of tumors through the blockade of transforming growth factor-beta signalling in T cells. *Nat Med* **7**, 1118–1122 (2001).
72. C.-P. Mao, C.-F. Hung, T.-C. Wu. Immunotherapeutic strategies employing RNA interference technology for the control of cancers. *J Biomedical Science* DOI 10.1007/s11373-006-9131-5 (2006).
73. S. Sakaguchi. Naturally arising CD4+ regulatory T cells for immunologic self-tolerance and negative control of immune responses. *Annu Rev Immunol* **22**, 531–562 (2004)
74. M.J. Turk, J.A. Guevara-Patino, G.A. Rizzuto, M.E. Engelhorn, S. Sakaguchi, A.N. Houghton. Concomitant tumor immunity to a poorly immunogenic melanoma is prevented by regulatory T cells. *J Exp Med* **200**, 771–782 (2004).
75. F.M. Foss. Interleukin-2 fusion toxin: targeted therapy for cutaneous T cell lymphoma. *Ann NY Acad Sci* **941**, 166–176 (2001).
76. J. Dannull, et al. Enhancement of vaccine-mediated antitumor immunity in cancer patients after depletion of regulatory T cells. *J Clin Invest* **115**, 3623–3633 (2005).
77. I. Lee, L. Wang, A.D. Wells, M.E. Dorf, E. Ozkaynak, W.W. Hancock. Recruitment of Foxp3+ T regulatory cells mediating allograft tolerance depends on the CCR4 chemokine receptor. *J Exp Med* **201**, 1037–1044 (2005).
78. V.A. Boussiotis, et al. Prevention of T cell anergy by signalling through the gamma c chain of the IL-2 receptor. *Science* **266**, 1039–1042 (1994).
79. R.M. Teague, et al. Interleukin-15 rescues tolerant CD8+ T cells for use in adoptive immunotherapy of established tumors. *Nat Med* **12**, 335–341 (2006).
80. H.L. Kaufman, et al. Targeting the local tumor microenvironment with vaccinia virus expressing B7.1 for the treatment of melanoma. *J Clin Invest* **115**, 1903–1912 (2005).

81. H.L. Kaufmann, S. Cohen, K. Cheung, G. DeRaffele, J. Mitcham, D. Moroziewicz, J. Schlom, C. Hesdorffer. Local delivery of vaccinia virus expressing multiple costimulatory molecules for the treatment of established tumors. *Hum Gene Ther* **17**(2), 239–244 (2006).
82. H.L. Kaufman, G. DeRaffele, J. Divito, H. Horig, D. Lee, D. Panicali, M. Voulo. A phase I trial of intralesional rV-Tricom vaccine in the treatment of malignant melanoma. *Hum Gene Ther* **12**(11), 1459–1480 (2001).
83. C. Haas, M. Lulei, P. Fournier, A. Arnold, V. Schirrmacher. T-cell triggering by CD3- and CD28-binding molecules linked to a human virus-modified tumor cell vaccine. *Vaccine* **23**, 2439–2453 (2005).
84. C. Haas, M. Lulei, P. Fournier, A. Arnold, V. Schirrmacher. A tumor vaccine containing anti-CD3 and anti-CD28 bispecific antibodies triggers strong and durable anti-tumor activity in human lymphocytes. *Int J Cance*r **118**(3) (Mar 1), 658–667 (2005).
85. G.G. Parmiani, C. Castelli, P. Dalerba, et al. Cancer immunotherapy with peptide-based vaccines: what have we achieved? Where are we going? *J Natl Cancer Inst* **94**, 805–818 (2002).
86. G.G. Zeng, Y. Li, M. El-Gamil, et al. Generation of NY-ESO-1-specific CD4+ and CD8+ T cells by a single peptide with dual MHC class I and class II specificities: a new strategy for vaccine design. *Cancer Res* **62**, 3630–3535 (2002).
87. Y.Y. Kawarada, R. Ganss, N. Garbi, T. Sacher, B. Arnold, G.J. Hämmerling. NK and CD8+ T cell mediated eradication of established tumors by peritumoral injections of CpG-oligodeoxynucleotides. *J Immunol* **167**, 5247–5253 (2001).
88. C.C. Fiola, B. Peeters, P. Fournier, A. Arnold, M. Bucur, V. Schirrmacher. Tumor selective replication of Newcastle Disease Virus: Association with defects of tumor cells in antiviral defence. *Int J Cancer* **119**, 328–338 (2006).
89. P.P. Fournier, J. Zeng, V. Schirrmacher. Two ways to induce innate immune responses in human PBMCs: paracrine stimulation of IFN-a responses by viral protein or dsRNA. *Int J Oncol* **23**, 673–680 (2003).
90. A.A. Pichlmair, O. Schulz, C.P. Tan, T.I. Näslund, P. Liljeström, Weber Friedemann, C. Reis e Sousa. RIG-I-mediated antiviral responses to single-stranded RNA bearing 5 -phosphates. *Science* **314**, 997–1001 (2006).
91. D.D. Stetson, R. Metzhitov. Type I interferons in host defense. *Immunity* **25**, 373–381 (2006).
92. M.M. Mohty, A. Viall-Castellano, J.A. Nunes, D. Isnardon, D. Olive, B. Gaugler. IFN-alpha skews monocyte differentiation into Toll-like receptor 7-expressing dendritic cells with potent functional activities. *J Immunol* **171**(7), 3385–3393 (2003).
93. M.M.T. Spiotto, H. Schreiber. Rapid destruction of the tumor microenvironment by CTLs recognizing cancer-specific antigens cross-presented by stromal cells. *Cancer Immun* **5**, 8 (2005).
94. B.B. Zhang, N.A. Bowerman, J.K. Salama, et al. Induced sensitization of tumor stroma leads to eradication of established cancer by T cells. *J Exp Med* **204**, 49–55 (2007).
95. S.S. Nair, D. Boczkowski, B. Moeller, M. Dewhirst, J. Vieweg, E. Gilboa. Synergy between tumor immunotherapy and anti-angiogenic therapy. *Blood* **102**, 964–971 (2003).
96. T.T. Kelly. Fibroblast activation protein-alpha and dipeptidyl peptidase IV (CD26): cell-surface proteases that activate cell signalling and are potential targets for cancer therapy. *Drug Resist Updat* **8**, 51–58 (2005).
97. S.S. Mocellin, C.R. Rossi, D. Nitti. Cancer vaccine development: on the way to break immune tolerance to malignant cells. *Exp Cell Res* **299**, 267–278 (2004).
98. C.C. Uyttenhove, et al. Evidence for a tumoral immune resistance mechanism based on tryptophan degradation by indoleamine 2, 3 dioxygenase. *Nat Med* **9**, 1269–1274 (2003).
99. P.P.C. Rodriguez, et al. Arginase I production in the tumor microenvironment by mature myeloid cells inhibits T-cell receptor expression and antigen-specific T-cell responses. *Cancer Res* **64**, 5839–5849 (2004).
100. B.B. Ahn, H. Ohshima. Suppression of intestinal polyposis in Apc (Min/+) mice by inhibiting nitric oxide production. *Cancer Res* **61**, 8357–8360 (2001).

101. S.S.S. Agarwala, M.H. Sabbagh. Histamine dihydrochloride: inhibiting oxidants and synergising IL-2 mediated immune activation in the tumor microenvironment. *Expert Opin Biol Ther* **1**, 869–879 (2001).
102. C.C. Peyssonnaux, R.S. Johnson. An Unexpected Role for Hypoxic Response.*Cell Cycle* **3**, 168–171 (2004).
103. V.V. Bocci, A. Larini, V. Micheli. Restoration of normoxia by ozone therapy may control neoplastic growth: a review and a working hypothesis. *J Altern Complement Med* **11**(2), 257–265 (2005).
104. N.N. Senzer, S. Mani, A. Rosemurgy, et al. TNFerade biologic, an adenovector with a radiation-inducible promoter, carrying the human tumor necrosis factor alpha gene: a phase I study in patients with solid tumors. *J Clin Oncol* **22**, 592–601 (2004).
105. J.J.L. Gulley, P.M. Arlen, N. Bastian, et al. Combining a recombinant cancer vaccine with standard definitive radiotherapy in patients with localized prostate cancer. *Clin Cancer Res* **11**, 3353–3362 (2005).
106. C.C. Lurquin, B. Lethe, E. DePlaen, et al. Constrasting frequencies of antitumor and anti-vaccine T cells in metastases of a melanoma patient vaccinated with a MAGE tumor antigen. *J Exp Med* **201**, 249–257 (2005).
107. S.S. Demaria, N. Bhardwaj, W.H. McBride, S.C. Formenti. Combining radiotherapy and immunotherapy: a revived partnership. *Int J Radiation Oncology Biol Phys* **63**(3), 655–666 (2005).
108. E.E. Vorothnikova, R. Ivkov, A. Foreman, M. Tries, S.J. Braunhut. The magnitude and time-dependence of the apoptotic response of normal and malignant cells subjected to ionizing radiation versus hyperthermia. *Int J Radiat Biol* **82**, 549–559 (2006).
109. Q.Q. Chen, DT Fisher, K.A. Clancy, J.M. Gauguet, W.C. Wang, E. Unger, S. Rose-John, U.H. von Andrian, H. Baumann, S.S. Evans. Fever-range thermal stress promotes lymphocyte trafficking across high endothelial venules via an interleukin 6 trans-signaling mechanism. *Nat Immunol* **7**(12), 1299–1308 (2006).
110. B.B. Tang, L. Li, Z. Jiang, Y. Luan, D. Li, W. Zhang, E. Reed, Q.Q. Li. Characterization of the mechanisms of electrochemotherapy in an in vitro model for human cervical cancer. *Int J Oncol* **26**(3), 703–711 (2005).
111. Y.Y. Xin, F. Xue, B. Ge, F. Zhao, B. Shi, W. Zhang. Electrochemical treatment of lung cancer, *Bioelectromagnetics* **18**, 8–13 (1997).
112. A.A. Loskog, H. Dzojic, S. Vikman, C. Ninalga, M. Essand, O. Korsgren, T.H. Totterman. Adenovirus CD40 ligand gene therapy counteracts immune escape mechanisms in the tumor microenvironment. *J Immunol* **172**, 7200–7205 (2004).
113. I.I. Van Bruggen, D.J. Nelson, A.J. Currie, C. Jackaman, B.W.S. Robinson. Intratumoral Poly-N-acetyl glucosamine-based polymer matrix provokes a prolonged local inflammatory response that, when combined with IL-2, induces regression of malignant mesothelioma in a murine model. *J Immunother* **28**, 359–367 (2005).
114. T.T.R. Sana, M.J. Janatpour, M. Sathe, L.M. McEvoy, T.K. McClanahan. Microarray analysis of primary endothelial cells challenged with different inflammatory and immune cytokines. *Cytokine* **29**, 256–269 (2005).
115. T.T. Lichtor, R.P. Glick. Cytokine immuno-gene therapy for treatment of brain tumors. *J Neuro-Oncol* **65**, 247–259 (2003).
116. E.E. Wang, L.D. Miller, G.A. Ohnmacht, et al. Prospective molecular profiling of melanoma metastases suggests classifiers of immune responsiveness. *Cancer Res* **62**, 3381–3386 (2002).

Chapter 4
Insights into Mechanisms of Immune Resistance in the Tumor Microenvironment through Molecular Profiling

Thomas F. Gajewski

Abstract Many patients with melanoma show spontaneous T cell responses against tumor antigens, and induction or amplification of these T cells responses can frequently be achieved through vaccination or adoptive T cell transfer. However, tumor responses as measured by tumor shrinkage remain infrequent. These observations have argued for analysis of the tumor microenvironment in metastatic melanoma for potential mechanisms of resistance to immune effector function at the level of the target tumor site. This review discusses two categories of regulation at the level of the tumor microenvironment, chemokine-mediated migration of effector T cells and active immune suppression, that have been identified through gene expression profiling of human specimens. Melanoma cell-intrinsic apoptosis also is discussed. The identification of these mechanisms points toward new strategies of intervention to consider for improving the clinical efficacy of T cell-based immunotherapy for cancer, and also suggest that molecular profiling of tumors might be used as a strategy for stratifying patients enrolled on immunotherapy clinical trials.

Keywords Melanoma, gene expression profiling, chemokines, T cells, immune suppression

4.1 Introduction

The large body of accumulated evidence indicating that many, if not most, cancers express antigens that can be recognized by T cells of the immune system has led to a critical question in the field of antitumor immunity. Namely, why are antigen-expressing tumors not spontaneously eliminated through a T cell-mediated immune response? While there are numerous potential mechanisms that could explain this failure, much focus during the past decade has been on the presumption of immunologic ignorance. If a growing tumor fails to prime a tumor antigen-specific T cell response de novo, then active immunization of patients, or adoptive therapy of tumor antigen-specific T cells, should bypass this putative block and achieve tumor

Department of Pathology and Department of Medicine, Section of Hematology/Oncology
University of Chicago, Chicago, IL, USA

E. Yefenof (ed.), *Innate and Adaptive Immunity in the Tumor Microenvironment.*

regression. However, these approaches have met with limited therapeutic success, and it has become clear that increasing the frequencies of functional antitumor T cells is not always sufficient to obtain a clinical response [1, 2]. In addition, spontaneous priming of an antitumor immune response has been observed in patients with melanoma and other cancers, as detected either at the T cell level using peptide/HLA tetramer staining or specific IFN-γ production [1, 3, 4], or at the humoral level by measuring antibodies specific for human tumor cells [5, 6]. Collectively, these and other similar observations have suggested that resistance to an antitumor immune response may frequently occur downstream from immune recognition and initial lymphocyte priming, likely at the level of the tumor microenvironment. Gaining a thorough understanding of the tumor microenvironment as it relates to the host immune response has thus become a significant priority. Gene expressing profiling of metastatic tumor sites in patients with melanoma has provided a rich source of information that is beginning to be confirmed mechanistically in preclinical and clinical contexts. Additional detailed understanding of molecular features of the tumor cells themselves is providing another level of information that highlights alternative mechanisms of immune resistance. These analyses are now shedding light on new potential interventions designed to overcome these mediators of immune escape at the level of the tumor microenvironment, and also is suggesting features that could guide patient selection in the setting for eligibility of T cell-based immunotherapy clinical trials.

4.2 Melanoma Microenvironment Analysis Through Gene Expression Profiling

One strategy that has been pursued to gain a broad array of information regarding the melanoma tumor microenvironment and potential associations with clinical outcome to immunotherapies is gene expression profiling. Marincola and colleagues were the first to investigate whether patterns of transcripts in the melanoma tumor site might correlate with clinical response in patients. Using cDNA arrays applied to retrospective fine needle aspirate samples, they identified a set of transcripts present in pretreatment lesions that appeared to be linked with a subsequent response to therapy. These included multiple immunologically relevant genes, including EBI3, TIA1, IRF2, and IFI27, suggesting features of interferon (IFN) responsiveness and cytolytic potential [7]. The expression of interferon-response genes suggests the possibility that innate immune signals mediated through host type I IFNs might contribute to generating a supportive tumor microenvironment for an adaptive immune response. This hypothesis is currently being explored in preclinical models and is consistent with prior mouse model experiments indicated the importance of host Stat1 signaling for the induction of antitumor CTL [8].

In our own laboratory we have utilized the Affymetrix oligonucleotide array platform to analyze a series of melanoma biopsies and cell lines. We opted to study samples obtained by excisional or core biopsy (as opposed to fine needle aspiration) in an attempt to reflect more accurately the totality of the microenvironment. The melanoma cell lines were included to obtain inferential information regarding the likely expression of genes in the melanoma tumor cells themselves versus the stromal compartment. Non-supervised hierarchical clustering revealed two major groups, with a second major division within one group, generating three major clusters. Interestingly, the differences between these three groups were largely accounted for by differential expression of immunologically relevant transcripts. In particular, one cluster of samples expressed transcripts unique to T cells and B cells, a second cluster expressed some inflammatory genes but lacked a lymphocyte signature, and the third cluster was characterized by absence of these inflammatory transcripts (Harlin et al., manuscript submitted). Thus, not all melanoma metastases are the same, and a major distinction seems to be the presence or absence of lymphocyte recruitment.

A second critical observation in this study was a transcriptional profile suggesting an immunosuppressive microenvironment in the tumors that do contain T cells. The tryptophan-catabolizing enzyme indoleamine-2,3-dioxygenase (IDO) was most frequently found in the T cell-rich sample cluster. Interestingly, a second amino acid catabolizing gene, Arginase I, was found in a second group of tumors, indicating that these two factors are often expressed in a mutually exclusive set of tumors. We also saw expression of the inhibitory ligand, PD-L1/B7-H1, and the presence of FoxP3$^+$ regulatory T cells (Tregs) in the T cell-containing tumors. These observations suggest that a rich network of immunosuppressive factors operates in concert to control the effector phase of the antitumor immune response in melanoma patients. Also noteworthy is the relative absence of expression of the T cell costimulatory ligands B7-1 and B7-2, suggesting the possibility of an anergy-promoting environment as well. These features of the metastatic melanoma microenvironment that are most relevant to the host immune response will be discussed in more detail below.

4.3 Regulation of Migration into Tumor Metastases

A striking difference between clustered melanoma metastases was the presence or absence of a lymphocyte gene expression signature. It is critical to perform confirmatory assays to determine whether this difference is truly reflected at the level of T cell involvement. In this case, differential presence of CD8$^+$ T cells was confirmed by immunohistochemistry. In some instances, T cells were extremely abundant and were uniformly distributed throughout the tumor mass. This result suggests that some tumors have the capability to recruit T cells and others do not. One might imagine that even if a patient developed circulating tumor antigen-specific T cells as measured in the blood, if homing into tumor sites did not occur then tumor regression would be unlikely to follow. Another implication of this result is, assuming a subset of these T cells is antigen specific as we and others have observed in similar cases previously [9–11], then it seems likely that the tumor microenvironment has rendered the T cells dysfunctional.

Fig. 4.1 Chemokines found differentially expressed in melanoma metastases that contain or lack CD8⁺ T cells. This list was extracted from gene expression profiling data as being statistically different between the three tumor categories. Group 1 contained T cell-specific transcripts indicating the presence of T cells in the tumor specimen.

- Group 1
  - CCL2, CCL3, CCL4, CCL5, CCL19, CCL21
  - CXCL9, CXCL10, CXCL11, CXCL13
  - CXCL8, CXCL12
- Group 2
  - CXCL14
  - CXCL8, CXCL12
- Group 3
  - CXCL8, CXCL12

It was of interest to understand in more detail the possible mechanism for differential migration of T cells into melanoma metastases. Correlating with the presence of T cells was expression of a broad array of chemokine genes (Fig. 4.1). These data were confirmed by real-time RT-PCR and for a subset of them by protein array on tumor lysates. Based on the chemokine receptors upregulated on $CD8^+$ effector T cells and known chemokine-receptor interactions, the list of candidate chemokines was narrowed to 6 (MIP-1α, MIP-1β, MCP-1, Mig, IP-10, and RANTES). Each of these chemokines was found to be sufficient for recruiting $CD8^+$ effector T cells in a transwell system *in vitro*. Thus, it seems likely that a cooperative activity of these 6 chemokines helps to support recruitment of activated $CD8^+$ T cells into tumor sites *in vivo*. Absence of these chemokines might therefore represent a barrier that precludes T cell recruitment and implies that inadequate T cell migration may be defined as an immune evasion mechanism.

Previous studies of primary melanoma lesions support a chemokine correlation with T cell infiltrates. It is well known that primary cutaneous melanomas sometimes have a brisk T cell infiltrate yet in other cases show a complete absence of T cells. Immunohistochemical staining has shown an association of T cell infiltration with expression of the chemokines Mig/CXCL9 and IP-10/CXCL10 [12]. Interestingly, these factors appeared to be produced by monocyte/macrophage-like cells in the tumor microenvironment, suggesting that the phenotype of this stromal cell component might help dictate the nature of the inflammatory infiltrate in melanoma tumors.

Mouse models have been utilized to explore whether introduction of chemokines into the tumor microenvironment may improve T cell migration *in vivo*. Expression of CCL2/MCP-1, CCL3/MIP-1α, CCL5/RANTES, CCL21/SLC, and CXCL10/IP-10 by tumor cells have each been found to improve antitumor immunity in various mouse tumor models [13, 14]. An interesting alternative to consider is the TNF superfamily member LIGHT. In binding to the LTβR, LIGHT triggers expression of a broad range of chemokines from stromal cells, including CCL21 and CCL3, and drives generation of a secondary lymphoid-like structure [15]. In tumor models, expression of LIGHT has been shown to promote recruitment of both naïve and activated $CD8^+$ T cells and support vigorous tumor rejection *in vivo* [16]. For clinical translation, use of a viral vector to transfer expression of LIGHT in tumor sites following direct injection might be considered. Preclinical experiments using an adenoviral vector encoding murine LIGHT have shown promising results, yielding not only increased T cell recruitment in the injected tumor but also control of micrometastatic disease (Yu et al, J Immunol., In Press). Thus, use of LIGHT to modify the tumor microenvironment is attractive to consider for clinical translation.

4.4 Inhibitory Mechanisms in the Tumor Microenvironment

Migration of activated T cells into tumor sites may not always be sufficient, as infiltrating T cells must retain their effector functions at the tumor site in order to be efficacious. The finding that T cells in some cases are already present at tumor sites, and even penetrated well into the tumor mass, raises the key question of why the tumor cells are not then killed through expected cytotoxicity mechanisms. One possibility is that the T cells are rendered dysfunctional through regulatory signals delivered *in situ* within the tumor microenvironment. Recent data from several groups including our own have demonstrated that T cells in tumor sites do have an activated surface phenotype, and include $CD8^+$ T cell reactive against defined tumor antigen peptide/HLA tetramers [9–11]. However, in several cases these same T cells have been shown not to express granules containing granzyme B or perforin, indicating a deficiency in cytolytic potential. In addition, cytokine production and proliferation upon analysis ex vivo also has been found to be defective. Similar results have been seen in murine tumor models and argue that T cell dysfunction at tumor sites may be a frequent occurrence [17, 18]. Of note, the expansion of so-called tumor-infiltrating lymphocytes (TIL) for adoptive transfer *in vivo* is based upon the existence of specific T cells that have localized to tumor sites [19]. These cells clearly can recover functionality following expansion *in vitro*, arguing that the apparent dysfunction is reversible. This recovery includes cytokine producing capability but also acquisition of cytotoxic granules. We recently have observed that the re-acquisition of granzyme B-containing granules in antigen experienced $CD8^+$ T cells can be driven by CD28 costimulation and is dependent upon cell cycle progression [20]. Thus, supporting the proliferation of TILs may be central in helping them to regain critical effector functions.

In our gene expression profiling studies, the T cell-rich metastatic melanoma subset was noted to have frequent expression of IDO. Confirmatory immunohistochemistry revealed expression of IDO protein in endothelial cells and also in infiltrating macrophage/dendritic-like cells. In concert with IDO, those tumors also contained PD-L1 expressed by the tumor cells themselves, as well as FoxP3 expressed by infiltrating lymphocytes. Other groups have also observed abundant Tregs in melanoma metastases [21]. Lieping Chen's laboratory has reported that 100% of melanomas examined express PD-L1 [22], supporting these results. Thus, multiple cell lineages appear to contribute to the immunosuppressive nature of the metastatic melanoma microenvironment. This observation also suggests that combined elimination of multiple negative regulatory mechanisms may be necessary to reveal optimal T cell effector function and gain maximal immune-mediated tumor control.

Preclinical models have been utilized to explore the mechanistic contribution of these potential barriers to tumor resistance, and at the same time to examine strategies to overcome them in an attempt to improve tumor rejection. For inhibition of

IDO, the inhibitor 1-methyltryptophan has received the most attention. Administration of 1-methyltryptophan *in vivo* has been shown to improve immune-mediated control of tumors transfected to express IDO [23]. Although IDO expression in melanoma tumors appears to be by stromal cells rather than by tumor cells, this strategy is expected to be effective in the physiologic scenario as well. For interfering with PD-L1/PD-1 interactions, several model experimental approaches have been taken. In our own laboratory, we have utilized PD-1-deficient mice to study tumor rejection. T cells obtained from transgenic mice expressing a TCR specific for a model antigen crossed onto the PD-1-deficient background were superior at tumor rejection *in vivo* [24]. This was associated with improved cytokine production and cytolysis directly at the effector phase. Neutralizing antibodies against PD-L1 also have been examined and have been found to improve T cell-mediated tumor control in mice [25].

T cell anergy is a third potential mechanism of immune escape to be explored in murine systems. Three general categories of approach have been evaluated to prevent or reverse anergy: manipulation of the T cells, alteration of the host environment, or modification of the tumor microenvironment. For influencing T cell biology directly, a better understanding of the molecular mechanisms of the anergic state has been required. Our group had previously shown that anergic T cells showed defective activation of the Ras/MAP kinase pathway in response to TCR/coreceptor ligation [26]. More recently, this defect has been correlated with upregulated expression of the lipid kinase DGK, which blunts RasGRP-mediated Ras activation in anergic T cells [27]. Thus, one attractive target to manipulate in an attempt to restore function of anergic cells is DGK. To extend this line of reasoning, inhibition of other negative regulatory signaling molecules in T cells also could be considered for potentiation of T cell function in the tumor context. Ongoing investigations are testing these notions in mouse tumor models *in vivo*.

As an approach to alter the host environment to counter T cell anergy, lymphopenia-induced homeostatic proliferation has been explored. Previous work had shown that T cells rendered anergic *in vitro* could recover function after proliferation in response to IL-7 [28]. Creating a lymphopenic environment liberates endogenous IL-7 and IL-15, cytokines which support so-called homeostatic proliferation of transferred T cells. Indeed, we recently have observed that homeostatic proliferation in lymphopenic recipients could restore and/or maintain $CD8^+$ T cell function and promote tumor rejection [29]. These data are consistent with those of Turka and colleagues who found that induction of $CD4^+$ T cell tolerance *in vivo* was prevented in the context of homeostatic proliferation [30]. Direct activity of exogenous IL-15 also has been reported to recover function of *in vivo* tolerized $CD8^+$ T cells [31].

As a strategy to manipulate the tumor microenvironment directly to prevent and/or reverse T cell anergy, perhaps the most straightforward approach is to introduce expression of B7 costimulatory ligands. Transfection of tumor cells to express B7-1 has been tested by multiple investigators and shown to support rejection of multiple different tumor cell types in mice [32, 33]. While the initial presumption was that this effect was mediated through more efficient induction of antitumor T cells, subsequent data supported the notion that B7-1 expression also acts at the effector phase of the antitumor T cell response [34], presumably to support maintenance of T cell function.

Finally, interfering with the suppressive effect of FoxP3+ Tregs in tumor models also has been investigated. Depletion with anti-CD25 mAbs alone has been shown to slow growth of several tumors, including B16 melanoma [35, 36]. Ex vivo depletion of CD25+ T cells from T cell products prior to adoptive transfer *in vivo* also can be considered. Like the anergy story, additional manipulations of Tregs may become possible as the details mechanisms by which they suppress effector T cells are uncovered. For example, Rongfu Wang's laboratory has reported that human CD4+ CD25+ Tregs express TLR8 and that their suppressive activity can be reversed upon exposure to a TLR8 ligand comprised of a poly-guanidine oligonucleotide [37]. All together, these preclinical experiments have helped to validate the inhibitory capability of IDO, PD-1, anergy, and Tregs on the antitumor immune response, and also point towards potential strategies to counter them for clinical translation.These putative mechanisms of tumor escape from immune attack are represented diagrammatically in Figure 4.2.

There are striking similarities between the negative regulatory mechanisms at play in the tumor microenvironment and those involved in limiting inflammation during

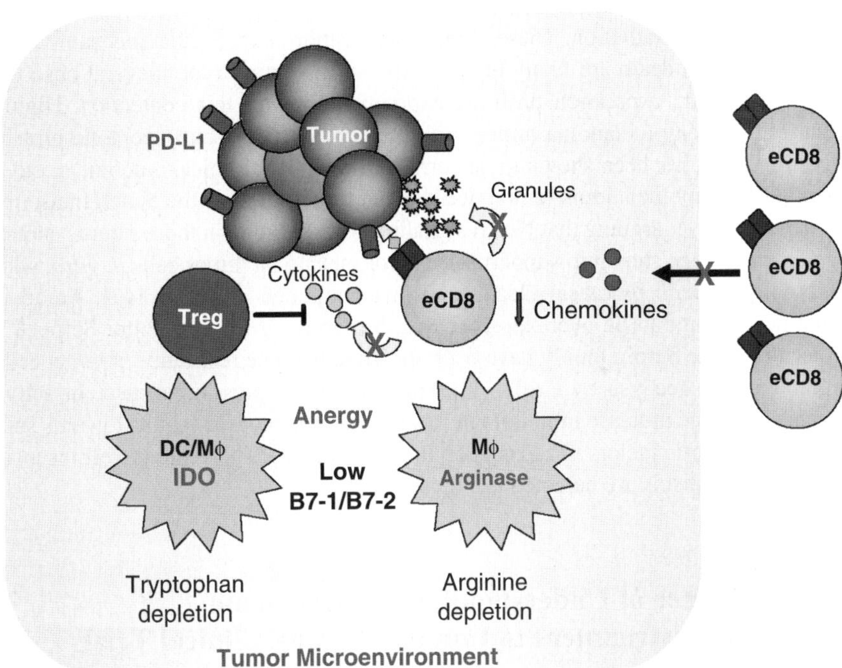

Fig. 4.2 Identified putative mechanisms of resistance in melanoma metastases that contain CD8+ T cells. The tumor microenvironment can contain low chemokine expression which may decrease migration of T cells that could recognize and kill tumor cells. The microenvironment can also contain cells that express IDO or arginase, which deplete critical amino acids for T cell function. APCs in the microenvironment also may be low for B7 expression, which would promote T cell anergy. Presence of FoxP3+ Tregs, and tumor cell expression of PD-L1, also may contribute to inhibition of T cell effector function.

chronic infection. In both HIV and CMV models, chronically stimulated $CD8^+$ T cells express PD-1 and are functionally limited by PD-1 ligation [38–40]. An abundance of $FoxP3^+$ Tregs has been observed in HIV-infected lymph nodes [41], and HIV also can induce expression of IDO [42]. Thus, it is plausible that the involvement of these negative regulatory pathways in the tumor context may be immune-intrinsic and become recruited into action in many different situations of chronic antigen exposure. Another reference model for natural immune suppression preventing tissue rejection is the placenta microenvironment. Each of the above described mechanisms present in tumor sites has been found to contribute to maternal tolerance to the semi-allogeneic fetus [43–45]. These parallels support the notion that much of the immune escape in tumor sites is likely a consequence of normal immune regulatory mechanisms rather than by factors unique to malignant tumors per se.

4.5 Resistance at the Level of the Tumor Cells Themselves

Melanoma in general is quite resistant to killing by cytotoxic chemotherapy and conventional doses of radiation. These clinical observations suggest that mechanisms of resistance to cell death are likely highly active in this tumor type. Several observations in our gene expression profiling experiment support this contention. Highly expressed in many melanoma tumors was the gene encoding anti-apoptotic protein survivin, which has been shown to be correlated with poor clinical outcome in other series [46]. Many melanoma tumors were found to overexpress the Notch transcriptional target Hey1, arguing that Notch signaling is constitutive in those tumors. Notch signaling has been shown to support survival of melanoma tumor cells *in vitro*, with inhibition of Notch processing leading to melanoma cell apoptosis [47]. An additional factor found to be overexpressed was the serine protease inhibitor SerpinA3. Molecules in the Serpin family have been shown to mediate resistance of target cells to granule-mediated lysis by T cells [48]. Thus, elevated expression of survivin, active Notch, and serine protease inhibitors in fresh melanoma biopsies points towards specific anti-apoptotic factors to focus upon in future studies. These tumor cell-intrinsic survival mechanisms are depicted in Figure 4.3.

4.6 Relevance of Understanding the Melanoma Microenvironment to Immunotherapy Clinical Trials

There are two principal implications of the results uncovered by studies of the metastatic melanoma tumor microenvironment. First, it is plausible to perform pretreatment tumor biopsies to examine features of the tumor microenvironment that may correlate with clinical response to melanoma vaccines. Identifying a molecular signature linked to clinical outcome should ultimately be able to guide patient selection for such studies. We have performed such a prospective analysis

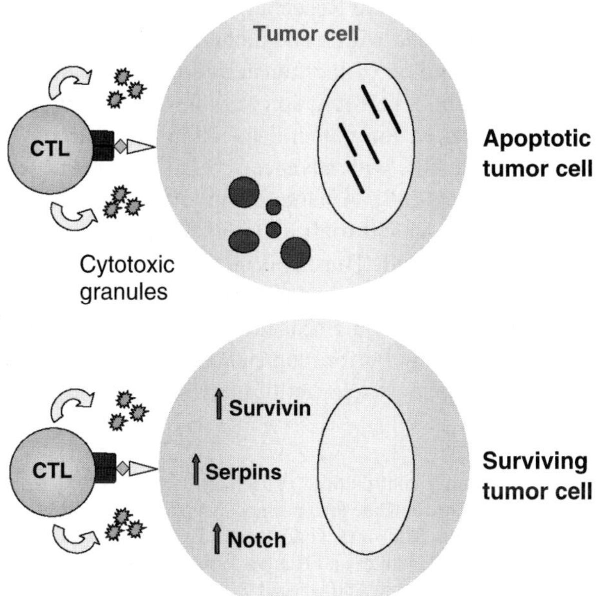

Fig. 4.3 Tumor cell-intrinsic resistance to T cell-mediated apoptosis. CTL-mediated lysis of tumor cells can be vigorous in the absence of potent anti-apoptotic pathways (upper panel). It is speculated that upregulated expression of survivin, serpins, and Notch signaling may generate melanoma cells that are relatively resistant to apoptosis triggered by CTL granule exocytosis (lower panel).

with 19 metastatic melanoma patients participating in a multipeptide melanoma vaccine trial. A small subset of transcriptional differences in the tumor biopsies was found to distinguish clinical responders from non-responders. Interestingly, among these was a set of chemokine genes more highly expressed in the responding patients (Gajewski et al., unpublished data). While this is a small sample size that will require confirmation, it is beginning to connect the biology of melanoma and the antitumor immune response with clinically relevant information.

The second implication of such gene expression profiling is to identify the relevant barriers that can be overcome with new immunotherapeutic interventions. Ultimately, one can envision an individualized patient therapy in which a tumor is biopsied to determine major potential barriers to effective antitumor immunity (e.g. absence of chemokines, over overexpression of specific negative regulatory mechanisms) then instituting the appropriate therapy to overcome those barriers in concert with active immunization or adoptive T cell transfer. Some such combination strategies are already being explored clinically, albeit with less selected patients. Vieweg and colleagues were the first to report on the use of Ontak (an IL-2-diptheria toxin fusion protein) to decrease the numbers of $CD4^+$ $CD25^+$ $FoxP3^+$ Tregs in patients in combination with a dendritic cell-based vaccine [49].

The TIL adoptive transfer approach of Rosenberg and colleagues has been found to be most effective when combined with a chemotherapy conditioning regimen that supports homeostatic proliferation of the transferred cells [19], which preclinical data argues helps to maintain T cell function and prevent anergy induction [29]. Howard Kaufman has explored intratumoral injection of a vaccine virus vector encoding human B7-1 in patients with advanced melanoma, and observed clinical activity in a subset of patients [50]. A neutralizing anti-PD-1 mAb developed by Medarex is entering phase I clinical trials soon, and clinical-grade 1-methyltryptophan also is being manufactured. Thus, it is conceivable that antagonizing PD-L1/PD-1 interactions and IDO activity in combination with vaccines or adoptive T cell transfer also may be possible in the near future. Individualizing such therapies should ultimately be possible by performing molecular analyses on pretreatment tumor biopsies to identify the most relevant tumor resistance mechanisms in individual cases.

Acknowledgments The laboratory contributions of Helena Harlin, Yuru Meng, Mark McKee, Craig Slingluff, Ian Brown, and Justin Kline are enormously appreciated. I also acknowledge the technical contributions of Alpana Sahu and Todd Kuna. This work relied heavily upon the University of Chicago Functional Genomics and Human Immunologic Monitoring core facilities, and was supported by NIH grants R01 CA90575 and P01 CA97296, a Translational Research Award from the Burroughs Wellcome Fund, and as a grant from the Ludwig Trust.

References

1. Peterson, A. C., H. Harlin, and T. F. Gajewski. 2003. Immunization with Melan-A peptide-pulsed peripheral blood mononuclear cells plus recombinant human interleukin-12 induces clinical activity and T-cell responses in advanced melanoma. *J Clin Oncol 21:2342–2348.*
2. Rosenberg, S. A., R. M. Sherry, K. E. Morton, W. J. Scharfman, J. C. Yang, S. L. Topalian, R. E. Royal, U. Kammula, N. P. Restifo, M. S. Hughes, D. Schwartzentruber, D. M. Berman, S. L. Schwarz, L. T. Ngo, S. A. Mavroukakis, D. E. White, and S. M. Steinberg. 2005. Tumor progression can occur despite the induction of very high levels of self/tumor antigen-specific CD8+ T cells in patients with Melanoma. *J Immunol 175:6169–6176.*
3. Valmori, D., D. Lienard, G. Waanders, D. Rimoldi, J. C. Cerottini, and P. Romero. 1997. Analysis of MAGE-3-specific cytolytic T lymphocytes in human leukocyte antigen-A2 melanoma patients. *Cancer Res 57:735–741.*
4. Valmori, D., V. Dutoit, V. Rubio-Godoy, C. Chambaz, D. Lienard, P. Guillaume, P. Romero, J. C. Cerottini, and D. Rimoldi. 2001. Frequent cytolytic T-cell responses to peptide MAGE-A10 (254–262) in melanoma. *Cancer Res 61:509–512.*
5. Jager, E., E. Stockert, Z. Zidianakis, Y. T. Chen, J. Karbach, D. Jager, M. Arand, G. Ritter, L. J. Old, and A. Knuth. 1999. Humoral immune responses of cancer patients against "Cancer-Testis" antigen NY-ESO-1: correlation with clinical events. *Int J Cancer 84:506–510.*
6. Chen, Y. T., A. O. Gure, S. Tsang, E. Stockert, E. Jager, A. Knuth, and L. J. Old. 1998. Identification of multiple cancer/testis antigens by allogeneic antibody screening of a melanoma cell line library. *Proc Natl Acad Sci USA 95:6919–6923.*
7. Wang, E., L. D. Miller, G. A. Ohnmacht, S. Mocellin, A. Perez-Diez, D. Petersen, Y. Zhao, R. Simon, J. I. Powell, E. Asaki, H. R. Alexander, P. H. Duray, M. Herlyn, N. P. Restifo, E. T. Liu, S. A. Rosenberg, and F. M. Marincola. 2002. Prospective molecular profiling of melanoma metastases suggests classifiers of immune responsiveness. *Cancer Res 62:3581–3586.*

8. Fallarino, F., and T. F. Gajewski. 1999. Cutting edge: differentiation of antitumor CTL in vivo requires host expression of Stat1. *J Immunol 163:4109–4113.*
9. Harlin, H., T. V. Kuna, A. C. Peterson, Y. Meng, and T. F. Gajewski. 2006. Tumor progression despite massive influx of activated CD8(+) T cells in a patient with malignant melanoma ascites. *Cancer Immunol Immunother:1–13.*
10. Zippelius, A., P. Batard, V. Rubio-Godoy, G. Bioley, D. Lienard, F. Lejeune, D. Rimoldi, P. Guillaume, N. Meidenbauer, A. Mackensen, N. Rufer, N. Lubenow, D. Speiser, J. C. Cerottini, P. Romero, and M. J. Pittet. 2004. Effector function of human tumor-specific CD8 T cells in melanoma lesions: a state of local functional tolerance. *Cancer Res 64:2865–2873.*
11. Mortarini, R., A. Piris, A. Maurichi, A. Molla, I. Bersani, A. Bono, C. Bartoli, M. Santinami, C. Lombardo, F. Ravagnani, N. Cascinelli, G. Parmiani, and A. Anichini. 2003. Lack of terminally differentiated tumor-specific CD8+ T cells at tumor site in spite of antitumor immunity to self-antigens in human metastatic melanoma. *Cancer Res 63:2535–2545.*
12. Kunz, M., A. Toksoy, M. Goebeler, E. Engelhardt, E. Brocker, and R. Gillitzer. 1999. Strong expression of the lymphoattractant C-X-C chemokine Mig is associated with heavy infiltration of T cells in human malignant melanoma. *J Pathol 189:552–558.*
13. Feldman, A. L., J. Friedl, T. E. Lans, S. K. Libutti, D. Lorang, M. S. Miller, E. M. Turner, S. M. Hewitt, and H. R. Alexander. 2002. Retroviral gene transfer of interferon-inducible protein 10 inhibits growth of human melanoma xenografts. *Int J Cancer 99:149–153.*
14. van Deventer, H. W., J. S. Serody, K. P. McKinnon, C. Clements, W. J. Brickey, and J. P. Ting. 2002. Transfection of macrophage inflammatory protein 1 alpha into B16 F10 melanoma cells inhibits growth of pulmonary metastases but not subcutaneous tumors. *J Immunol 169:1634–1639.*
15. Wang, J., and Y. X. Fu. 2004. The role of LIGHT in T cell-mediated immunity. *Immunol Res 30:201–214.*
16. Yu, P., Y. Lee, W. Liu, R. K. Chin, J. Wang, Y. Wang, A. Schietinger, M. Philip, H. Schreiber, and Y. X. Fu. 2004. Priming of naive T cells inside tumors leads to eradication of established tumors. *Nat Immunol 5:141–149.*
17. Koneru, M., D. Schaer, N. Monu, A. Ayala, and A. B. Frey. 2005. Defective proximal TCR signaling inhibits CD8+ tumor-infiltrating lymphocyte lytic function. *J Immunol 174:1830–1840.*
18. Radoja, S., M. Saio, D. Schaer, M. Koneru, S. Vukmanovic, and A. B. Frey. 2001. CD8(+) tumor-infiltrating T cells are deficient in perforin-mediated cytolytic activity due to defective microtubule-organizing center mobilization and lytic granule exocytosis. *J Immunol 167:5042–5051.*
19. Rosenberg, S. A., and M. E. Dudley. 2004. Cancer regression in patients with metastatic melanoma after the transfer of autologous antitumor lymphocytes. *Proc Natl Acad Sci USA.*
20. Meng, Y., H. Harlin, J. P. O'Keefe, and T. F. Gajewski. 2006. Induction of cytotoxic granules in human memory CD8+ T cell subsets requires cell cycle progression. *J Immunol 177:1981–1987.*
21. Viguier, M., F. Lemaitre, O. Verola, M. S. Cho, G. Gorochov, L. Dubertret, H. Bachelez, P. Kourilsky, and L. Ferradini. 2004. Foxp3 expressing CD4+CD25(high) regulatory T cells are overrepresented in human metastatic melanoma lymph nodes and inhibit the function of infiltrating T cells. *J Immunol 173:1444–1453.*
22. Dong, H., and L. Chen. 2003. B7-H1 pathway and its role in the evasion of tumor immunity. *J Mol Med 81:281–287.*
23. Uyttenhove, C., L. Pilotte, I. Theate, V. Stroobant, D. Colau, N. Parmentier, T. Boon, and B. J. Van Den Eynde. 2003. Evidence for a tumoral immune resistance mechanism based on tryptophan degradation by indoleamine 2,3-dioxygenase. *Nat Med 9:1269–1274.*
24. Blank, C., I. Brown, A. C. Peterson, M. Spiotto, Y. Iwai, T. Honjo, and T. F. Gajewski. 2004. PD-L1/B7H-1 inhibits the effector phase of tumor rejection by T cell receptor (TCR) transgenic CD8+ T cells. *Cancer Res 64:1140–1145.*

25. Iwai, Y., M. Ishida, Y. Tanaka, T. Okazaki, T. Honjo, and N. Minato. 2002. Involvement of PD-L1 on tumor cells in the escape from host immune system and tumor immunotherapy by PD-L1 blockade. *Proc Natl Acad Sci USA 99:12293–12297.*
26. Fields, P. E., T. F. Gajewski, and F. W. Fitch. 1996. Blocked Ras activation in anergic CD4+ T cells [see comments]. *Science 271:1276–1278.*
27. Zha, Y., R. Marks, A. W. Ho, A. C. Peterson, S. Janardhan, I. Brown, K. Praveen, S. Stang, J. C. Stone, and T. F. Gajewski. 2006. T cell anergy is reversed by active Ras and is regulated by diacylglycerol kinase-alpha. *Nat Immunol 7:1166–1173.*
28. Boussiotis, V. A., D. L. Barber, T. Nakarai, G. J. Freeman, J. G. Gribben, G. M. Bernstein, A. D. D'Andrea, J. Ritz, and L. M. Nadler. 1994. Prevention of T cell anergy by signaling through the gamma c chain of the IL-2 receptor. *Science 266:1039–1042.*
29. Brown, I. E., C. Blank, J. Kline, A. K. Kacha, and T. F. Gajewski. 2006. Homeostatic proliferation as an isolated variable reverses CD8+ T cell anergy and promotes tumor rejection. *J Immunol 177:4521–4529.*
30. Wu, Z., S. J. Bensinger, J. Zhang, C. Chen, X. Yuan, X. Huang, J. F. Markmann, A. Kassaee, B. R. Rosengard, W. W. Hancock, M. H. Sayegh, and L. A. Turka. 2004. Homeostatic proliferation is a barrier to transplantation tolerance. *Nat Med 10:87–92.*
31. Teague, R. M., B. D. Sather, J. A. Sacks, M. Z. Huang, M. L. Dossett, J. Morimoto, X. Tan, S. E. Sutton, M. P. Cooke, C. Ohlen, and P. D. Greenberg. 2006. Interleukin-15 rescues tolerant CD8+ T cells for use in adoptive immunotherapy of established tumors. *Nat Med 12:335–341.*
32. Townsend, S. E., and J. P. Allison. 1993. Tumor rejection after direct costimulation of CD8+ T cells by B7- transfected melanoma cells [see comments]. *Science 259:368–370.*
33. Chen, L., S. Ashe, W. A. Brady, I. Hellstrom, K. E. Hellstrom, J. A. Ledbetter, P. McGowan, and P. S. Linsley. 1992. Costimulation of antitumor immunity by the B7 counterreceptor for the T lymphocyte molecules CD28 and CTLA-4. *Cell 71:1093–1102.*
34. Bai, X. F., J. Bender, J. Liu, H. Zhang, Y. Wang, O. Li, P. Du, P. Zheng, and Y. Liu. 2001. Local costimulation reinvigorates tumor-specific cytolytic T lymphocytes for experimental therapy in mice with large tumor burdens. *J Immunol 167:3936–3943.*
35. Jones, E., M. Dahm-Vicker, A. K. Simon, A. Green, F. Powrie, V. Cerundolo, and A. Gallimore. 2002. Depletion of CD25+ regulatory cells results in suppression of melanoma growth and induction of autoreactivity in mice. *Cancer Immun 2:1.*
36. Turk, M. J., J. A. Guevara-Patino, G. A. Rizzuto, M. E. Engelhorn, and A. N. Houghton. 2004. Concomitant tumor immunity to a poorly immunogenic melanoma is prevented by regulatory T cells. *J Exp Med 200:771–782.*
37. Peng, G., Z. Guo, Y. Kiniwa, K. S. Voo, W. Peng, T. Fu, D. Y. Wang, Y. Li, H. Y. Wang, and R. F. Wang. 2005. Toll-like receptor 8-mediated reversal of CD4+ regulatory T cell function. *Science 309:1380–1384.*
38. Day, C. L., D. E. Kaufmann, P. Kiepiela, J. A. Brown, E. S. Moodley, S. Reddy, E. W. Mackey, J. D. Miller, A. J. Leslie, C. DePierres, Z. Mncube, J. Duraiswamy, B. Zhu, Q. Eichbaum, M. Altfeld, E. J. Wherry, H. M. Coovadia, P. J. Goulder, P. Klenerman, R. Ahmed, G. J. Freeman, and B. D. Walker. 2006. PD-1 expression on HIV-specific T cells is associated with T-cell exhaustion and disease progression. *Nature 443:350–354.*
39. Radziewicz, H., C. C. Ibegbu, M. L. Fernandez, K. A. Workowski, K. Obideen, M. Wehbi, H. L. Hanson, J. P. Steinberg, D. Masopust, E. J. Wherry, J. D. Altman, B. T. Rouse, G. J. Freeman, R. Ahmed, and A. Grakoui. 2007. Liver-infiltrating lymphocytes in chronic human hepatitis C virus infection display an exhausted phenotype with high levels of PD-1 and low levels of CD127 expression. *J Virol 81:2545–2553.*
40. Barber, D. L., E. J. Wherry, D. Masopust, B. Zhu, J. P. Allison, A. H. Sharpe, G. J. Freeman, and R. Ahmed. 2006. Restoring function in exhausted CD8 T cells during chronic viral infection. *Nature 439:682–687.*
41. Andersson, J., A. Boasso, J. Nilsson, R. Zhang, N. J. Shire, S. Lindback, G. M. Shearer, and C. A. Chougnet. 2005. The prevalence of regulatory T cells in lymphoid tissue is correlated with viral load in HIV-infected patients. *J Immunol 174:3143–3147.*

42. Grant, R. S., H. Naif, S. J. Thuruthyil, N. Nasr, T. Littlejohn, O. Takikawa, and V. Kapoor. 2000. Induction of indolamine 2,3-dioxygenase in primary human macrophages by human immunodeficiency virus type 1 is strain dependent. *J Virol 74:4110–4115.*
43. Munn, D. H., M. Zhou, J. T. Attwood, I. Bondarev, S. J. Conway, B. Marshall, C. Brown, and A. L. Mellor. 1998. Prevention of allogeneic fetal rejection by tryptophan catabolism. *Science 281:1191–1193.*
44. Petroff, M. G., L. Chen, T. A. Phillips, D. Azzola, P. Sedlmayr, and J. S. Hunt. 2003. B7 family molecules are favorably positioned at the human maternal-fetal interface. *Biol Reprod 68:1496–1504.*
45. Zenclussen, A. C., K. Gerlof, M. L. Zenclussen, S. Ritschel, A. Zambon Bertoja, S. Fest, S. Hontsu, S. Ueha, K. Matsushima, J. Leber, and H. D. Volk. 2006. Regulatory T cells induce a privileged tolerant microenvironment at the fetal-maternal interface. *Eur J Immunol 36:82–94.*
46. Takeuchi, H., D. L. Morton, D. Elashoff, and D. S. Hoon. 2005. Survivin expression by metastatic melanoma predicts poor disease outcome in patients receiving adjuvant polyvalent vaccine. *Int J Cancer 117:1032–1038.*
47. Qin, J. Z., L. Stennett, P. Bacon, B. Bodner, M. J. Hendrix, R. E. Seftor, E. A. Seftor, N. V. Margaryan, P. M. Pollock, A. Curtis, J. M. Trent, F. Bennett, L. Miele, and B. J. Nickoloff. 2004. p53-independent NOXA induction overcomes apoptotic resistance of malignant melanomas. *Mol Cancer Ther 3:895–902.*
48. Medema, J. P., J. de Jong, L. T. Peltenburg, E. M. Verdegaal, A. Gorter, S. A. Bres, K. L. Franken, M. Hahne, J. P. Albar, C. J. Melief, and R. Offringa. 2001. Blockade of the granzyme B/perforin pathway through overexpression of the serine protease inhibitor PI-9/SPI-6 constitutes a mechanism for immune escape by tumors. *Proc Natl Acad Sci USA 98:11515–11520.*
49. Dannull, J., Z. Su, D. Rizzieri, B. K. Yang, D. Coleman, D. Yancey, A. Zhang, P. Dahm, N. Chao, E. Gilboa, and J. Vieweg. 2005. Enhancement of vaccine-mediated antitumor immunity in cancer patients after depletion of regulatory T cells. *J Clin Invest 115:3623–3633.*
50. Kaufman, H. L., G. Deraffele, J. Mitcham, D. Moroziewicz, S. M. Cohen, K. S. Hurst-Wicker, K. Cheung, D. S. Lee, J. Divito, M. Voulo, J. Donovan, K. Dolan, K. Manson, D. Panicali, E. Wang, H. Horig, and F. M. Marincola. 2005. Targeting the local tumor microenvironment with vaccinia virus expressing B7.1 for the treatment of melanoma. *J Clin Invest 115:1903–1912.*

Chapter 5
Tumor Antigens as Modulators of the Tumor Microenvironment

Katja Engelmann and Olivera J. Finn

Abstract The tumor microenvironment regulates tumor growth by providing cancer cells with extracellular matrix molecules and stromal cells, specific growth factors and cytokines, and a supply of nutrients through increased angiogenesis. In addition, it supports tumor growth by actively suppressing immune surveillance. Tumor-associated antigens (TAAs) play a pivotal role in the ability of the immune system to detect and possibly eradicate nascent tumors. Some TAAs, however, appear also to influence the tumor microenvironment through increased or aberrant cell signaling and modifications of cell-cell or cell-stroma interactions. Many TAAs are self-molecules that have become antigenic due to overexpression or aberrant expression on tumor cells compared to their normal counterparts, and therefore are able to elicit a host immune response. The recently proposed cancer stem cell theory of tumor development begs the question of whether some of these same TAAs are expressed on stem cells and what is their role in shaping their microenvironment. This chapter will focus on what it known about tumor antigens in the microenvironment of mature tumors. We will also discuss tumor antigens and other molecules expressed on stem cells since by influencing the microenvironment at the earliest stages of tumor development these molecules could be deciding factors in tumor growth and metastasis.

Keywords Cancer immunotheraphy, immune targeting, immune prevention, cancer stemcells, MUC1

5.1 Introduction

Cancer is a result of multiple genetic mutations that present the immune system with numerous new antigens, either products of the mutated genes or normal proteins that are aberrantly expressed due to oncogenic mutations. There are many examples of molecules expressed in tumors that are also expressed in normal cells, however on

Department of Immunology, University of Pittsburgh School of Medicine, Pittsburgh, PA 15232, USA

E. Yefenof (ed.), *Innate and Adaptive Immunity in the Tumor Microenvironment.*

tumors they are found either at a much higher concentration, in the wrong location (in the cytoplasm instead of the nucleus, on the entire cell surface instead of polarized), or in a differentially processed form (e.g. tumor-specific glycosylation, phosphorylation, lipidation, etc.). These molecules are collectively known as tumor associated antigens (TAA). This distinguishes them from tumor specific antigens present on cancer cells but not detectable on the normal cell counterparts. They are products of DNA-mutations that cause altered protein structures and often influence their function. In the last two decades, research in the mechanisms of immune recognition of tumors centred in great part on the molecular characterization of tumor antigens and revealed that a large number of these molecules can be recognized by the immune system. However, while this work provided proof that tumors are highly antigenic, it also showed that in spite of often simultaneous expression of multiple antigens on tumors, the immune responses generated against them are weak and do not result in tumor rejection. In the case of TAA that closely resemble self-molecules, weak immune responses are considered to be due to self-tolerance. Indeed, experiments in animal models have shown that immunization against tumor-associated antigens generates much stronger immune responses in wild type mice compared to mice transgenic for these antigens [1–4]. Nevertheless, even in transgenic mice that show evidence of tolerance against these antigens, it is possible to induce immunity strong enough to reject a tumor challenge. There is also accumulating evidence that in people, immune responses that are generated against self-tumor associated antigens may be protective against tumor occurrence [5–8]. Furthermore, the presence of immune responses against several of these antigens in cancer patients is positively correlated with better prognosis.

Many different techniques have been utilized to identify tumor antigens: transfection of recombinant tumor cDNA libraries and human leukocyte antigen (HLA) molecules into target cells [9, 10], isolation of peptides from the binding cleft of major histocompatibility complex (MHC) class I molecules expressed on tumor cells [11, 12], deduction of immunogenic peptide sequences from suspected tumor antigens (such as known oncogenes, tumor suppressor genes or other tumor-associated proteins using computer algorithms based on HLA-anchor motifs) [13, 14], and serological expression cloning of recombinant tumor cDNA expression libraries (SEREX) [15, 16]. They have led to the identification of a large number of molecules that are clearly recognized by the immune system and by that definition are targets of immune surveillance. The question that is of special interest to tumor immunologists is why the immune response against these molecules is not sufficient to prevent tumor outgrowth. It is clearly possible, although difficult to demonstrate, that in most individuals responses against tumor antigens are protective, and some evidence for this is beginning to accumulate [5, 17]. Cancer may only win when it succeeds to escape immune surveillance and this can happen through the loss of specific antigens or gain of immunosuppressive properties.

In parallel to the work on tumor antigen discovery, work has been done on the reasons why tumor antigen-specific immunity fails to control disease. This has proven to be a very rich area for discovery. Numerous immunosuppressive mechanisms have been unveiled and described that are unique to a tumor microenvironment. Tumor microenvironment is an altered local environment in which the incipient neoplasia develops, progresses, and begins to metastasize. Fibroblasts, epithelial cells, adipocytes, smooth muscle cells, inflammatory cells, as well as resident and recruited immune and vascular cells, constitute the stroma that communicates with tumor cells and co-evolves with them over time. In addition, extracellular matrix molecules dynamically interact with tumor cells to provide nutrients and stimuli necessary for tumor proliferation, angiogenesis, invasion and insulation from the host's immunosurveillance. Interaction of cancer cells with their microenvironment is mediated through many cell surface molecules, some of which have been also defined by the immune system as tumor antigens.

In this chapter we will focus on immunosuppressive mechanisms that are promoted by the expression of some well-known tumor antigens that appear to play a dual role reminiscent of the Dr. Jekyll and Mr. Hyde story. Under certain circumstances they can activate effective immunity and serve as good targets for tumor rejection, thus behaving like the good Dr. Jekyll. Unexpectedly and only recently appreciated, these same molecules can contribute to the creation of an immunosuppressive microenvironment such that immune effector cells that recognize these same antigens, and potentially many other tumor antigens, are prevented from destroying the tumor. That is their bad Mr. Hyde behavior. The success of immunotherapy based on these antigens will depend on the ability to take advantage of their Dr. Jekyll side and prevent them from showing their Mr. Hyde side.

As a cardinal example of a Dr. Jekyll and Mr. Hyde antigen, we will present in detail the tumor antigen MUC1, which has been extensively studied as a tumor antigen on over 80% of all human solid tumors, as well as multiple myelomas and some B cell lymphomas. It was realized only recently that MUC1 could also be used by the tumor for survival and immune evasion, if it is not targeted efficiently by the immune system for tumor destruction. We will describe also recent realization that MUC1 is expressed on cancer stem cells, where it can be a target for therapy but also a facilitator of their development into mature tumor cells. While much less is known about such dual behavior of other molecules studied as tumor antigens, we will review evidence that at least some of them, such as CEA, tumor glycolipids and gp100, share this dual behavior with MUC1.

5.2 Mr. Hyde MUC1 Promotes Tumorigenesis

MUC1 is the first described and best known member of the mucin family that currently has 20 members [18]. This family is characterized by transmembrane-associated and secreted glycoproteins, which show a dense O-glycosylation linked to serine-, threonine- and proline-rich tandem repeat domain called VNTR (variable number of tandem repeats). The VNTR domain of MUC1 is comprised of 20–120 tandemly repeated 20 amino acid segments whose sequence includes 5 potential O-glycosylation sites per repeat [19]. In its normal expression, MUC1 is found at low amounts on the apical cell surface of most glandular epithelial cells [20, 21]. There, it performs a variety of biological functions, such as lubrication and hydration of epithelia and, protection against colonization by pathogens. It is also involved in cell-cell (via E-cadherin) and cell-matrix (via integrins) interactions as a part of both adhesion and anti-adhesion mechanisms. Adhesion effects are mediated by carbohydrate structures on the extracellular repeat domain, whereas anti-adhesive effects can be mediated by the glycosylation-stabilized rigid conformation of the several hundred-nanometer long mucin molecules [22]. Its highly conserved cytoplasmic domain can be phosphorylated at seven phosphorylation sites by, among other signaling proteins, glycogen synthase kinase 3β, c-src and protein kinase Cδ, and thus, regulate further cell signaling, such as through interaction with β-catenin [23–26]. MUC1 furthermore associates with members of the ErbB family of receptor tyrosine kinases and FGFR3 [27].

During tumorigenesis, cells lose their polarity and MUC1 expression is found at very high levels on the entire cell surface [28, 29]. The extracellular tandem repeat region can act as a ligand for intercellular adhesion molecule (ICAM) 1, involved in metastatic intravasation [30, 31]. The O-glycosylated extended ectodomain can receive signals from selectin binding or conceivably trigger signaling events. Overexpression of MUC1 regulates binding to Wnt effector β-catenin competitively for E-cadherin, modulates the localization of β-catenin to the cytoplasm and subverts E-cadherin-mediated cell adhesion in epithelial cells, leading to destabilization of intercellular junctions, which benefits tumor cell migration [23]. MUC1 has been shown to bind to the signaling mediators Grb2/SOS upon phosphorylation [32] and to mediate activation of numerous receptor tyrosine kinases. Activation of the ras signaling pathway phosphorylates Raf, MEK, and ERK1/2, with the latter translocating to the nucleus and inducing transcription of genes involved in mitogenesis, differentiation, apoptosis, and quiescence [33]. Expression of high levels of MUC1 has been shown to elicit EGF-dependent activation of ERK1/2 MAPK [34]. Moreover, MUC1 can mediate resistance to apoptosis in a survival response to oxidative stress [35] by suppression of H_2O_2-induced accumulation of reactive oxygen species (ROS) and in response to DNA damaging agents, suggesting its important contribution to the resistance of cancer cells to genotoxic agents [36].

5.3 Dr. Jekyll MUC1 is a Tumor Antigen Targeted by Immune Surveillance

In normal epithelia MUC1 expression is found in small amounts and restricted to the apical cell surface. However, in >80% of most premalignant lesions that are precursors to cancer and other adenocarcinomas [28, 37], MUC1 expression is highly up-regulated; for instance in breast cancer it is expressed up to 10-fold higher than in normal cells [38]. MUC1 expressed on tumor cells displays striking alterations with respect to shortened glycan chains (Tn-antigen (GalNAc1-O-S/T) and Thomsen-Friedenreich (TF) or T antigen (Galβ1-3GalNAcα1-O-S/T)), increased sialylation (mono- and di-sialylated TF-antigen, mono-sialyl Tn-antigen, sialyl-core1), and a shift in the carbohydrate core-type [39–41]. As an example, normal lactating mammary gland expresses MUC1 with primarily long polylactosamine-type chains based on the elongation of the core2 structure (GlcNAcβ1–6[Galβ1-3] GalNAcα1-O), and is dominated by neutral glycans with linear and branched backbone structures that can comprise up to 16 monosaccharide units [42, 43]. During malignant transformation the core2 forming β6-glucosaminyltransferase level is low or not expressed and leads to the accumulation of core1 structures [41] that now serve as substrates of Gal-specific α3-sialyltransferase and GalNAc-specific β6-sialyltransferases. In addition, the increased presence of CMP-NeuAc:Galβ1–3GalNAcα3-sialyltransferase may compete with the core2 enzyme for substrate [44]. Thus, the length of O-linked glycans on tumor MUC1 is restricted by a high degree of sialic acid that dominates over neutral carbohydrates [19, 45]. This underglycosylation results in unmasking of previously cryptic protein core epitopes and short carbohydrate epitopes. This is important for the humoral and cellular immune response, because the peptide core is more accessible for peptide specific anti-MUC1 antibodies and for enzymatic processing into peptides for presentation to T cells, compared to MUC1 on normal epithelia.

The new epitopes on MUC1 are tumor-specific targets for both T and B-cells. The DTR sequence within the tandemly repeated icosapeptide sequence is characterized as the immunodominant epitope [46] recognized by tumor-specific anti-MUC1 antibodies. Because the VNTR domain of MUC1 contains 25–125 tandem repeats, the immunodominant DTR epitope is presented many times by every MUC1 molecule. Although chemical analyses have shown that the threonines in the DTR-motif can be glycosylated [47, 48], it appears to be a low affinity site for glycosyltranferases and thus it is usually presented in its non-glycosylated form as a knob-like structure [49]. This self-stabilizing and tandemly repeated structure [50] is assumed to favor immune recognition, especially for non-MHC restricted T cells [51]. Other studies revealed that the presence of a GalNAc moiety on the threonine in the DTR motif increases antigenicity [52, 53]. Vaccination studies in mice also showed that a 60mer MUC1 glycopeptide composed of three tandem repeats with GalNAc (Tn) or NeuAc-GalNAc (sialyl-Tn) moieties in each of the five glycosylation sites per repeat, elicited the strongest MUC1-specific antibody response compared to less glycosylated or unglycosylated MUC1 peptides [54]. Interestingly, in up to 50% of the tandem repeats the DTR motif is replaced by an ESR motif, often in concert with a Pro to Ala replacement in position +10 relative to the DT or ES in the repeat (P*DT/ES*RPAPGSTAP*P/A*AHGVTSA). These variant repeats often occur in clusters [55]. The amino acid changes revealed a higher conformational flexibility of ES/P peptides and reduced glycosylation density of ES/A sequence-variant repeats compared to DT/P peptides. Moreover, sera of healthy controls, pregnant woman and patients with benign tumors contained IgGs preferentially directed to variant repeat clusters, whereas sera of cancer patients showed predominant specificity to the invariant DTR peptide [56].

The most unique aspect of the DTR motif is that due to its stable structure identically repeated multiple times, it is recognized by human T-lymphocytes as an unprocessed epitope on the native MUC1 protein expressed on the surface of tumor cells [57, 58]. This MHC-unrestricted recognition of native MUC1 epitopes induces calcium mobilization, phosphorylation of ZAP-70 and proliferation of T cells (CTL) [59]. MUC1 processing by DCs is influenced by O-glycosylation such that a high degree of complex O-glycosylation on MUC1 is unable to prime strong helper T cell responses when presented by dendritic cells *in vitro* [60]. This is due to the high density of O-glycans that protect the molecule from proteolytic degradation. In contrast, less complex O-linked glycans, like core-type structures, are not removed from MUC1 when processed by DCs and presented on MHC class II molecules [61]. Efficient processing of MUC1 glycopeptides is also a matter of site-specific glycosylation [62]. Elongated glycans proximate to the minor cleavage sites of the tandem repeats, AH↓GV and PG↓ST (↓represents the cleavage site), can sterically impede the access of the protease (Cathepsin L) and glycosylation adjacent to the major cleavage site VT↓SA prevents cleavage completely [62].

Considering the undesirable Mr. Hyde tumor promoting characteristics of MUC1, it was gratifying to discover that in its tumor form this molecule is under immune surveillance. Because immune responses against MUC1 can be found in cancer patients, an expectation can be made that if these responses were boosted, they may eliminate or control the growth of cancer cells that express MUC1. Work to date has shown that MUC1 is a promising target for immunotherapeutic strategies to treat cancer in humans [63–65]. Several phase I studies have been completed so far and demonstrated that MUC1 vaccines are safe and in a small number of cases also effective in controlling disease. In one study, a vaccine consisting of a non-glycosylated synthetic MUC1 peptide corresponding to five 20-amino acid long repeats and SB-AS2 adjuvant, was tested for safety, toxicity, and ability to elicit or boost MUC1-specific immune responses in patients with resected or locally advanced pancreatic cancer without prior chemotherapy or radiotherapy [66]. Another clinical phase I study used the same MUC1 peptide and incomplete Freund's adjuvant as a vaccine for patients with pancreatic or bile duct cancer. This study showed again safety of the vaccine [67]. In a yet another phase I trial, autologous mature dendritic cells pulsed with HLA-A2-binding MUC1 peptides were used in patients with metastatic renal cell carcinoma (RCC). This study showed that MUC1 peptide-pulsed dendritic cells could induce clinical and immunologic responses in some patients [68]. Moreover, the study revealed epitope spreading, whereby some patients developed T-cell responses to antigens not used for vaccinations, such as adipophilin, telomerase, or oncofetal antigen.

The MUC1 peptide from the tandem repeat region was the chosen antigen in all these vaccine formulations because this region differs the most between normal epithelial cells and tumor cells and thus the immune response induced against this form would not be expected to target normal tissues. There is another form of MUC1 that is equally promising in its immunogenicity and safety and that is the MUC1 glycopeptide that bears tumor specific carbohydrates. This form induces strong helper T cell responses that are able to help generation of MUC1-specific antibodies, preferentially of IgG_1 isotype and ADCC, and/or effective T cell mediated cytotoxicity [69].

5.4 Mr. Hyde MUC1 Manipulates Tumor Microenvironment

Tumor-associated forms of MUC1 have been shown to be chemotactic for circulating immature human myeloid DCs [70]. This is mediated by the peptide epitopes in the tandem repeat region of the underglycosylated MUC1. Moreover, glycopeptides from the same region provide a maturation/activation signal for the DCs that migrate to the tumor site. These DCs that have matured under the influence of MUC1 produce IL-6 and TNF-α, cytokines that have been implicated in tumor metastasis and progression. They also fail to promote a type 1 response [70]. Hence, instead of promoting adaptive immunity for efficient immune surveillance, these DCs promote inflammation that mediates immunosuppression at the tumor site and later increased tumor invasion within the tumor microenvironment. A striking example of the "wrong" immune response that exists in this MUC1 modified tumor microenvironment is induction by the DCs at the tumor site of T cells that secrete IL-13 [70], a cytokine involved in suppression of tumor specific CTL and direct promotion of growth of tumor cells that bear IL-13 receptor [71].

As mentioned above, MUC1 has been linked to several signaling pathways via its cytoplasmic domain that has multiple phosphorylation sites and binding motifs for multiple kinases. In addition, it was reported recently that MUC1 regulates p53 transcription by binding to its regulatory domain [72]. P53 is a tumor suppressor gene whose function is often inactivated either through mutation or overexpression in a variety of cancers. Overexpression of MUC1 by tumor cells can lead to aberrant overexpression of p53. MUC1 expression also leads to decreased apoptosis of tumor cells under oxidative stress that occurs naturally during tumor expansion or in response to genotoxic agents [35]. This might be related to the ability of MUC1 to modulate transcription of the nuclear factor kB (NF-kB) as well as to regulate the ERK1/2 pathway and AP-1 mediated transcription [73]. The involvement of MUC1 in these networks reveals its key role in regulating tumorigenesis.

The impact of MUC1 on shaping the tumor microenvironment is well illustrated by a study in which MUC1 expression on tumor cells was suppressed by RNA interference. The siRNA treated pancreatic cancer cell line with low level of MUC1 expression showed a greatly reduced proliferative and metastatic capacity [74]. High levels of sialyl-Tn antigen, a carbohydrate structure decorating MUC1 on tumor cells, has been linked to increased tumor cell migration *in vitro* and decreased cell adhesion to the extracellular matrix molecules collagen type I, collagen type IV and fibronectin [75]. Moreover, MUC1 has been identified as a potential target in the colonization of metastasizing tumor cells by interacting with ICAM1 [31] or with E-selectin [76].

5.5 Other Dr. Jekyll and Mr. Hyde-Like Tumor Antigens

5.5.1 Dr. Jekyll CEA

The carcinoembryonic antigen (CEA), a heterogeneous cell surface and secreted glycoprotein (~50% glycosylation), member of the immunoglobulin supergene family, performs a function as an intercellular adhesion molecule. During malignant transformation its expression is largely upregulated, accompanied by changes in its glycosylation.

Because CEA is recognized as a self-antigen, individuals are immunologically tolerant to CEA. Although, immune responses to CEA have been observed in healthy and cancer bearing individuals, these are weak and have to be boosted to become protective against or to reject CEA positive tumors. Serum titers for CEA are used for diagnosis of a variety of tumors, especially of the colon, pancreas, breast and lung.

Using human CEA-transgenic mouse models immune tolerance has been successfully overcome by certain vaccination strategies that generate CEA-specific, MHC-restricted CTLs, T cell proliferation, CD4+ T cell responses and anti-CEA antibody generation, which resulted in tumor rejection in these mice [77–81]. Very recently it was shown that genetically modified DCs that express CEA and Th1-type cytokines IL-12 and GM-CSF, can overcome peripheral immune tolerance to CEA and induce strong cellular antitumor activity in transgenic mice without creating autoimmunity [82, 83]. Similar results were achieved with a vaccine based on anti-Id antibody pulsed DCs, which induced both humoral and cellular (CTLs, helper and memory T cells) immune responses in the vaccinated CEA-transgenic mice and provided protective antitumor immunity. The anti-Id antibody was raised against anti-CEA antibody and mimics a single protective epitope of human CEA. This might be of advantage in comparison to CEA protein or gene based vaccines, that contain multiple different CEA epitopes, some of which are protective but others could initiate immunosuppressive or autoimmune responses [81]. In general, humoral responses against CEA have been seen in response to vaccination with recombinant CEA protein and CEA anti-idiotype antibodies, whereas T cell responses have been preferentially generated by dendritic cells loaded with immunogenic epitopes of CEA [84].

In a recently performed clinical study, generation of CEA-specific T cell responses was seen in patients with advanced CEA-expressing tumors who received recombinant vaccinia and subsequently avipox vectors encoding CEA and costimulatory molecules (e.g. B7.1). Additional supplementation with GM-CSF and low-dose IL-2 increased the number of CEA-specific T cell precursors. Some patients also developed anti-CEA antibodies [85]. In another vaccination study, several patients with advanced colorectal cancer developed responses to CEA after intradermally and intravenously administered vaccination with CEA-loaded matured monocyte-derived DCs [86]. CEA-specific CD8+ T cells were detected in post-treatment delayed type hypersensitivity biopsies and in one resected abdominal draining lymph node of one patient. These T cells produced large amounts of IFN-gamma and IL-2, but no IL-4 or IL-10 in response to CEA-peptide loaded target cells. Majority of the patients showed preexisting immunity against CEA (induration of the vaccination site), although this immunity did not yield outgrowth of T cells from the injection site. A larger clinical study was performed with recombinant CEA and with or without low-dose GM-CSF given with each immunization [87]. All patients generated CEA-specific T cells and anti-CEA antibodies of IgG isotype. The immune responses persisted for more than 2 years and correlated with increased survival. Of the patients that received only recombinant CEA vaccine and no GM-CSF, only about 70% developed humoral and cellular immune responses against CEA. In 2004 a clinical study was conducted on chemotherapy resistant colorectal cancer patients using synthetic CEA peptides that bind to either MHC class I HLA-A 0201 or HLA-A2402 [88]. Peptide-pulsed dendritic cells were injected into the inguinal lymph node. About 70% of the patients showed significant increase in IFN-gamma producing CEA-specific CTLs. However, most patients still progressed and only two showed disease stabilization for 12 weeks. Similarly challenging combination therapy, chemotherapy with concurrent immunotherapy, against CEA was performed in patients with metastatic colorectal cancer [89]. The target antigen was the immunodominant epitope CAP-1, HLA-A2-restricted nonamer peptide. Approximately 47% of the patients showed significant increases in CAP-1-specific CTLs. Multiple adjuvants (CpG, GM-CSF or DCs) were tested in this trial without any major differences observed between them. The overall clinical response in this study was 35%.

5.5.2 *Mr. Hyde CEA*

Like MUC1, CEA has also been shown to play an important role in cancer cell metastasis as well as to be chemotactic for immature DCs [90]. Neo-expression and over-expression of distinct carbohydrate epitopes on the tumor-associated CEA seems to be responsible for this effect. The degree of expression of fucosylated structures as well as the increased levels of Lewis^x carbohydrate and the *de novo* expression of Lewis^y has been associated with recognition by immature DC. Immature DCs express the specific intercellular adhesion molecule (ICAM)-3–grabbing nonintegrin (DC-SIGN) [91], a C-type lectin that displays high affinity for nonsialylated Lewis^x and Lewis^y, carbohydrate moieties found abundantly expressed on the tumor forms of CEA. Mature DCs do not interact with tumor forms of CEA, whereas immature DCs do not bind to CEA on normal cells. Besides ICAM-2 on endothelial cells [92] and ICAM-3 on resting T cells [91], which represent additional ligands for DC-SIGN, the glycosylation-dependent interaction between immature DCs and the secreted tumor cell derived CEA is suggested to play a central role in tumor tolerance through suppression of dendritic cell functions [90]. It has also been shown in colorectal cancer that binding of CEA to Kupffer cells in the liver can stimulate the secretion of cytokines, e.g. TNF-alpha and interleukin 1-beta [93] that stimulate endothelial cells to increase their expression of adhesion molecules such as E-selectin. The binding of E-selectin to sialyl-Lewis antigen, carbohydrate structure on CEA and MUC1 on tumor cells, promotes tumor cell extravasation and metastasis to the liver [94, 95].

5.5.3 *Dr. Jekyll Tumor Glycolipids*

Many glycolipids have been extensively studied as tumor-associated antigens and potential targets of anti-cancer therapies, primarily on melanomas. The GD3 ganglioside has a restricted distribution on normal brain cells, connective tissue, and a small population of T cells, and is considered to be only weakly immunogenic. In comparison,

GM2 is expressed in the brain and at secretory borders of all epithelial tissues and has been shown to be highly immunogenic. Fucosyl-GM1 (Fucα1–2Galβ1–3GalNAcα1–4[NeuAcα2–3]-Galβ1–4Glcβ1–1Cer) is only sparsely expressed in the thymus, spleen, small intestine, and islet cells of the pancreas [96], and seems to be the most immunogenic of all three. The presence of serum antibodies against ganglioside GM2 is associated with prolonged disease-free survival in melanoma patients [97]. Similarly, induction of anti-GD3 immunity through vaccination with an anti-idiotypic monoclonal antibody that mimics GD3 correlated with prolonged survival of small cell lung cancer patients [97]. Also in small cell lung cancer patients, who were previously treated with chemotherapy and radiation therapy, Fucosyl-GM1 immunity was induced by vaccination with bovine Fucosyl-GM1 conjugated to the carrier protein keyhole limpet hemocyanin (KLH) and mixed with QS-21 as adjuvant. All patients showed humoral responses, IgM and IgG_1 isotype, specific to Fucosyl-GM1 [97] that demonstrated complement binding activity and complement dependent cytotoxicity of tumor cells *in vitro*. A similar vaccination study was performed to target GM2 in previously untreated melanoma patients. Immune response was seen in form of GM2-specific IgM, IgG_1 and IgG_3 antibodies [98, 99]. In a more recent study, synthetic fucosyl-GM1 injected with QS-21 to patients with SCLC induced only Ig-specific antibodies that were unable to mediate antibody-dependent cellular cytotoxicity [100].

GM2, GD2 and GD3 have also been used as targets of passive immunotherapy. A murine IgG_3 monoclonal antibody specific for GD2 was administered to patients with metastatic neuroblastoma or melanoma [101]. This antibody could activate human complement and mediate ADCC *in vitro*. Two patients (12%) showed complete tumor regression and four patients had a partial or mixed antitumor response. A human IgM monoclonal antibody specific for GD2 has been administered intralesionally in patients with metastatic melanoma. Regression was seen in almost all patients, with evidence of tumor degeneration, fibrosis, free melanin, and some degree of lymphocyte or macrophage infiltration [102]. Administration of a GD3-specific murine monoclonal IgG_3 antibody produced lymphocyte and mast cell infiltration, mast cell degranulation, and complement deposition in the injected lesion. In 25% of the patients major tumor regression was observed [103].

5.5.4 Mr. Hyde Tumor Glycolipids

GD2 is aberrantly expressed in small cell lung cancer and is considered to be a key factor in cancer cell proliferation and invasion. Upregulated expression of GD3 in melanomas has been reported to enhance phosphorylation of two adaptor molecules, p130Cas and paxillin, which regulate cell growth and cell adhesion [104]. Similarly, polysialic acid (polySA) is involved in cell-cell interactions and cell migration [105]. Based on its large size and negative charge of the carbohydrate side chain, polySA has an inhibitory effect on cell adhesion and thus promotes cell motility [106]. This has been confirmed in animal studies, where its expression correlated with tumor invasiveness, metastases, and increased growth rate [107].

5.5.5 *Dr. Jekyll gp100*

Gp100 is a melanocyte differentiation antigen that is often used in anti-melanoma therapies as an antigen and a target. Several gp100 derived peptides are restricted by HLA-A2, -A3, and -A24 molecules and have been shown to be immunogenic in peptide-based vaccines. Vaccination with gp100 derived epitopes and IL-2 has induced tumor regression in patients with advanced disease [108]. A recently performed immunization study in patients with metastatic disease compared vaccines based on HLA-A2-restricted peptides of MART-1_{27-35}, gp$100_{209-217}$, and tyrosinase$_{368-376}$ administered in combination with a low dose of GM-CSF. The results established that the HLA-A2 restricted peptide gp-$100_{209-217}$ is more immunogenic than the other melanoma peptide antigens [109]. One patient of 25 showed an overall survival of 3-year post-completed vaccination without disease progression. In another study performed on patients with unresectable or metastatic melanoma, immunization with a slightly modified HLA-A2-restricted gp100 (G280-9V) peptide loaded on autologous DCs, resulted in uniform immunogenicity. High-avidity CTLs, generated against the modified peptide, were capable of killing melanoma cells that co-expressed $A^{*}0201$ and gp100 [110]. A study performed in a human-mouse chimeric model showed that the modification of the naturally occurring gp100 sequence enhanced the immunogenicity of the peptide epitope. The modified gp100 peptide showed a much slower dissociation rate from the HLA-A2-peptide complex than the bound natural peptide [111]. As an extension of the modified peptide strategy, PBMCs from vaccinated patients were expanded *in vitro* and used for adoptive therapy. While there was no clear clinical response, two patients developed autoimmunity against their melanocytes showing antigen-specific activity of the transferred cells [112]. Best results were derived from trials that recruited patients with early or minimal disease. A vaccination study in HLA-$A2^{+}$ patients with completely resected non-metastatic melanoma, using a modified melanoma peptide, gp100_{209-2M} and adjuvant, demonstrated significant increases in the frequency of IFN-gamma producing $CD8^{+}$ T cells in 33 of 35 patients and durable presence of gp100_{209-2M}-specific memory $CD8^{+}$ T cells with high proliferative potential [113].

Overall, vaccination with gp100 melanoma peptides can mount a significant antigen-specific $CD8^{+}$ T-cell immune response, but translation into a clinical response appears to be compromised by the immunosuppressive environment of an advanced disease. As in the case of the tumor antigens discussed previously, gp100 in its Mr. Hyde role may be in part responsible for creating this immunosuppressed state.

5.5.6 *Mr. Hyde gp100*

Gp100 is a melanogenic enzyme that is expressed at low levels in normal melanocytes and displays aberrantly expression in melanoma cells. This tumor-associated gp100 exhibits chemotactic properties to immature DCs, monocytes and T cells

[114] and thus, contributes to the leucocyte infiltration in the tumor tissue. This function appears to be mediated through its interaction with the chemokine receptor 2 (CCR2) that is expressed on these immune cells. However, recruiting of immune cells will not result in induction of effective tumor immunity, instead, the cellular infiltrate promotes tumor progression through the action of gp100. Furthermore, the efficiency of T cells in recognition and eradication of the melanoma antigen gp100 depends on the number of gene copies expressed. The threshold level that is necessary for T cell stimulation has been reported to be 500 mRNA copies per 10^8 copies of rRNA [115]. In addition, a soluble form of gp100, ME20-S, circulates in the serum and might be involved in tumor protection, mediated by antibody quenching and inhibition of humoral immune responses, e.g. complement mediated toxicities expressed at low levels in quiescent adult melanocytes but overexpressed by proliferating neonatal melanocytes and during tumor growth. Release of the soluble form, ME20-S, could protect tumor cells from antibody mediated immunity.

5.6 Cancer Stem Cells and their Dr. Jekyll and Mr. Hyde Antigens

The cancer stem cell hypothesis of tumor development has recently been proposed for many types of cancer [116] and therefore has replaced the older stochastic model of cancer development. It provides a likely explanation for failed cancer treatments and recurrences that are now considered to be due to therapy resistant cancer stem cells. Furthermore, this model is supported by the observations that cancer may arise from embryonic cells, as observed in childhood tumors, or through activation of hormone-sensitive stem cells in the case of breast cancer [117, 118].

Normal adult epithelial stem cells are a minor population of mostly resting cells that are defined by their long life span and capacity to self-renew, which allows them to maintain the stem-cell compartment as well as to generate differentiated lineage-restricted progeny with the ability to form mature cell types in specific tissues [119, 120]. In general, the slow dividing stem cells are responsible for tissue renewal and repair of tissue damage. The integrity and cell divisions of stem cells is regulated within a local environment, called niche, where stem cells interact with niche cells and extracellular matrix proteins according to external and internal stimuli. Stem cell divisions can be asymmetrical to simultaneously self-renew and generate a differentiated progeny or symmetrical resulting in two identical daughter stem cells, which will expand the stem cell pool, or two progeny, which will deplete the stem cell population [121].

Tissue committed stem cells or stem-like cells have been identified in both normal and cancerous human tissues such as mammary gland and breast cancer, skeletal muscle, lung, liver, epidermis and melanoma, forebrain and gliomas, testis, heart, kidney, limbal epithelium, gastrointestinal tract and prostate, as well as in several long-term tumor cell lines [122–135]. As an example, the existence of adult breast stem cells was experimentally shown by studies of X-chromosome inactivation, retroviral tagging, and transplantation studies [136–138]. During mammary gland development, pregnancy, lactation and involution, tissue stem cells undergo a differentiation process into functionally ductal-alveolar structures, whereupon the stem cell pool is reduced. The mammary gland in humans is composed of three lineages that generate the lobulo-alveolar structure: contractile myoepithelial cells forming the basal layer of ducts and alveoli, ductal epithelial cells lining the lumen of ducts, and alveolar epithelial cells initiating milk production [139]. More than 90% of the adult mammary gland is composed of differentiated luminal and myoepithelial cells and only few cells occur with undifferentiated morphology, localized in the terminal end buds [133]. It is suggested that stem cells are partially located in the Terminal Ductal Lobulo-Alveolar Units (TDLUs) that are lined by a layer of luminal epithelial cells surrounded by a basal layer of myoepithelial cells. Interestingly, this area has been described as origin for more than 90% of all (luminal epithelial) breast cancers [140]. In 2003, the propagation of breast stem cells was achieved *in vitro* [141] and potential breast cancer stem cells were isolated and characterized from patients [142]. In 2005, candidate breast cancer stem cells were successfully propagated *in vitro* as well [143].

Stem cells have also been identified from acute myelogenous leukaemia (AML). They express similar cell surface molecules as normal hematopoietic stem cells and thus are thought to arise from normal stem cells [144]. In contrast, blast crisis chronic myelogenous leukemia (CML) seems to derive from a committed granulocyte-macrophage progeny that may have undergone mutations and regained stem cell-like features [144].

5.7 Candidate Cancer Stem Cell Markers

Cancer stem cells represent a very small percentage of highly malignant cells, probably less than 1% within the bulk of mature, differentiated tumor cells. Self-renewal and multipotency have been used as the "gold standard" assay to identify cancer stem cells. Overexpression of multi drug-resistance proteins in cancer stem cells and therefore their capability to efflux the fluorescent DNA-binding dyes Hoechst 33342 or rhodamine 123 that can be monitored by flow cytometry, has been used as complementary approach to identify and enrich a candidate cancer stem cell population [145]. Proliferation assays with 5-Bromo-2-deoxy-uridine (BrdU) have also been used to detect the slowly dividing (cancer) stem cells that retain the label longer than the more rapidly cycling progeny and mature cells. Identification of surface markers is an important goal that will help their isolation and the study of their role in tumor development, cancer recurrence and tumor cell resistance to conventional cancer therapies. It is already clear that new therapies that can kill the mostly quiescent and drug-resistant tumor stem cells will need to be developed. One therapeutic approach that can be predicted to be successful is immunotherapy. As long as specific antigens are present on the cancer stem cell, the immune response can be directed against it and its efficiency should not depend on the proliferative state of the target cells.

Tumor antigens have been studied to date exclusively on mature tumor cells. Their expression needs to be confirmed on cancer stem cells or new antigens need to be identified that are expressed on cancer stem cells. With the exception of MUC1, to date there is no information regarding the expression on cancer stem (like) cells of the many already known tumor antigens. The field has focused on identifying specific markers on these cells, some of which may very well in the future play a role of an antigen or a specific target. Several candidate (cancer) stem cell markers have been identified in hematologic and solid tumors. In several forms (M0, M1, M2, M4, M5) of human acute myeloid leukemias (AML) cancer stem cells were detected [146] based on their capacity to engraft in non-obese diabetic/severe combined immunodeficiency (NOD-scid) mice, to reinitiate tumor development in recipient animals throughout serial transplantation, and to reestablish tumor cell heterogeneity that suggested their self-renewal capacity and high potency. Only cells defined as $CD34^{+}$ /$CD38^{neg}$ were able to induce AML in immunodeficient animals by serial transplantation, confirming self-renewal capacity [147]. Importantly, the $CD34^{+}$ /$CD38^{neg}$ cells could differentiate into $CD38^{+}$ /$lineage^{+}$ cells and thus reproduce a population hierarchy seen in normal cell differentiation [148]. While the cancer stem cell population displays a similar cell surface marker profile $CD34^{+}$ / $CD38^{neg}$/$CD71^{-}$/HLA-DR^{-} as the normal hematopoietic stem cells (HSC) [149], only AML stem cells are $CD90^{-}$, $CD117^{-}$ and $CD123^{+}$, a surface pattern not described for normal HSC [146]. $CD34^{+}$ /$CD38^{neg}$ stem cells were also successfully isolated from acute lymphoblastic leukaemia (ALL) patients carrying the BCR/ABL translocation and only these cells were able to engraft in the marrow of immunodeficient mice [146].

In brain tumors (glioblastomas and medullablastomas), cancer stem cells were characterized by the $CD133^{+}$ /$nestin^{+}$ phenotype [150]. Like in the case of AML stem cells, CD133 (Prominin) transmembrane glycoprotein and nestin, which belongs to the family of intermediary filaments, are also markers of normal neural [151] and normal prostate stem cells [152], as well as other non-stem cells in many tissues, like pancreas, kidney, liver, lung, heart and CD34 hematopoietic stem/progenitor cells in adult and fetal bone marrow [152, 153].

Breast cancer stem cells have been defined by their capacity to efflux the DNA-binding dye Hoechst 33342 by virtue of overexpression of ABCG2 drug resistance proteins. These Hoechst stain negative cells are a minor population in a tumor cell preparation and have thus been named "Side Population" or SP. The normal breast stem cell population is supposed to lack the epithelial membrane antigen (EMA) and CD10 (myoepithelial marker), but may be enriched in cells expressing cytokeratin 18, a luminal marker, and cytokeratin 14, a myoepithelial marker [154]. Normal breast stem cells can be estrogen receptor positive (ER) and cytokeratin 19 positive, and might up-regulate other putative stem cell markers, e.g. the cell cycle regulator $p21^{CIP/WAF1}$ and RNA-binding protein Musashi homolog 1 (Msi1), which is involved in asymmetric stem cell division [155]. Breast stem cells cultured as non-adherent mammospheres were shown to express CD49f (alpha 6 integrin), cytokeratin 5 and CD10, but only a few cells were positive for the epithelial specific antigen (ESA) and cytokeratin 14 [141]. In 2003, candidate breast cancer stem cells were reported to reside within a $CD44^+/CD24^{low/neg}$ cell population, which was confirmed by *in vivo* tumor cell transplantation studies [131]. In 2005, *in vitro* propagation of $CD44^+/CD24^{low/neg}/Oct\text{-}4^+$/connexin 43^{neg} (Cx43) tumor-initiating breast cancer stem cells was reported [143]. However, a recently performed evaluation showed no correlation between the number of $CD44^+/CD24^{low/neg}$ cells in primary breast cancer samples and breast tumor progression, tumor recurrence, and survival of these patients [156]. Thus definitive markers for breast cancer stem cells are still missing.

The situation is similar in prostate cancer. Stem cells account for approximately 0.1% of the prostate tumor mass and display a $CD44^+/\ \alpha_2\beta_1{}^{hi}/CD133^+$ phenotype. Only the $CD133^+$ cells show capability for self-renewal and immortality *in vitro*. Prostate cancer cells that do not express CD133, but are still $CD44^+\ /\ \alpha_2\beta_1{}^{hi}$ are identified as transiently amplifying progeny. However, no correlation could be found between the number of $CD44^+\ /\alpha_2\beta_1{}^{hi}/CD133^+$ cells and tumor grade [134]. In another study, the $CD44^+$ cells were characterized as more tumorigenic, metastatic and having higher mRNA levels of "stemness" genes [157]. CD44 glycoprotein is an adhesion molecule that is involved in cell signaling and occurs in multiple splice variants. Like the CD133 molecule, CD44 is also found on normal prostate stem cells as well as on mature tumor cells and normal tissues, thus its importance to cancer stem cell biology and function is still poorly understood.

5.8 The Stem Cell Niche as the Tumor Microenvironment

Normal adult tissue specific stem cells reside in a local microenvironment called "niche" [158]. Via adhesive molecules, cadherins and integrins, stem cells anchor in the basement membrane that separates niche cells from stem cells. Extracellular and intrinsic factors, e.g. cytokines, growth factors, neural inputs or physical stimuli, regulate self-renewal, cell division and daughter cell fate, proliferation, and migration of stem cells [159]. Secretion of certain growth factors, such as fibroblast growth factor (FGF), and mobilization of matrix molecules by niche cells stimulates activity and growth of stem cells [121]. The niche cells also release growth inhibitory molecules, such as TGF-β, bone morphogenic proteins (BMPs) and cell cycle inhibitors to suppress stem cell differentiation [160], [161]. Vascular endothelial cells may also provide external signals and growth factors. For normal tissue development the number of niches may vary in response to the developmental needs, e.g. the expansion of the stem cell pool or stromal cells after injury [162]. In addition, niches can modify their regulatory properties in response to stem cell activity [163]. Deregulation of the complex signaling pathways within the niche (cells) may lead to significant changes in stem cell number and stem cell fates, including remodeling the niche to favor malignant transformation. By the same token, malignant transformation of the stem cells can change completely their niche by affecting secretion of cytokines and growth factors through aberrant expression of cell surface molecules leading to uncontrolled cell signaling. Change in the intensity and nature of infiltrating immune cells and release of inflammatory cytokines may contribute to further alteration of the niche and progression of the tumor to a more mature state [144].

5.9 Signaling Pathways that Control Cancer Stem Cell Function

Many important regulatory signaling pathways have been identified that contribute to stem cell survival and proliferation, malignant transformation as well as cell-cell and cell-matrix communication. For example, Wnt, a secreted highly hydrophobic signaling molecule controls stem cell activity and daughter cell fate through regulation of β-catenin [164–166]. In the absence of Wnt signaling, phosphorylated β-catenin is degraded by a complex of adenomatous polyposis coli (APC), Axin and glycogen synthase kinase-3β proteins via the ubiquitin pathway. Abnormal activation of Wnt signaling and accumulation of nuclear β-catenin in cancer stem cells can lead to a continuous targeting of a large number of downstream genes, such as the proto-oncogene c-myc, which allows undefined expansion of stem cells through inhibition of $p21^{cip-1/WAF}$, a cell-cycle inhibitor [167], as well as their migration out of the niche and accumulation of mutations that result in tumor progression [168, 169].

Another example is overexpression of membrane-associated ABC-transporters as regulators in stem and cancer stem cell survival. MDR1, the multidrug resistance P-glycoprotein (P-gp), also called ABCB1, was the first identified drug transporter that effluxes a wide range of diverse substrates [170]. Among the large family of ABC-transporters, the breast cancer resistance protein (BCRP or ABCG2) has been characterized as a novel stem cell-related transporter [171]. ABC transporters can also protect cells from apoptosis induced by a variety of factors, such as TNF-alpha or UV-irradiation, as well as modulate signal transduction pathways [170] that may have a key role in cancer stem cell survival. Bone morphogenetic proteins (BMPs) and Wnt signaling act synergistically in decisions about stem cell fate by restricting their activation, self-renewal and maintenance of their multipotency as shown during neurogenesis on neural crest stem cells [172].

5.10 Cancer Stem Cell Antigens and the Tumor Microenvironment

As was discussed earlier for mature tumor cells, expression of certain tumor-associated antigens on cancer stem cells can profoundly influence their microenvironment. Evidence is beginning to accumulate that some of the tumor antigens described on mature tumor cells are also present on cancer stem (like) cells. One of these antigens is MUC1. Because a lot is known about its immunoregulatory and oncogenic functions in mature tumor cells, we will present a hypothetical picture of its impact on the microenvironment of a cancer stem cell.

A tissue specific adult stem cell that has accumulated a significant number of oncogenic mutations over its long life might eventually turn malignant if it successfully passes through the many survival checkpoints. In comparison to normal stem cells that reside in their niche and depend on interactions with the surrounding stromal cells that regulate survival and stem cell divisions, a transformed stem cell becomes independent from the niche. During this process we suspect that the malignant stem cell will start expressing modified tumor forms of some self-antigens. Overexpression of the tumor form of MUC1 could at that point serve to protect the cancer stem cell from apoptosis, increase its proliferation and perhaps signal differentiation into tumor progenitor cells. At the same time, it will send a chemotactic signal to the cells of the innate immune system to survey the emerging tumor site, but also provide a supportive cytokine milieu for further growth of the nascent tumor. The host immune system eventually becomes aware of the changes in the cancer stem cell niche in part through the recognition of the modified self-antigens. In the very early stage of tumor growth, the local microenvironment is not yet tumor supportive and it is not immunogenic either. The immune system recognizes the tumor-induced changes in the self-molecules but this recognition is still below the threshold of full activation. During the time when a few cancer stem and progenitor cells are tolerated by the immune system, they have the opportunity to shape their microenvironment to support their further growth. This includes promotion of an immunosuppressive environment that will inhibit antitumor immunity once it is formed and arrives to the tumor site. Expression of MUC1, for example, can influence and manipulate cell signaling via phosphorylation of its C-terminal domain and thus stimulate gene transcription and influence uncontrolled cell proliferation. Overexpression of phosphorylated MUC1 can compete with E-cadherin expressed by neighbouring cells, for binding to β-catenin and stimulating anchorage independent growth. The abundance of sialic acid moieties on tumor MUC1 can mediate repulsion from a united cell structure and eventually help migration of the tumor cells. This is furthermore facilitated by the extended rigid structure of MUC1. MUC1 has been reported to compromise complement-dependent cytotoxicity on mucin expressing tumor cells. When cells are coated with MUC1, complement is activated very far away from the cancer cell surface and thus, the membrane attack complex that forms pores in the cell membrane, can not be initiated or could not reach the cell membrane [173].

As stem cells convert to transient amplifying cells they become more and more independent from contact inhibitory growth and growth factors and turn more aggressive, eventually generating a heterogeneous mature tumor cell population that forms the tumor mass. Throughout this process, MUC1 expression is high and its glycosylation is low. This serves to attract immature DCs to the nascent tumor microenvironment. Immature DCs express various C-type lectin receptors that can bind and internalize MUC1. Immature DCs can also take up other tumor antigens through their active endocytosis. This would at first appear as a very beneficial process that could favor development of effective tumor immunosurveillance.

However, just the opposite may happen. In order to effectively present MUC1 and other tumor antigens to T cells, DCs need to fully mature into cells with high stimulatory capacity promoted by high levels of co-stimulatory molecules and cytokine production, preferably IL-12. This does not appear to happen after their interaction with tumor MUC1. Immature DCs attracted to the site on MUC1-expressing tumor cells mature into regulatory DCs that secrete high concentration of IL-10, but not IL-12, and express very low levels of co-stimulatory molecules [174, 175]. These DCs also produce proinflammatory cytokines IL-6 and TNF-alpha and promote generation of T cells that produce IL-13. IL-13 can have a dual function in the tumor microenvironment. On one hand it can suppress CTLs, and induces the generation of regulatory T cells, thus preventing effective immune surveillance of the tumor. On the other hand, it serves as a paracrine growth factor for tumor cells that express IL-13 receptor. Another important component of the tumor microenvironment is represented by the stromal fibroblasts, which become activated and begin to secrete TGF-β and thus, further inhibit antitumor immunity. All of these events, in part driven by MUC1, are known to happen in the environment of a mature tumor. The likelihood is high that these processes also influence the cancer stem cell niche and are responsible for tumor development and/or recurrence.

5.11 Targeting Tumor Antigens to Change Tumor Microenvironment and Restore Immunosurveillance

Most of the events described above that happen in the tumor microenvironment are not observed if strong adaptive immunity exists against the aberrant forms of various tumor antigens. In highlighting the good, Dr. Jekyll side of these antigens, we discussed evidence of low immunity against these antigens could prevent tumor outgrowth and lead to long-term protective antitumor memory responses. Vaccination directed against certain TAAs can induce strong immune response that may mediate clearance of existing cancer foci and moreover, might be protective against new malignant transformation starting from a cancer stem cell. The presence of strong adaptive immunity in the tumor microenvironment can restore the balance between the proinflammatory responses of innate immune cells, that can be immunosuppressive and tumor promoting, and antitumor effector cells such as activated macrophage, T cells, NK cells and their cytokines.

Strong antitumor effector responses can be induced through vaccination with specific antigens and should be expected to change the tumor microenvironment such that it no longer supports tumor growth. The problem has been, that vaccines have been administered in the therapeutic setting where the tumor microenvironment is already fully established, immunosuppressive and tumor growth supportive. Even when effective immunity is generated, the effector cells do not function well once they arrive at the tumor site. The potential solution to this problem is to try to alter the tumor microenvironment prior to the application of immunotherapy. This can be accomplished by delivery of certain cytokines and chemokines [176–178] as well as by depletion of specific cell populations, such as Tregs [179]. Alternatively, induction of antitumor immune responses can be generated prophylactically in people at high risk for developing certain forms of cancer. These memory responses would get activated faster in response to early changes, such as new antigens on cancer stem cells and eliminate these cells before they have a chance to establish a tumor-promoting environment.

Abbreviations

ALL	acute lymphoblastic leukaemia
AML	acute myeloid leukemias
CEA	carcinoembryonic antigen
CML	chronic myelogenous leukemia
CTL	cytotoxic T cells
DCs	dendritic cell(s)
ERK1/2	extracellular signal-regulated kinase 1/2
FGFR3	fibroblast growth factor receptor 3
GM-CSF	granulocyte macrophage colony stimulating factor
HLA	human leukocyte antigen
HSC	hematopoietic stem cells
ICAM	intercellular adhesion molecule 1
IL	interleucin
M-CSF	macrophage colony stimulating factor
MHC	major histocompatibility complex
MUC1	mucin1
NOD/scid	non-obese diabetic/severe combined immunodeficiency
ROS	reactive oxygen species
SEREX	serological expression cloning
SP	side Population
TAA	tumor Associated Antigen(s)
TF	Thomsen-Friedenreich antigen
TGF-ß	transforming growth factor ß
Th1	T helper cells type I
Treg	regulatory $CD4^+/CD25^+/FOXP3^+$ T cells
VEGF	vascular endothelial growth factor
VNTR	variable number of tandem repeats

References

1. Kass, E., Schlom, J., Thompson, J., Guadagni, F., Graziano, P., and Greiner, J. W. Induction of protective host immunity to carcinoembryonic antigen (CEA), a self-antigen in CEA transgenic mice, by immunizing with a recombinant vaccinia-CEA virus. Cancer Res, *59*: 676–683, 1999.

2. Satoi, J., Murata, K., Lechmann, M., Manickan, E., Zhang, Z., Wedemeyer, H., Rehermann, B., and Liang, T. J. Genetic immunization of wild-type and hepatitis C virus transgenic mice reveals a hierarchy of cellular immune response and tolerance induction against hepatitis C virus structural proteins. J Virol, *75*: 12121–12127, 2001.
3. Bos, R., van Duikeren, S., van Hall, T., Kaaijk, P., Taubert, R., Kyewski, B., Klein, L., Melief, C. J., and Offringa, R. Expression of a natural tumor antigen by thymic epithelial cells impairs the tumor-protective CD4+ T-cell repertoire. Cancer Res, *65*: 6443–6449, 2005.
4. Muders, M., Ghoreschi, K., Suckfuell, M., Zimmermann, W., and Enders, G. Studies on the immunogenicity of hCEA in a transgenic mouse model. Int J Colorectal Dis, *18*: 153–159, 2003.
5. Cramer, D. W., Titus-Ernstoff, L., McKolanis, J. R., Welch, W. R., Vitonis, A. F., Berkowitz, R. S., and Finn, O. J. Conditions associated with antibodies against the tumor-associated antigen MUC1 and their relationship to risk for ovarian cancer. Cancer Epidemiol Biomarkers Prev, *14*: 1125–1131, 2005.
6. Toubi, E. and Shoenfeld, Y. Protective autoimmunity in cancer (review). Oncol Rep, *17*: 245–251, 2007.
7. Vollmers, H. P. and Brandlein, S. The "early birds": natural IgM antibodies and immune surveillance. Histol Histopathol, *20*: 927–937, 2005.
8. Brandlein, S., Pohle, T., Ruoff, N., Wozniak, E., Muller-Hermelink, H. K., and Vollmers, H. P. Natural IgM antibodies and immunosurveillance mechanisms against epithelial cancer cells in humans. Cancer Res, *63*: 7995–8005, 2003.
9. Brichard, V., Van Pel, A., Wolfel, T., Wolfel, C., De Plaen, E., Lethe, B., Coulie, P., and Boon, T. The tyrosinase gene codes for an antigen recognized by autologous cytolytic T lymphocytes on HLA-A2 melanomas. J Exp Med, *178*: 489–495, 1993.
10. Coulie, P. G., Brichard, V., Van Pel, A., Wolfel, T., Schneider, J., Traversari, C., Mattei, S., De Plaen, E., Lurquin, C., Szikora, J. P., Renauld, J. C., and Boon, T. A new gene coding for a differentiation antigen recognized by autologous cytolytic T lymphocytes on HLA-A2 melanomas. J Exp Med, *180*: 35–42, 1994.
11. Hunt, D. F., Michel, H., Dickinson, T. A., Shabanowitz, J., Cox, A. L., Sakaguchi, K., Appella, E., Grey, H. M., and Sette, A. Peptides presented to the immune system by the murine class II major histocompatibility complex molecule I-Ad. Science, *256*: 1817–1820, 1992.
12. Hunt, D. F., Henderson, R. A., Shabanowitz, J., Sakaguchi, K., Michel, H., Sevilir, N., Cox, A. L., Appella, E., and Engelhard, V. H. Characterization of peptides bound to the class I MHC molecule HLA-A2.1 by mass spectrometry. Science, *255*: 1261–1263, 1992.
13. Fisk, B., Blevins, T. L., Wharton, J. T., and Ioannides, C. G. Identification of an immunodominant peptide of HER-2/neu protooncogene recognized by ovarian tumor-specific cytotoxic T lymphocyte lines. J Exp Med, *181*: 2109–2117, 1995.
14. Blake, J., Johnston, J. V., Hellstrom, K. E., Marquardt, H., and Chen, L. Use of combinatorial peptide libraries to construct functional mimics of tumor epitopes recognized by MHC class I-restricted cytolytic T lymphocytes. J Exp Med, *184*: 121–130, 1996.
15. Tureci, O., Sahin, U., and Pfreundschuh, M. Serological analysis of human tumor antigens: molecular definition and implications. Mol Med Today, *3*: 342–349, 1997.
16. Kao, H., Marto, J. A., Hoffmann, T. K., Shabanowitz, J., Finkelstein, S. D., Whiteside, T. L., Hunt, D. F., and Finn, O. J. Identification of cyclin B1 as a shared human epithelial tumor-associated antigen recognized by T cells. J Exp Med, *194*: 1313–1323, 2001.
17. Terry, K. L., Titus-Ernstoff, L., McKolanis, J. R., Welch, W. R., Finn, O. J., and Cramer, D. W. Incessant ovulation, mucin 1 immunity, and risk for ovarian cancer. Cancer Epidemiol Biomarkers Prev, *16*: 30–35, 2007.
18. Gendler, S. J., Lancaster, C. A., Taylor-Papadimitriou, J., Duhig, T., Peat, N., Burchell, J., Pemberton, L., Lalani, E. N., and Wilson, D. Molecular cloning and expression of human tumor-associated polymorphic epithelial mucin. J Biol Chem, *265*: 15286–15293, 1990.
19. Hanisch, F. G. and Muller, S. MUC1: the polymorphic appearance of a human mucin. Glycobiology, *10*: 439–449, 2000.
20. Taylor-Papadimitriou, J., Burchell, J., Miles, D. W., and Dalziel, M. MUC1 and cancer. Biochim Biophys Acta, *1455*: 301–313, 1999.

21. Gendler, S. J. MUC1, the renaissance molecule. J Mammary Gland Biol Neoplasia, *6*: 339–353, 2001.
22. O'Connor, J. C., Julian, J., Lim, S. D., and Carson, D. D. MUC1 expression in human prostate cancer cell lines and primary tumors. Prostate Cancer Prostatic Dis, *8*: 36–44, 2005.
23. Li, Y., Bharti, A., Chen, D., Gong, J., and Kufe, D. Interaction of glycogen synthase kinase 3beta with the DF3/MUC1 carcinoma-associated antigen and beta-catenin. Mol Cell Biol, *18*: 7216–7224, 1998.
24. Ren, J., Li, Y., and Kufe, D. Protein kinase C delta regulates function of the DF3/MUC1 carcinoma antigen in beta-catenin signaling. J Biol Chem, *277*: 17616–17622, 2002.
25. Li, Y., Kuwahara, H., Ren, J., Wen, G., and Kufe, D. The c-Src tyrosine kinase regulates signaling of the human DF3/MUC1 carcinoma-associated antigen with GSK3 beta and beta-catenin. J Biol Chem, *276*: 6061–6064, 2001.
26. Li, Y., Ren, J., Yu, W., Li, Q., Kuwahara, H., Yin, L., Carraway, K. L., 3rd, and Kufe, D. The epidermal growth factor receptor regulates interaction of the human DF3/MUC1 carcinoma antigen with c-Src and beta-catenin. J Biol Chem, *276*: 35239–35242, 2001.
27. Ren, J., Raina, D., Chen, W., Li, G., Huang, L., and Kufe, D. MUC1 oncoprotein functions in activation of fibroblast growth factor receptor signaling. Mol Cancer Res, *4*: 873–883, 2006.
28. Girling, A., Bartkova, J., Burchell, J., Gendler, S., Gillett, C., and Taylor-Papadimitriou, J. A core protein epitope of the polymorphic epithelial mucin detected by the monoclonal antibody SM-3 is selectively exposed in a range of primary carcinomas. Int J Cancer, *43*: 1072–1076, 1989.
29. Burchell, J., Durbin, H., and Taylor-Papadimitriou, J. Complexity of expression of antigenic determinants, recognized by monoclonal antibodies HMFG-1 and HMFG-2, in normal and malignant human mammary epithelial cells. J Immunol, *131*: 508–513, 1983.
30. Kam, J. L., Regimbald, L. H., Hilgers, J. H., Hoffman, P., Krantz, M. J., Longenecker, B. M., and Hugh, J. C. MUC1 synthetic peptide inhibition of intercellular adhesion molecule-1 and MUC1 binding requires six tandem repeats. Cancer Res, *58*: 5577–5581, 1998.
31. Regimbald, L. H., Pilarski, L. M., Longenecker, B. M., Reddish, M. A., Zimmermann, G., and Hugh, J. C. The breast mucin MUCI as a novel adhesion ligand for endothelial intercellular adhesion molecule 1 in breast cancer. Cancer Res, *56*: 4244–4249, 1996.
32. Pandey, P., Kharbanda, S., and Kufe, D. Association of the DF3/MUC1 breast cancer antigen with Grb2 and the Sos/Ras exchange protein. Cancer Res, *55*: 4000–4003, 1995.
33. Olayioye, M. A., Neve, R. M., Lane, H. A., and Hynes, N. E. The ErbB signaling network: receptor heterodimerization in development and cancer. Embo J, *19*: 3159–3167, 2000.
34. Schroeder, J. A., Thompson, M. C., Gardner, M. M., and Gendler, S. J. Transgenic MUC1 interacts with epidermal growth factor receptor and correlates with mitogen-activated protein kinase activation in the mouse mammary gland. J Biol Chem, *276*: 13057–13064, 2001.
35. Yin, L., Huang, L., and Kufe, D. MUC1 oncoprotein activates the FOXO3a transcription factor in a survival response to oxidative stress. J Biol Chem, *279*: 45721–45727, 2004.
36. Ren, J., Agata, N., Chen, D., Li, Y., Yu, W. H., Huang, L., Raina, D., Chen, W., Kharbanda, S., and Kufe, D. Human MUC1 carcinoma-associated protein confers resistance to genotoxic anticancer agents. Cancer Cell, *5*: 163–175, 2004.
37. Jerome, K. R., Barnd, D. L., Bendt, K. M., Boyer, C. M., Taylor-Papadimitriou, J., McKenzie, I. F., Bast, R. C., Jr., and Finn, O. J. Cytotoxic T-lymphocytes derived from patients with breast adenocarcinoma recognize an epitope present on the protein core of a mucin molecule preferentially expressed by malignant cells. Cancer Res, *51*: 2908–2916, 1991.
38. Gaemers, I. C., Vos, H. L., Volders, H. H., van der Valk, S. W., and Hilkens, J. A stat-responsive element in the promoter of the episialin/MUC1 gene is involved in its overexpression in carcinoma cells. J Biol Chem, *276*: 6191–6199, 2001.
39. Engelmann, K., Kinlough, C. L., Muller, S., Razawi, H., Baldus, S. E., Hughey, R. P., and Hanisch, F. G. Transmembrane and secreted MUC1 probes show trafficking-dependent changes in O-glycan core profiles. Glycobiology, *15*: 1111–1124, 2005.
40. Agrawal, B., Gendler, S. J., and Longenecker, B. M. The biological role of mucins in cellular interactions and immune regulation: prospects for cancer immunotherapy. Mol Med Today, *4*: 397–403, 1998.

41. Brockhausen, I., Yang, J. M., Burchell, J., Whitehouse, C., and Taylor-Papadimitriou, J. Mechanisms underlying aberrant glycosylation of MUC1 mucin in breast cancer cells. Eur J Biochem, *233*: 607–617, 1995.
42. Hanisch, F. G., Peter-Katalinic, J., Egge, H., Dabrowski, U., and Uhlenbruck, G. Structures of acidic O-linked polylactosaminoglycans on human skim milk mucins. Glycoconj J, *7*: 525–543, 1990.
43. Hanisch, F. G., Uhlenbruck, G., Peter-Katalinic, J., Egge, H., Dabrowski, U., and Dabrowski, J. Unbranched polylactosamino-O-glycans on human skim milk mucins exhibit Gal beta(1–4)GlcNAc beta(1–6) repeating units. Symp Soc Exp Biol, *43*: 155–162, 1989.
44. Whitehouse, C., Burchell, J., Gschmeissner, S., Brockhausen, I., Lloyd, K. O., and Taylor-Papadimitriou, J. A transfected sialyltransferase that is elevated in breast cancer and localizes to the medial/trans-Golgi apparatus inhibits the development of core-2-based O-glycans. J Cell Biol, *137*: 1229–1241, 1997.
45. Lloyd, K. O., Burchell, J., Kudryashov, V., Yin, B. W., and Taylor-Papadimitriou, J. Comparison of O-linked carbohydrate chains in MUC-1 mucin from normal breast epithelial cell lines and breast carcinoma cell lines. Demonstration of simpler and fewer glycan chains in tumor cells. J Biol Chem, *271*: 33325–33334, 1996.
46. Price, M. R., Rye, P. D., Petrakou, E., Murray, A., Brady, K., Imai, S., Haga, S., Kiyozuka, Y., Schol, D., Meulenbroek, M. F., Snijdewint, F. G., von Mensdorff-Pouilly, S., Verstraeten, R. A., Kenemans, P., Blockzjil, A., Nilsson, K., Nilsson, O., Reddish, M., Suresh, M. R., Koganty, R. R., Fortier, S., Baronic, L., Berg, A., Longenecker, M. B., Hilgers, J., et al. Summary report on the ISOBM TD-4 Workshop: analysis of 56 monoclonal antibodies against the MUC1 mucin. San Diego, Calif., November 17–23, 1996. Tumour Biol, *19 Suppl. 1*: 1–20, 1998.
47. Muller, S., Alving, K., Peter-Katalinic, J., Zachara, N., Gooley, A. A., and Hanisch, F. G. High density O-glycosylation on tandem repeat peptide from secretory MUC1 of T47D breast cancer cells. J Biol Chem, *274*: 18165–18172, 1999.
48. Muller, S., Goletz, S., Packer, N., Gooley, A., Lawson, A. M., and Hanisch, F. G. Localization of O-glycosylation sites on glycopeptide fragments from lactation-associated MUC1. All putative sites within the tandem repeat are glycosylation targets in vivo. J Biol Chem, *272*: 24780–24793, 1997.
49. Fontenot, J. D., Gatewood, J. M., Mariappan, S. V., Pau, C. P., Parekh, B. S., George, J. R., and Gupta, G. Human immunodeficiency virus (HIV) antigens: structure and serology of multivalent human mucin MUC1-HIV V3 chimeric proteins. Proc Natl Acad Sci USA, *92*: 315–319, 1995.
50. Kirnarsky, L., Nomoto, M., Ikematsu, Y., Hassan, H., Bennett, E. P., Cerny, R. L., Clausen, H., Hollingsworth, M. A., and Sherman, S. Structural analysis of peptide substrates for mucin-type O-glycosylation. Biochemistry, *37*: 12811–12817, 1998.
51. Finn, O. J., Jerome, K. R., Henderson, R. A., Pecher, G., Domenech, N., Magarian-Blander, J., and Barratt-Boyes, S. M. MUC-1 epithelial tumor mucin-based immunity and cancer vaccines. Immunol Rev, *145*: 61–89, 1995.
52. Karsten, U., Diotel, C., Klich, G., Paulsen, H., Goletz, S., Muller, S., and Hanisch, F. G. Enhanced binding of antibodies to the DTR motif of MUC1 tandem repeat peptide is mediated by site-specific glycosylation. Cancer Res, *58*: 2541–2549, 1998.
53. Karsten, U., Serttas, N., Paulsen, H., Danielczyk, A., and Goletz, S. Binding patterns of DTR-specific antibodies reveal a glycosylation-conditioned tumor-specific epitope of the epithelial mucin (MUC1). Glycobiology, *14*: 681–692, 2004.
54. Sorensen, A. L., Reis, C. A., Tarp, M. A., Mandel, U., Ramachandran, K., Sankaranarayanan, V., Schwientek, T., Graham, R., Taylor-Papadimitriou, J., Hollingsworth, M. A., Burchell, J., and Clausen, H. Chemoenzymatically synthesized multimeric Tn/STn MUC1 glycopeptides elicit cancer-specific anti-MUC1 antibody responses and override tolerance. Glycobiology, *16*: 96–107, 2006.
55. Engelmann, K., Baldus, S. E., and Hanisch, F. G. Identification and topology of variant sequences within individual repeat domains of the human epithelial tumor mucin MUC1. J Biol Chem, *276*: 27764–27769, 2001.
56. von Mensdorff-Pouilly, S., Kinarsky, L., Engelmann, K., Baldus, S. E., Verheijen, R. H., Hollingsworth, M. A., Pisarev, V., Sherman, S., and Hanisch, F. G. Sequence-variant repeats

of MUC1 show higher conformational flexibility, are less densely O-glycosylated and induce differential B lymphocyte responses. Glycobiology, *15*: 735–746, 2005.
57. Barnd, D. L., Lan, M. S., Metzgar, R. S., and Finn, O. J. Specific, major histocompatibility complex-unrestricted recognition of tumor-associated mucins by human cytotoxic T cells. Proc Natl Acad Sci USA, *86*: 7159–7163, 1989.
58. Jerome, K. R., Domenech, N., and Finn, O. J. Tumor-specific cytotoxic T cell clones from patients with breast and pancreatic adenocarcinoma recognize EBV-immortalized B cells transfected with polymorphic epithelial mucin complementary DNA. J Immunol, *151*: 1654–1662, 1993.
59. Magarian-Blander, J., Ciborowski, P., Hsia, S., Watkins, S. C., and Finn, O. J. Intercellular and intracellular events following the MHC-unrestricted TCR recognition of a tumor-specific peptide epitope on the epithelial antigen MUC1. J Immunol, *160*: 3111–3120, 1998.
60. Hiltbold, E. M., Ciborowski, P., and Finn, O. J. Naturally processed class II epitope from the tumor antigen MUC1 primes human CD4+ T cells. Cancer Res, *58*: 5066–5070, 1998.
61. Vlad, A. M., Muller, S., Cudic, M., Paulsen, H., Otvos, L., Jr., Hanisch, F. G., and Finn, O. J. Complex carbohydrates are not removed during processing of glycoproteins by dendritic cells: processing of tumor antigen MUC1 glycopeptides for presentation to major histocompatibility complex class II-restricted T cells. J Exp Med, *196*: 1435–1446, 2002.
62. Hanisch, F. G. Design of a MUC1-based cancer vaccine. Biochem Soc Trans, *33*: 705–708, 2005.
63. Vlad, A. M., Kettel, J. C., Alajez, N. M., Carlos, C. A., and Finn, O. J. MUC1 immunobiology: from discovery to clinical applications. Adv Immunol, *82*: 249–293, 2004.
64. Gilewski, T., Adluri, S., Ragupathi, G., Zhang, S., Yao, T. J., Panageas, K., Moynahan, M., Houghton, A., Norton, L., and Livingston, P. O. Vaccination of high-risk breast cancer patients with mucin-1 (MUC1) keyhole limpet hemocyanin conjugate plus QS-21. Clin Cancer Res, *6*: 1693–1701, 2000.
65. Reddish, M., MacLean, G. D., Koganty, R. R., Kan-Mitchell, J., Jones, V., Mitchell, M. S., and Longenecker, B. M. Anti-MUC1 class I restricted CTLs in metastatic breast cancer patients immunized with a synthetic MUC1 peptide. Int J Cancer, *76*: 817–823, 1998.
66. Ramanathan, R. K., Lee, K. M., McKolanis, J., Hitbold, E., Schraut, W., Moser, A. J., Warnick, E., Whiteside, T., Osborne, J., Kim, H., Day, R., Troetschel, M., and Finn, O. J. Phase I study of a MUC1 vaccine composed of different doses of MUC1 peptide with SB-AS2 adjuvant in resected and locally advanced pancreatic cancer. Cancer Immunol Immunother, *54*: 254–264, 2005.
67. Yamamoto, K., Ueno, T., Kawaoka, T., Hazama, S., Fukui, M., Suehiro, Y., Hamanaka, Y., Ikematsu, Y., Imai, K., Oka, M., and Hinoda, Y. MUC1 peptide vaccination in patients with advanced pancreas or biliary tract cancer. Anticancer Res, *25*: 3575–3579, 2005.
68. Wierecky, J., Mueller, M., and Brossart, P. Dendritic cell-based cancer immunotherapy targeting MUC-1. Cancer Immunol Immunother, *55*: 63–67, 2006.
69. von Mensdorff-Pouilly, S., Petrakou, E., Kenemans, P., van Uffelen, K., Verstraeten, A. A., Snijdewint, F. G., van Kamp, G. J., Schol, D. J., Reis, C. A., Price, M. R., Livingston, P. O., and Hilgers, J. Reactivity of natural and induced human antibodies to MUC1 mucin with MUC1 peptides and n-acetylgalactosamine (GalNAc) peptides. Int J Cancer, *86*: 702–712, 2000.
70. Carlos, C. A., Dong, H. F., Howard, O. M., Oppenheim, J. J., Hanisch, F. G., and Finn, O. J. Human tumor antigen MUC1 is chemotactic for immature dendritic cells and elicits maturation but does not promote Th1 type immunity. J Immunol, *175*: 1628–1635, 2005.
71. Terabe, M., Park, J. M., and Berzofsky, J. A. Role of IL-13 in regulation of anti-tumor immunity and tumor growth. Cancer Immunol Immunother, *53*: 79–85, 2004.
72. Wei, X., Xu, H., and Kufe, D. Human MUC1 oncoprotein regulates p53-responsive gene transcription in the genotoxic stress response. Cancer Cell, *7*: 167–178, 2005.
73. Thompson, E. J., Shanmugam, K., Hattrup, C. L., Kotlarczyk, K. L., Gutierrez, A., Bradley, J. M., Mukherjee, P., and Gendler, S. J. Tyrosines in the MUC1 cytoplasmic tail modulate transcription via the extracellular signal-regulated kinase 1/2 and nuclear factor-kappaB pathways. Mol Cancer Res, *4*: 489–497, 2006.
74. Tsutsumida, H., Swanson, B. J., Singh, P. K., Caffrey, T. C., Kitajima, S., Goto, M., Yonezawa, S., and Hollingsworth, M. A. RNA interference suppression of MUC1 reduces the

growth rate and metastatic phenotype of human pancreatic cancer cells. Clin Cancer Res, *12*: 2976–2987, 2006.

75. Julien, S., Lagadec, C., Krzewinski-Recchi, M. A., Courtand, G., Le Bourhis, X., and Delannoy, P. Stable expression of sialyl-Tn antigen in T47-D cells induces a decrease of cell adhesion and an increase of cell migration. Breast Cancer Res Treat, *90*: 77–84, 2005.
76. Zhang, K., Baeckstrom, D., Brevinge, H., and Hansson, G. C. Secreted MUC1 mucins lacking their cytoplasmic part and carrying sialyl-Lewis a and x epitopes from a tumor cell line and sera of colon carcinoma patients can inhibit HL-60 leukocyte adhesion to E-selectin-expressing endothelial cells. J Cell Biochem, *60*: 538–549, 1996.
77. Morse, M. A., Nair, S. K., Boczkowski, D., Tyler, D., Hurwitz, H. I., Proia, A., Clay, T. M., Schlom, J., Gilboa, E., and Lyerly, H. K. The feasibility and safety of immunotherapy with dendritic cells loaded with CEA mRNA following neoadjuvant chemoradiotherapy and resection of pancreatic cancer. Int J Gastrointest Cancer, *32*: 1–6, 2002.
78. Greiner, J. W., Zeytin, H., Anver, M. R., and Schlom, J. Vaccine-based therapy directed against carcinoembryonic antigen demonstrates antitumor activity on spontaneous intestinal tumors in the absence of autoimmunity. Cancer Res, *62*: 6944–6951, 2002.
79. Hodge, J. W., Grosenbach, D. W., Aarts, W. M., Poole, D. J., and Schlom, J. Vaccine therapy of established tumors in the absence of autoimmunity. Clin Cancer Res, *9*: 1837–1849, 2003.
80. Saha, A., Chatterjee, S. K., Foon, K. A., and Bhattacharya-Chatterjee, M. Anti-idiotype antibody induced cellular immunity in mice transgenic for human carcinoembryonic antigen. Immunology, *118*: 483–496, 2006.
81. Saha, A., Chatterjee, S. K., Foon, K. A., Primus, F. J., Sreedharan, S., Mohanty, K., and Bhattacharya-Chatterjee, M. Dendritic cells pulsed with an anti-idiotype antibody mimicking carcinoembryonic antigen (CEA) can reverse immunological tolerance to CEA and induce antitumor immunity in CEA transgenic mice. Cancer Res, *64*: 4995–5003, 2004.
82. Ojima, T., Iwahashi, M., Nakamura, M., Matsuda, K., Nakamori, M., Ueda, K., Naka, T., Ishida, K., Primus, F. J., and Yamaue, H. Successful cancer vaccine therapy for carcinoembryonic antigen (CEA)-expressing colon cancer using genetically modified dendritic cells that express CEA and T helper-type 1 cytokines in CEA transgenic mice. Int J Cancer, *120*: 585–593, 2007.
83. Ojima, T., Iwahashi, M., Nakamura, M., Matsuda, K., Naka, T., Nakamori, M., Ueda, K., Ishida, K., and Yamaue, H. The boosting effect of co-transduction with cytokine genes on cancer vaccine therapy using genetically modified dendritic cells expressing tumor-associated antigen. Int J Oncol, *28*: 947–953, 2006.
84. Berinstein, N. L. Carcinoembryonic antigen as a target for therapeutic anticancer vaccines: a review. J Clin Oncol, *20*: 2197–2207, 2002.
85. Marshall, J. L., Hoyer, R. J., Toomey, M. A., Faraguna, K., Chang, P., Richmond, E., Pedicano, J. E., Gehan, E., Peck, R. A., Arlen, P., Tsang, K. Y., and Schlom, J. Phase I study in advanced cancer patients of a diversified prime-and-boost vaccination protocol using recombinant vaccinia virus and recombinant nonreplicating avipox virus to elicit anti-carcinoembryonic antigen immune responses. J Clin Oncol, *18*: 3964–3973, 2000.
86. Lesterhuis, W. J., de Vries, I. J., Schuurhuis, D. H., Boullart, A. C., Jacobs, J. F., de Boer, A. J., Scharenborg, N. M., Brouwer, H. M., van de Rakt, M. W., Figdor, C. G., Ruers, T. J., Adema, G. J., and Punt, C. J. Vaccination of colorectal cancer patients with CEA-loaded dendritic cells: antigen-specific T cell responses in DTH skin tests. Ann Oncol, *17*: 974–980, 2006.
87. Ullenhag, G. J., Frodin, J. E., Jeddi-Tehrani, M., Strigard, K., Eriksson, E., Samanci, A., Choudhury, A., Nilsson, B., Rossmann, E. D., Mosolits, S., and Mellstedt, H. Durable carcinoembryonic antigen (CEA)-specific humoral and cellular immune responses in colorectal carcinoma patients vaccinated with recombinant CEA and granulocyte/macrophage colony-stimulating factor. Clin Cancer Res, *10*: 3273–3281, 2004.
88. Liu, K. J., Wang, C. C., Chen, L. T., Cheng, A. L., Lin, D. T., Wu, Y. C., Yu, W. L., Hung, Y. M., Yang, H. Y., Juang, S. H., and Whang-Peng, J. Generation of carcinoembryonic antigen (CEA)-specific T-cell responses in HLA-A*0201 and HLA-A*2402 late-stage colorectal cancer patients after vaccination with dendritic cells loaded with CEA peptides. Clin Cancer Res, *10*: 2645–2651, 2004.

89. Weihrauch, M. R., Ansen, S., Jurkiewicz, E., Geisen, C., Xia, Z., Anderson, K. S., Gracien, E., Schmidt, M., Wittig, B., Diehl, V., Wolf, J., Bohlen, H., and Nadler, L. M. Phase I/II combined chemoimmunotherapy with carcinoembryonic antigen-derived HLA-A2-restricted CAP-1 peptide and irinotecan, 5-fluorouracil, and leucovorin in patients with primary metastatic colorectal cancer. Clin Cancer Res, 11: 5993–6001, 2005.
90. van Gisbergen, K. P., Aarnoudse, C. A., Meijer, G. A., Geijtenbeek, T. B., and van Kooyk, Y. Dendritic cells recognize tumor-specific glycosylation of carcinoembryonic antigen on colorectal cancer cells through dendritic cell-specific intercellular adhesion molecule-3-grabbing nonintegrin. Cancer Res, 65: 5935–5944, 2005.
91. Geijtenbeek, T. B., Torensma, R., van Vliet, S. J., van Duijnhoven, G. C., Adema, G. J., van Kooyk, Y., and Figdor, C. G. Identification of DC-SIGN, a novel dendritic cell-specific ICAM-3 receptor that supports primary immune responses. Cell, 100: 575–585, 2000.
92. Geijtenbeek, T. B., Krooshoop, D. J., Bleijs, D. A., van Vliet, S. J., van Duijnhoven, G. C., Grabovsky, V., Alon, R., Figdor, C. G., and van Kooyk, Y. DC-SIGN-ICAM-2 interaction mediates dendritic cell trafficking. Nat Immunol, 1: 353–357, 2000.
93. Gangopadhyay, A., Bajenova, O., Kelly, T. M., and Thomas, P. Carcinoembryonic antigen induces cytokine expression in Kuppfer cells: implications for hepatic metastasis from colorectal cancer. Cancer Res, 56: 4805–4810, 1996.
94. Phillips, M. L., Nudelman, E., Gaeta, F. C., Perez, M., Singhal, A. K., Hakomori, S., and Paulson, J. C. ELAM-1 mediates cell adhesion by recognition of a carbohydrate ligand, sialyl-Lex. Science, 250: 1130–1132, 1990.
95. Takada, A., Ohmori, K., Yoneda, T., Tsuyuoka, K., Hasegawa, A., Kiso, M., and Kannagi, R. Contribution of carbohydrate antigens sialyl Lewis A and sialyl Lewis X to adhesion of human cancer cells to vascular endothelium. Cancer Res, 53: 354–361, 1993.
96. Brezicka, F. T., Olling, S., Nilsson, O., Bergh, J., Holmgren, J., Sorenson, S., Yngvason, F., and Lindholm, L. Immunohistological detection of fucosyl-GM1 ganglioside in human lung cancer and normal tissues with monoclonal antibodies. Cancer Res, 49: 1300–1305, 1989.
97. Dickler, M. N., Ragupathi, G., Liu, N. X., Musselli, C., Martino, D. J., Miller, V. A., Kris, M. G., Brezicka, F. T., Livingston, P. O., and Grant, S. C. Immunogenicity of a fucosyl-GM1-keyhole limpet hemocyanin conjugate vaccine in patients with small cell lung cancer. Clin Cancer Res, 5: 2773–2779, 1999.
98. Helling, F., Zhang, S., Shang, A., Adluri, S., Calves, M., Koganty, R., Longenecker, B. M., Yao, T. J., Oettgen, H. F., and Livingston, P. O. GM2-KLH conjugate vaccine: increased immunogenicity in melanoma patients after administration with immunological adjuvant QS-21. Cancer Res, 55: 2783–2788, 1995.
99. Livingston, P. O., Adluri, S., Helling, F., Yao, T. J., Kensil, C. R., Newman, M. J., and Marciani, D. Phase 1 trial of immunological adjuvant QS-21 with a GM2 ganglioside-keyhole limpet haemocyanin conjugate vaccine in patients with malignant melanoma. Vaccine, 12: 1275–1280, 1994.
100. Krug, L. M., Ragupathi, G., Hood, C., Kris, M. G., Miller, V. A., Allen, J. R., Keding, S. J., Danishefsky, S. J., Gomez, J., Tyson, L., Pizzo, B., Baez, V., and Livingston, P. O. Vaccination of patients with small-cell lung cancer with synthetic fucosyl GM-1 conjugated to keyhole limpet hemocyanin. Clin Cancer Res, *10*: 6094–6100, 2004.
101. Cheung, N. K., Lazarus, H., Miraldi, F. D., Abramowsky, C. R., Kallick, S., Saarinen, U. M., Spitzer, T., Strandjord, S. E., Coccia, P. F., and Berger, N. A. Ganglioside GD2 specific monoclonal antibody 3F8: a phase I study in patients with neuroblastoma and malignant melanoma. J Clin Oncol, *5*: 1430–1440, 1987.
102. Irie, R. F. and Morton, D. L. Regression of cutaneous metastatic melanoma by intralesional injection with human monoclonal antibody to ganglioside GD2. Proc Natl Acad Sci USA, *83*: 8694–8698, 1986.
103. Houghton, A. N., Mintzer, D., Cordon-Cardo, C., Welt, S., Fliegel, B., Vadhan, S., Carswell, E., Melamed, M. R., Oettgen, H. F., and Old, L. J. Mouse monoclonal IgG3 antibody detecting GD3 ganglioside: a phase I trial in patients with malignant melanoma. Proc Natl Acad Sci USA, *82*: 1242–1246, 1985.

104. Furukawa, K., Hamamura, K., and Aixinjueluo, W. Biosignals modulated by tumor-associated carbohydrate antigens: novel targets for cancer therapy. Ann NY Acad Sci, 1086: 185–198, 2006.
105. Acheson, A., Sunshine, J. L., and Rutishauser, U. NCAM polysialic acid can regulate both cell-cell and cell-substrate interactions. J Cell Biol, 114: 143–153, 1991.
106. Rutishauser, U. Polysialic acid at the cell surface: biophysics in service of cell interactions and tissue plasticity. J Cell Biochem, 70: 304–312, 1998.
107. Daniel, L., Trouillas, J., Renaud, W., Chevallier, P., Gouvernet, J., Rougon, G., and Figarella-Branger, D. Polysialylated-neural cell adhesion molecule expression in rat pituitary transplantable tumors (spontaneous mammotropic transplantable tumor in Wistar-Furth rats) is related to growth rate and malignancy. Cancer Res, 60: 80–85, 2000.
108. Slingluff, C. L., Jr., Petroni, G. R., Yamshchikov, G. V., Barnd, D. L., Eastham, S., Galavotti, H., Patterson, J. W., Deacon, D. H., Hibbitts, S., Teates, D., Neese, P. Y., Grosh, W. W., Chianese-Bullock, K. A., Woodson, E. M., Wiernasz, C. J., Merrill, P., Gibson, J., Ross, M., and Engelhard, V. H. Clinical and immunologic results of a randomized phase II trial of vaccination using four melanoma peptides either administered in granulocyte-macrophage colony-stimulating factor in adjuvant or pulsed on dendritic cells. J Clin Oncol, 21: 4016–4026, 2003.
109. Markovic, S. N., Suman, V. J., Ingle, J. N., Kaur, J. S., Pitot, H. C., Loprinzi, C. L., Rao, R. D., Creagan, E. T., Pittelkow, M. R., Allred, J. B., Nevala, W. K., and Celis, E. Peptide vaccination of patients with metastatic melanoma: improved clinical outcome in patients demonstrating effective immunization. Am J Clin Oncol, 29: 352–360, 2006.
110. Linette, G. P., Zhang, D., Hodi, F. S., Jonasch, E. P., Longerich, S., Stowell, C. P., Webb, I. J., Daley, H., Soiffer, R. J., Cheung, A. M., Eapen, S. G., Fee, S. V., Rubin, K. M., Sober, A. J., and Haluska, F. G. Immunization using autologous dendritic cells pulsed with the melanoma-associated antigen gp100-derived G280–9V peptide elicits CD8+ immunity. Clin Cancer Res, 11: 7692–7699, 2005.
111. Yu, Z., Theoret, M. R., Touloukian, C. E., Surman, D. R., Garman, S. C., Feigenbaum, L., Baxter, T. K., Baker, B. M., and Restifo, N. P. Poor immunogenicity of a self/tumor antigen derives from peptide-MHC-I instability and is independent of tolerance. J Clin Invest, 114: 551–559, 2004.
112. Powell, D. J., Jr., Dudley, M. E., Hogan, K. A., Wunderlich, J. R., and Rosenberg, S. A. Adoptive transfer of vaccine-induced peripheral blood mononuclear cells to patients with metastatic melanoma following lymphodepletion. J Immunol, 177: 6527–6539, 2006.
113. Walker, E. B., Haley, D., Miller, W., Floyd, K., Wisner, K. P., Sanjuan, N., Maecker, H., Romero, P., Hu, H. M., Alvord, W. G., Smith, J. W., 2nd, Fox, B. A., and Urba, W. J. gp100(209–2M) peptide immunization of human lymphocyte antigen-A2+ stage I-III melanoma patients induces significant increase in antigen-specific effector and long-term memory CD8+ T cells. Clin Cancer Res, 10: 668–680, 2004.
114. Oppenheim, J. J., Dong, H. F., Plotz, P., Caspi, R. R., Dykstra, M., Pierce, S., Martin, R., Carlos, C., Finn, O., Koul, O., and Howard, O. M. Autoantigens act as tissue-specific chemoattractants. J Leukoc Biol, 77: 854–861, 2005.
115. Riker, A. I., Kammula, U. S., Panelli, M. C., Wang, E., Ohnmacht, G. A., Steinberg, S. M., Rosenberg, S. A., and Marincola, F. M. Threshold levels of gene expression of the melanoma antigen gp100 correlate with tumor cell recognition by cytotoxic T lymphocytes. Int J Cancer, 86: 818–826, 2000.
116. Reya, T., Morrison, S. J., Clarke, M. F., and Weissman, I. L. Stem cells, cancer, and cancer stem cells. Nature, 414: 105–111, 2001.
117. Dean, M. Cancer stem cells: redefining the Paradigm of Cancer Treatment Strategies. Mol Interv, 6: 140–148, 2006.
118. Houghton, J., Morozov, A., Smirnova, I., and Wang, T. C. Stem cells and cancer. Semin Cancer Biol, 2006.
119. Morrison, S. J., Shah, N. M., and Anderson, D. J. Regulatory mechanisms in stem cell biology. Cell, 88: 287–298, 1997.

120. Morrison, S. J., Shah, N. M., and Anderson, D. J. Regulatory mechanisms in stem cell biology. Cell, 88: 287–298, 1997.
121. Spradling, A., Drummond-Barbosa, D., and Kai, T. Stem cells find their niche. Nature, 414: 98–104, 2001.
122. Blanpain, C. and Fuchs, E. Epidermal Stem Cells of the Skin. Annu Rev Cell Dev Biol, 2005.
123. Collins, A. T., Berry, P. A., Hyde, C., Stower, M. J., and Maitland, N. J. Prospective identification of tumorigenic prostate cancer stem cells. Cancer Res, 65: 10946–10951, 2005.
124. Haraguchi, N., Utsunomiya, T., Inoue, H., Tanaka, F., Mimori, K., Barnard, G. F., and Mori, M. Characterization of a side population of cancer cells from human gastrointestinal system. Stem Cells, 24: 506–513, 2006.
125. Fiegel, H. C., Gluer, S., Roth, B., Rischewski, J., von Schweinitz, D., Ure, B., Lambrecht, W., and Kluth, D. Stem-like cells in human hepatoblastoma. J Histochem Cytochem, 52: 1495–1501, 2004.
126. Patrawala, L., Calhoun, T., Schneider-Broussard, R., Zhou, J., Claypool, K., and Tang, D. G. Side population is enriched in tumorigenic, stem-like cancer cells, whereas ABCG2+ and ABCG2- cancer cells are similarly tumorigenic. Cancer Res, 65: 6207–6219, 2005.
127. Fang, D., Nguyen, T. K., Leishear, K., Finko, R., Kulp, A. N., Hotz, S., Van Belle, P. A., Xu, X., Elder, D. E., and Herlyn, M. A tumorigenic subpopulation with stem cell properties in melanomas. Cancer Res, 65: 9328–9337, 2005.
128. Gibbs, C. P., Kukekov, V. G., Reith, J. D., Tchigrinova, O., Suslov, O. N., Scott, E. W., Ghivizzani, S. C., Ignatova, T. N., and Steindler, D. A. Stem-like cells in bone sarcomas: implications for tumorigenesis. Neoplasia, 7: 967–976, 2005.
129. Kim, C. F., Jackson, E. L., Woolfenden, A. E., Lawrence, S., Babar, I., Vogel, S., Crowley, D., Bronson, R. T., and Jacks, T. Identification of bronchioalveolar stem cells in normal lung and lung cancer. Cell, 121: 823–835, 2005.
130. Setoguchi, T., Taga, T., and Kondo, T. Cancer stem cells persist in many cancer cell lines. Cell Cycle, 3: 414–415, 2004.
131. Al-Hajj, M., Wicha, M. S., Benito-Hernandez, A., Morrison, S. J., and Clarke, M. F. Prospective identification of tumorigenic breast cancer cells. Proc Natl Acad Sci USA, 100: 3983–3988, 2003.
132. Singh, S. K., Clarke, I. D., Terasaki, M., Bonn, V. E., Hawkins, C., Squire, J., and Dirks, P. B. Identification of a cancer stem cell in human brain tumors. Cancer Res, 63: 5821–5828, 2003.
133. Stingl, J., Raouf, A., Emerman, J. T., and Eaves, C. J. Epithelial progenitors in the normal human mammary gland. J Mammary Gland Biol Neoplasia, 10: 49–59, 2005.
134. Collins, A. T., Berry, P. A., Hyde, C., Stower, M. J., and Maitland, N. J. Prospective identification of tumorigenic prostate cancer stem cells. Cancer Res, 65: 10946–10951, 2005.
135. Fang, D., Nguyen, T. K., Leishear, K., Finko, R., Kulp, A. N., Hotz, S., Van Belle, P. A., Xu, X., Elder, D. E., and Herlyn, M. A tumorigenic subpopulation with stem cell properties in melanomas. Cancer Res, 65: 9328–9337, 2005.
136. Smith, G. H. Experimental mammary epithelial morphogenesis in an in vivo model: evidence for distinct cellular progenitors of the ductal and lobular phenotype. Breast Cancer Res Treat, 39: 21–31, 1996.
137. Kordon, E. C. and Smith, G. H. An entire functional mammary gland may comprise the progeny from a single cell. Development, 125: 1921–1930, 1998.
138. Diallo, R., Schaefer, K. L., Poremba, C., Shivazi, N., Willmann, V., Buerger, H., Dockhorn-Dworniczak, B., and Boecker, W. Monoclonality in normal epithelium and in hyperplastic and neoplastic lesions of the breast. J Pathol, 193: 27–32, 2001.
139. Liu, S., Dontu, G., and Wicha, M. S. Mammary stem cells, self-renewal pathways, and carcinogenesis. Breast Cancer Res, 7: 86–95, 2005.
140. Gudjonsson, T. and Magnusson, M. K. Stem cell biology and the cellular pathways of carcinogenesis. Apmis, 113: 922–929, 2005.

141. Dontu, G., Abdallah, W. M., Foley, J. M., Jackson, K. W., Clarke, M. F., Kawamura, M. J., and Wicha, M. S. In vitro propagation and transcriptional profiling of human mammary stem/progenitor cells. Genes Dev, 17: 1253–1270, 2003.
142. Al-Hajj, M., Wicha, M. S., Benito-Hernandez, A., Morrison, S. J., and Clarke, M. F. Prospective identification of tumorigenic breast cancer cells. Proc Natl Acad Sci USA, 100: 3983–3988, 2003.
143. Ponti, D., Costa, A., Zaffaroni, N., Pratesi, G., Petrangolini, G., Coradini, D., Pilotti, S., Pierotti, M. A., and Daidone, M. G. Isolation and in vitro propagation of tumorigenic breast cancer cells with stem/progenitor cell properties. Cancer Res, 65: 5506–5511, 2005.
144. Clarke, M. F., Dick, J. E., Dirks, P. B., Eaves, C. J., Jamieson, C. H., Jones, D. L., Visvader, J., Weissman, I. L., and Wahl, G. M. Cancer Stem Cells – Perspectives on Current Status and Future Directions: AACR Workshop on Cancer Stem Cells. Cancer Res, 66: 9339–9344, 2006.
145. Leemhuis, T., Yoder, M. C., Grigsby, S., Aguero, B., Eder, P., and Srour, E. F. Isolation of primitive human bone marrow hematopoietic progenitor cells using Hoechst 33342 and Rhodamine 123. Exp Hematol, 24: 1215–1224, 1996.
146. Guzman, M. L. and Jordan, C. T. Considerations for targeting malignant stem cells in leukemia. Cancer Control, 11: 97–104, 2004.
147. Bonnet, D. and Dick, J. E. Human acute myeloid leukemia is organized as a hierarchy that originates from a primitive hematopoietic cell. Nat Med, 3: 730–737, 1997.
148. Dalerba, P., Cho, R. W., and Clarke, M. F. Annu Rev Med, 58: 267–284, 2007.
149. Hope, K. J., Jin, L., and Dick, J. E. Acute myeloid leukemia originates from a hierarchy of leukemic stem cell classes that differ in self-renewal capacity. Nat Immunol, 5: 738–743, 2004.
150. Singh, S. K., Hawkins, C., Clarke, I. D., Squire, J. A., Bayani, J., Hide, T., Henkelman, R. M., Cusimano, M. D., and Dirks, P. B. Identification of human brain tumour initiating cells. Nature, 432: 396–401, 2004.
151. Singh, S. K., Clarke, I. D., Terasaki, M., Bonn, V. E., Hawkins, C., Squire, J., and Dirks, P. B. Identification of a cancer stem cell in human brain tumors. Cancer Res, 63: 5821–5828, 2003.
152. Richardson, G. D., Robson, C. N., Lang, S. H., Neal, D. E., Maitland, N. J., and Collins, A. T. CD133, a novel marker for human prostatic epithelial stem cells. J Cell Sci, 117: 3539–3545, 2004.
153. Yin, A. H., Miraglia, S., Zanjani, E. D., Almeida-Porada, G., Ogawa, M., Leary, A. G., Olweus, J., Kearney, J., and Buck, D. W. AC133, a novel marker for human hematopoietic stem and progenitor cells. Blood, 90: 5002–5012, 1997.
154. Woodward, W. A., Chen, M. S., Behbod, F., and Rosen, J. M. On mammary stem cells. J Cell Sci, 118: 3585–3594, 2005.
155. Clarke, R. B., Spence, K., Anderson, E., Howell, A., Okano, H., and Potten, C. S. A putative human breast stem cell population is enriched for steroid receptor-positive cells. Dev Biol, 277: 443–456, 2005.
156. Abraham, B. K., Fritz, P., McClellan, M., Hauptvogel, P., Athelogou, M., and Brauch, H. Prevalence of CD44+/CD24-/low cells in breast cancer may not be associated with clinical outcome but may favor distant metastasis. Clin Cancer Res, 11: 1154–1159, 2005.
157. Patrawala, L., Calhoun, T., Schneider-Broussard, R., Li, H., Bhatia, B., Tang, S., Reilly, J. G., Chandra, D., Zhou, J., Claypool, K., Coghlan, L., and Tang, D. G. Highly purified CD44+ prostate cancer cells from xenograft human tumors are enriched in tumorigenic and metastatic progenitor cells. Oncogene, 25: 1696–1708, 2006.
158. Li, L. and Xie, T. Stem cell niche: structure and function. Annu Rev Cell Dev Biol, 21: 605–631, 2005.
159. Fuchs, E. and Segre, J. A. Stem cells: a new lease on life. Cell, 100: 143–155, 2000.
160. Tumbar, T., Guasch, G., Greco, V., Blanpain, C., Lowry, W. E., Rendl, M., and Fuchs, E. Defining the epithelial stem cell niche in skin. Science, 303: 359–363, 2004.
161. Clarke, M. F. and Fuller, M. Stem cells and cancer: two faces of eve. Cell, 124: 1111–1115, 2006.

162. Wright, N. A. Epithelial stem cell repertoire in the gut: clues to the origin of cell lineages, proliferative units and cancer. Int J Exp Pathol, 81: 117–143, 2000.
163. Taylor, G., Lehrer, M. S., Jensen, P. J., Sun, T. T., and Lavker, R. M. Involvement of follicular stem cells in forming not only the follicle but also the epidermis. Cell, 102: 451–461, 2000.
164. Liu, B. Y., McDermott, S. P., Khwaja, S. S., and Alexander, C. M. The transforming activity of Wnt effectors correlates with their ability to induce the accumulation of mammary progenitor cells. Proc Natl Acad Sci USA, 101: 4158–4163, 2004.
165. Li, Y., Welm, B., Podsypanina, K., Huang, S., Chamorro, M., Zhang, X., Rowlands, T., Egeblad, M., Cowin, P., Werb, Z., Tan, L. K., Rosen, J. M., and Varmus, H. E. Evidence that transgenes encoding components of the Wnt signaling pathway preferentially induce mammary cancers from progenitor cells. Proc Natl Acad Sci USA, 100: 15853–15858, 2003.
166. Merrill, B. J., Gat, U., DasGupta, R., and Fuchs, E. Tcf3 and Lef1 regulate lineage differentiation of multipotent stem cells in skin. Genes Dev, 15: 1688–1705, 2001.
167. Clarke, M. F. A self-renewal assay for cancer stem cells. Cancer Chemother Pharmacol, 56 Suppl. 1: 64–68, 2005.
168. Clevers, H. At the crossroads of inflammation and cancer. Cell, 118: 671–674, 2004.
169. Reya, T. and Clevers, H. Wnt signalling in stem cells and cancer. Nature, 434: 843–850, 2005.
170. Bunting, K. D. ABC transporters as phenotypic markers and functional regulators of stem cells. Stem Cells, 20: 11–20, 2002.
171. Zhou, S., Schuetz, J. D., Bunting, K. D., Colapietro, A. M., Sampath, J., Morris, J. J., Lagutina, I., Grosveld, G. C., Osawa, M., Nakauchi, H., and Sorrentino, B. P. The ABC transporter Bcrp1/ABCG2 is expressed in a wide variety of stem cells and is a molecular determinant of the side-population phenotype. Nat Med, 7: 1028–1034, 2001.
172. Kleber, M., Lee, H. Y., Wurdak, H., Buchstaller, J., Riccomagno, M. M., Ittner, L. M., Suter, U., Epstein, D. J., and Sommer, L. Neural crest stem cell maintenance by combinatorial Wnt and BMP signaling. J Cell Biol, 169: 309–320, 2005.
173. Ragupathi, G., Liu, N. X., Musselli, C., Powell, S., Lloyd, K., and Livingston, P. O. Antibodies against tumor cell glycolipids and proteins, but not mucins, mediate complement-dependent cytotoxicity. J Immunol, 174: 5706–5712, 2005.
174. Monti, P., Leone, B. E., Zerbi, A., Balzano, G., Cainarca, S., Sordi, V., Pontillo, M., Mercalli, A., Di Carlo, V., Allavena, P., and Piemonti, L. Tumor-derived MUC1 mucins interact with differentiating monocytes and induce IL-10highIL-12low regulatory dendritic cell. J Immunol, 172: 7341–7349, 2004.
175. Rughetti, A., Pellicciotta, I., Biffoni, M., Backstrom, M., Link, T., Bennet, E. P., Clausen, H., Noll, T., Hansson, G. C., Burchell, J. M., Frati, L., Taylor-Papadimitriou, J., and Nuti, M. Recombinant tumor-associated MUC1 glycoprotein impairs the differentiation and function of dendritic cells. J Immunol, 174: 7764–7772, 2005.
176. Kilinc, M. O., Aulakh, K. S., Nair, R. E., Jones, S. A., Alard, P., Kosiewicz, M. M., and Egilmez, N. K. Reversing tumor immune suppression with intratumoral IL-12: activation of tumor-associated T effector/memory cells, induction of T suppressor apoptosis, and infiltration of CD8+ T effectors. J Immunol, 177: 6962–6973, 2006.
177. Broderick, L., Yokota, S. J., Reineke, J., Mathiowitz, E., Stewart, C. C., Barcos, M., Kelleher, R. J., Jr., and Bankert, R. B. Human CD4+ effector memory T cells persisting in the microenvironment of lung cancer xenografts are activated by local delivery of IL-12 to proliferate, produce IFN-gamma, and eradicate tumor cells. J Immunol, 174: 898–906, 2005.
178. Nishimura, F., Dusak, J. E., Eguchi, J., Zhu, X., Gambotto, A., Storkus, W. J., and Okada, H. Adoptive transfer of type 1 CTL mediates effective anti-central nervous system tumor response: critical roles of IFN-inducible protein-10. Cancer Res, 66: 4478–4487, 2006.
179. Knutson, K. L., Dang, Y., Lu, H., Lukas, J., Almand, B., Gad, E., Azeke, E., and Disis, M. L. IL-2 immunotoxin therapy modulates tumor-associated regulatory T cells and leads to lasting immune-mediated rejection of breast cancers in neu-transgenic mice. J Immunol, 177: 84–91, 2006.

Chapter 6
Tumor Cell Resistance to Apoptosis by Infiltrating Cytotoxic Lymphocytes

Benjamin Bonavida

Abstract Cancer patients have been treated with conventional chemotherapy, radiation, and hormonal therapy with significant clinical responses. However, patients develop resistance and they no longer respond to the above therapies. Immunotherapy has been considered as an alternative approach to overcome resistance and several antibody-mediated new cell-mediated therapies have been introduced and yielded significant clinical responses. The development of cell mediated antitumor cytotoxic immunotherapy has also advanced significantly with the design of numerous strategies to generate a specific antitumor cytotoxic response. However, the clinical response with cytotoxic immunotherapy remains poor and several proposed mechanisms have been suggested for the poor response. One mechanism that received little attention is the development of tumor cell resistance to cytotoxic stimuli, including both antibody-mediated and cell-mediated. The resistance, inherent or acquired, primarily results from the dysregulation of apoptotic pathways in the tumor cells and thus rendering them refractory to apoptotic stimuli including chemotherapy, radiation, and immunotherapy. The development of cross-resistance to apoptosis may play an important role in the poor clinical response observed in patients with a strong cytotoxic immune system. The molecular understanding of immune resistance identified many targets whose modifications by selective sensitizing agents reversed resistance. Such sensitizing agents included low concentrations of chemotherapeutic drugs, chemical and pharmacological inhibitors of survival pathways, anti-receptor antibodies, siRNA, cytokines, etc. Such sensitizing agents when used in combination with immunotherapy reversed immune resistance. In addition, the tumor microenvironment is regulated by infiltrating cells and the secretion of several factors that regulate tumor cell sensitivity to killing. Agents that can interfere with these factors should potentiate the reversal of immune resistance.

Department of Microbiology, Immunology, and Molecular Genetics;
Jonsson Comprehensive Cancer, David Geffen School of Medicine, University of California, Los Angeles, CA 90095, USA

E. Yefenof (ed.), *Innate and Adaptive Immunity in the Tumor Microenvironment.*

Keywords Immune resistance, dysregulation of apoptosis, immunosensitizing agents, tumor microenvironment, cytotoxic lymphocytes, drugs-immune cross resistance

6.1 Introduction

From the early development of chemotherapeutic drugs and their application in the treatment of cancer significant advances have resulted on the underlying molecular mechanisms of their mode of cytotoxic action and the biology of cancer diseases. Although treatment with chemotherapeutic drugs resulted in significant clinical responses and overall survival, there remains the problem of the development and/or acquisition of tumor resistance to chemotherapy [1]. The failure to eradicate resistant tumors to conventional therapeutics has led to the introduction of immunotherapy. In practice, tumor immunotherapy is an ideal therapeutic approach because it offers several advantages over chemotherapy including low organ toxicity and high tumor selectivity. In immunotherapy, cytotoxic cells are derived from the host's own immune system such as lymphokine-activated killer (LAK) cells and interleukin 2 (IL-2)-activated tumor infiltrating lymphocytes (TIL) [2]. Also, based on the principle that most tumors have the capacity to trigger an immune response, immunotherapy can be used for the selective and specific recognition of tumor targets by the generation of specific antitumor cytotoxic T-cell responses [3]. There is a prevalent dogma that immunotherapeutic strategies under investigation can conquer chemoresistant tumors and assume that immunotherapy attacks tumor cells by different mechanisms of action than those achieved by chemotherapeutic drugs and may not be subjected to the same mechanisms underlying drug resistance. Despite the proposed advantages over chemotherapy, immunotherapy today fails to deliver a significant curative rate. Initially, immunotherapy or T-cell based immunotherapeutic approaches (e.g. LAK and TIL) have generated a great deal of excitement when they were shown to be effective in certain transplantable tumors in mice [4]. Subsequent studies with the LAK or TIL systems in clinical trials in patients with advanced diseases failed to demonstrate a significant clinical response rate [5]. Further, studies with cytokine gene transfer into tumor targets and pulsing CTLs with specific tumor peptides (that is tumor vaccines) still proved to be unsuccessful in long term cures as well [6,7]. Clearly, T-cell based immunotherapy has its own limitations. The lack of significant positive response by immunotherapy against drug resistant tumor cells suggests that the mere manipulation of the immune system may be not sufficient to restore a positive antitumor killing. Other aspects such as tumor manipulation of neo-antigens, MHC-1, suppressive factors, etc. might be involved in the failure to respond to immunotherapy. However, one aspect that has not been seriously considered is the tumor cell sensitivity to killing by cytotoxic lymphocytes and the development of resistance to killing by the cytotoxic lymphocytes. It is unclear if drug resistant tumors are initially sensitive to killing mediated by cytotoxic lymphocytes. It is possible that the poor effectiveness of immunotherapy is that, although immune cells could recognize chemoresistant tumors, chemoresistant tumors are also equally resistant to immune-mediated killing mechanisms. It follows that chemoresistant tumors are also equally resistant to immune-mediated killing mechanisms. Thus, the success of immunotherapy will ultimately be dictated by both the presence of antitumor CTLs and the sensitivity of tumor targets to the killing mediated by these cells. In addition, tumor chemoresistance may reflect an important part of the general tumor resistant mechanism to a common cytotoxic pathway mediated by various stimuli, namely apoptosis or programmed cell death, and such resistant scheme to a central cytotoxic pathway may also render the cells resistant to immune-mediated killing [8]. The definition of cross-resistance actually could go beyond the multi-drug resistance phenomena and encompasses other cytotoxic stimuli including the cytotoxic immune cells. Therefore, the ultimate goal for a successful antitumor therapy, be with chemotherapy, radiation, or immunotherapy is to overcome cross-resistance for the ultimate induction of apoptosis. Many physiological and external stimuli can induce apoptosis in susceptible cells including both chemotherapeutic drugs and host activated immune cells. However, not all tumor cells are intrinsically sensitive to apoptosis and most malignant cells develop resistance to apoptosis by dysregulating the apoptotic pathways that are triggered by drugs and immune cells. With the premise that chemoresistant tumors develop general mechanisms of resistance to apoptosis-mediated stimuli, our hypothesis and reported studies put forth are that for an effective antitumor therapeutic strategy it is essential to utilize complimentary pro-apoptotic signals to overcome tumor resistance to immune-mediated apoptosis.

6.2 Apoptosis as a Cytotoxic Mechanism Induced by Cytotoxic Lymphocytes

Apoptosis or programmed cell death plays an important role in the induction of tumor cell death [9]. Dysregulation of this process may lead to pathological diseases such as cancer and autoimmune diseases [10]. The molecular machinery for executing apoptosis is inherently in place in most cells including tumor cells but it is only triggered with the proper stimulus. There are two major cytotoxic mechanisms by which cytotoxic lymphocytes induce apoptosis, namely, the granule exocytosis pathway mediated by perforin and granzymes and the death receptor signaling pathways, which involve the TNF-related ligand family of proteins (FasL, TNFα, TRAIL) reviewed by [11]. The induction of the granule/exocytosis pathway requires direct lymphocytes to tumor cell contact and TCR/MHC engagement and recognition in order to initiate the release of cytotoxic granules containing perforin, granzymes and other cytotoxic constituents. Granzymes trigger apoptosis by either activating a caspase-dependent central apoptotic pathway or cleaving directly some of the substrates that are also cleaved by caspases. Perforin, however, induces necrotic cell death by causing membrane damage. The TNF apoptosis-inducing ligand members trigger death receptor-mediated apoptosis in tumor cells upon cross-linking with their death receptors on the cell surface reviewed by [12]. These ligands can also induce apoptosis in the appropriate aggregated soluble forms. Although the granule exocytosis pathway was originally considered the primary killing mechanism against tumor cells, a number of studies suggest that death receptor-mediated apotosis is an important mechanism of tumor cell killing by cytotoxic lymphocytes [13, 14]. Two major apoptotic signaling pathways have been recognized. One is mediated from the death receptors (extrinsic or type 1) and the other originates from the activation of the mitochondrial release of cytochrome c and other molecules see review by [15]. Central to both apoptotic pathways is the activation of caspases required for the induction of the final apoptotic phenotype. The death receptor pathway can also divert its signal through the mitochondrial pathway in some cases [16].

6.3 Inhibition of Apoptosis as a Mechanism of Cross Resistance

Many cancer cells are initially sensitive to drug/radiation/hormonal-mediated apoptosis. However, resistant variants and relapses result in cancer cells that are also resistant to apoptosis-induced by internal cellular cues. Such tumors also become cross-resistant to chemotherapy and immunotherapy. Numerous reports have demonstrated that resistance to chemotherapeutic drugs and immunocytotoxics are related since they induce apoptosis using a similar caspase-dependent pathway [17, 18]. A hierarchical pattern of tumor resistance to various apoptotic stimuli was documented previously using various cell lines [19, 20].

6.4 Mechanisms of Resistance to Cytotoxic Immune Cells

6.4.1 *Extrinsic*

CTL-mediated specific immunotherapy against tumor cells could be expected to kill all tumor cells expressing the appropriate tumor-specific associated antigens that come in contact with CTL, as long as the antigen induces significant and long lasting immune responses. Most if not all of the tumor cells express MHC/peptide complexes at sufficient levels to be recognized by CTL and the majority of tumor cells are sensitive to CTL. Infiltrate of cytotoxic lymphocytes in the tumor microenvironment is not indicative of an ongoing antitumor response and is not an indication of a favorable response. Tumor cells have developed several strategies to resist killing by cytotoxic immune lymphocytes. Resistance to preforin/granzyme-mediated cytotoxicity/apoptosis by tumor cells has been reported in several studies. For instance, [21] have reported that human leukemic cells were resistant to

NK-mediated cytotoxicity due to the failure of perforin to bind on the cell surface membrane of the tumor cells. Tumor cells may also be deficient in the mannose-6 phosphate receptor that binds and internalizes granzyme B and cell lines deficient in the receptor may be resistant to CTL-mediated DNA fragmentation [22].

6.4.2 Intrinsic

Tumor cells can acquire various intrinsic mechanisms to avoid killing by cytotoxic lymphocytes. For example, tumor cells may express serine protease inhibitors (Serpin/PI-9) that can irreversibly prevent granzyme-B-induced cell death as it has been determined both *in vitro* and *in vivo* [23, 24]. The expression level of death receptors establishes the initial stage of immune-mediated apoptosis. In the Fas system various resistance mechanisms that negatively regulate the initiation of death signal from the Fas receptor include overexpression of the receptor with mutations of the death domain region, loss of Fas expression and alternative splicing of the Fas receptor mRNA that generates a secreted soluble Fas [25–27]. Similarly, in the TRAIL receptor system, the overexpression of decoy receptors (DcR1 and DcR2), lacking the functional intracellular signaling domain or loss/mutation of the agonist receptors (DR4 and DR5) contribute to resistance to TRAIL- induced apoptosis [12].

Tumor cells undergo morphological changes and alterations in the cytoskeleton, leading to resistance to cytotoxic mechanisms [28]. In a recent study, [29] reported on the analysis of a CTL-derived resistant tumor cell variants and identified two gene products, namely ephrin-A1 and scuderi, that were overexpressed in the variants and regulated resistance to CTL-induced cytotoxicity. However, the exact role of each is not yet clear.

The tumor suppressor p53 acts as a stress sensor transcription factor that protects cells from DNA damage, oncogene activation, radiation, and chemotherapeutic drugs. p53 activates the transcription of genes involved in cell cycle arrest, DNA repair and apoptosis [30]. The activation of wild-type p53 leads to cell growth arrest and apoptosis. Most tumor cells express a mutated p53 and some cancers are deficient in p53 and thus, the vital function of wild-type p53 is compromised. Inactivation of p53 was shown to contribute negatively to chemotherapy drug-induced apoptosis [31]. The potential role of p53 in regulating cell-induced cytotoxicity is illustrated with resistance to Fas ligand [32].

Tumor cells develop several mechanisms to dysregulate the apoptotic pathways by modulating gene products involved, such as overexpression of anti-apoptotic gene products, underexpression of apoptotic gene products, mutations and deletions of gene products, etc. Examples of such gene modifications will be discussed below in the section of sensitization of immune resistant tumors. Since the tumors acquire resistance to cytotoxic cell-mediated killing, intervention by sensitizing agents, whose actions are to alter the dysregulated apoptotic gene products, should lead to sensitivity of the tumor to killing by cytotoxic cells.

6.5 Sensitization of Resistant Tumor Cells to Cytotoxic Lymphocytes/Factors-mediated Apoptosis

The modulation of the apoptotic pathways to reverse immune resistance presents a unique opportunity to improve the outcome of current anticancer treatment strategies. Even though immunotherapy has not been successful in killing tumor cells, due to the profound tumor immuneresistance to apoptosis, immunosensitizing agents that modulate signaling molecules involved in the immune-mediated apoptosis may enhance tumor killing and improve the efficacy of immunotherapy. Sensitizing agents of various kinds have been reported to sensitize tumor cells to death-ligand apoptosis such as chemotherapeutic drugs, pharmacologic inhibitors of survival pathways, cytokines, nitric oxide donors, specific inhibitors (chemicals or siRNA) of anti-apoptotic gene products, anti-receptor antibodies, etc. The description of these various sensitizing agents and their mechanisms of actions are outside the scope of this review. However, representative sensitizing agents that have been used by us will be briefly described below.

6.5.1 Chemosensitizing Drugs as Immunosensitizing Agents

Even though most advanced malignant tumor cells are resistant to chemotherapeutic drugs and cytotoxic immune cells, in many instances, low levels of the same or different drugs could sensitize resistant tumor cells to immune-mediated apoptosis [33–36]. Sensitization by chemotherapeutic drugs has also been documented in cells that express the drug efflux pump, p-glycoprotein [37]. These observations suggest that the sensitizing property of chemotherapeutic drugs may be distinct from their direct apoptosis-inducing effect. Several proposed mechanisms of immunosensitization have been reported, including transcriptional upregulation of pro-apoptotic proteins and downregulation of anti-apoptotic proteins [38]. The protein expression activity of signaling molecules and regulatory proteins involved in both the receptor-mediated proteins and the mitochondrial pathways have been suggested to be modulated by the drugs (see schematic diagram in Fig. 6.1). Other possible mechanisms for immunosensitization, such as post translational modification and protein translocation, have also been suggested [39]. Other molecular mechanisms of immunosensitization by chemotherapeutic drugs have also been examined and include the following: 1) Upregulation of death receptors. The expression level of death receptors establishes the initial stage for the control of immune-mediated apoptosis. Treatment of certain tumor cells with subtoxic concentrations of chemotherapeutic drugs (e.g. CDDP, ADR, VP16, 5FU, Camptothecin) upregulates the expression of death receptors (Fas, DR4, DR5) while other alternatively spliced variants for soluble receptors or decoy receptors remain unaltered [40–42]. The upregualtion of Fas and DR5 by chemotherapeutic drugs appears to be mediated by a p53-independent mechanism [43]. Recent studies

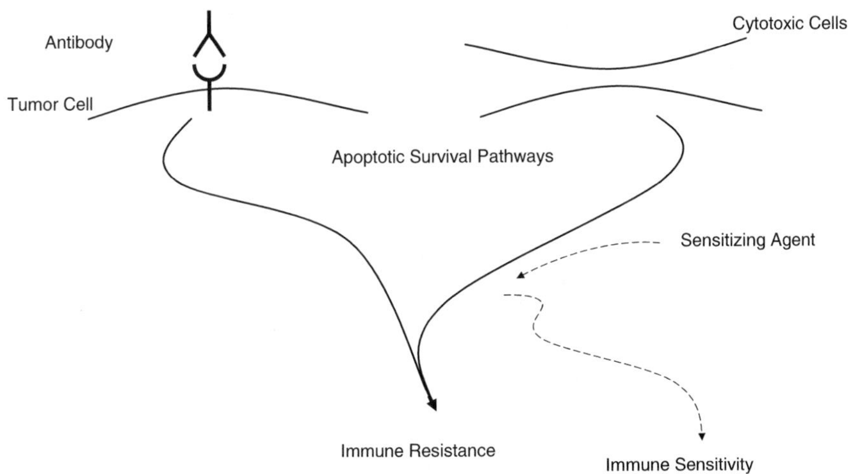

Fig. 6.1 Tumor Cells Resistance to Killing by Immunotherapy. Schematic diagram illustrating that tumor cells fail to respond to the cytotoxic activity of either antibody or cytotoxic cells. Immune resistance to cytotoxic stimuli by the tumor cells results from multiple mechanisms including hyperactivation of cell survival/anti-apoptotic pathways as well as by dysregulation of gene products that regulate the apoptotic pathways. Intervention by sensitizing agents that can modulate these various anti-apoptotic mechanisms can reverse immune resistance to an immune sensitive phenotype

examined the molecular mechanism by which death receptors are dysregulated and the findings revealed that the death receptors are negatively transcribed by the transcription factor Yin Yang 1 (YY1) and its inhibition by chemotherapeutic drugs or by YY1 siRNA sensitized the tumor cells to death ligand-induced apoptosis [36, 44].

6.5.1.1 Upregulation of FADD and Apaf-1 Adaptor Proteins

Following receptor trimerization, the recruitment of FADD protein to intracellular death domains of the death receptor to form DISC is required for initiation of caspase activation. The inability to activate the initiator caspase-8 has been proven to be involved in resistance [45]. Upregulation of FADD in drug-mediated sensitization has been reported in several tumor systems [46, 47]. Apaf-1 is the necessary link for mediating the mitochondrial apoptotic pathway [48]. Apaf-1 is upregulated by several chemotherapeutic drugs and sensitizes drugs to death ligands [37].

6.5.1.2 Downregulation of FLIP

FLIP (flice-inhibitory proteins) acts as a dominant negative for caspase-8 and inhibits the death receptor signaling by association with procaspase-8 in the DISC and prevents the recruitment and activation of caspase-8. Chemotherapeutic drugs sensitize immunoresistant tumor cells by downregulation of FLIP [38, 49].

6.5.1.3 Upregulation of Procaspases

Conventional chemotherapeutic drugs (e.g. etoposide, cisplatin, doxorubiricin and mitomycin) could sensitize tumor cells by selective induction of procaspases-8, 3 and 2 [46, 50]. The drug adriamycin increases the expression of caspase-9 in the drug resistant myeloma cells and sensitizes cells to TRAIL-induced apoptosis [37].

6.5.1.4 Regulation of Bcl-2 Family Proteins

The release of cytochrome c from the mitochondria into the cytoplasm is a crucial and decisive event in the induction of apoptosis [51]. Chemotherapeutic drugs induce the proapoptotic member, Bax via a p53-dependent transcription mechanism [52]. The upregulation of Bax and deregulation of Bcl-xL induced by chemotherapeutic drugs are associated with immunosensitization to death receptor-mediated apoptosis [33, 34].

IAP families of proteins (inhibitors of apoptosis) are a group of anti-apoptotic proteins that function by directly inhibiting certain caspases. The expression of IAP proteins inhibits the execution phase of death receptors-mediated apoptosis in the resistant tumor as they bind and inhibit the active forms of caspase-3 [53] and also bind to caspases-7 and 9 [54]. Several members of IAP proteins have been identified. Overexpressions of IAP proteins have been shown in a variety of chemoresistant tumors. Subtoxic levels of chemotherapeutic drugs have been shown to reduce the activity of IAP family proteins in several tumor cell systems [47, 55, 56].

The above examples have revealed that subtoxic concentrations of chemotherapeutic drugs can elicit a variety of regulatory signals at multiple levels of the apoptotic process. These differ from the direct apoptotic effects induced by drugs. Hence, certain chemotherapeutic drugs can sensitize resistant cancer cells to immune-mediated apoptosis by selectively downregulating anti-apoptotic proteins or upregulating pro-apoptotic proteins involved in the apoptotic pathways.

6.5.2 Nitric Oxide (NO) Donors as Immunosensitizing Agents

The discovery of NO donors as immunosensitizing agents was inadvertently found by examining the mechanism underlying the regulation of Fas expression by interferon-γ and sensitization of Fas ligand resistant tumors to Fas ligand-induced apoptosis. Treatment with interferon-γ induced the expression of NOSII and treatment with NO donors mimicked interferon-γ. Hence, we hypothesized that NO may inhibit a transcription repressor that regulates Fas transcription and resulting in upregulation of Fas and sensitization to Fas ligand-induced apoptosis. We have found that YY1 negatively regulates Fas transcription and that NO inhibits YY1 DNA-binding activity [44, 57, 58]. Further, we demonstrated that NO modified YY1 into an S-nitrosylated form that inhibited its DNA-binding activity [59]. Further studies identified a YY1-DNA binding activity in the silencer region of the Fas promoter [44]. Subsequent studies revealed that YY1 also regulates the transcription of the TRAIL-receptor DR5 and inhibition of YY1 by NO or siRNA upregulates DR5 and sensitized the tumor cells to TRAIL-induced apoptosis [36]. The role of NO donors as sensitizing agents for immunotherapy of cancer is an attractive approach provided that NO levels are appropriate for sensitization. Further, NO administration can also affect angiogenesis. Further details on the role of NO and cancer have been recently reviewed [60].

6.5.3 Antibody-mediated Immunosensitization

Antibodies directed against cell surface receptors on tumor cells have been recently used in the treatment of several cancers and autoimmune diseases. In cancer, such studies have been suggested to mediate the *in vivo* activities via ADCC, CDC and sometimes apoptosis. However, the antibody-mediated cell triggering and consequences on tumor cell viability and resistance have not been explored. We have initiated studies with rituximab (chimeric anti-human CD20 monoclonal antbody) approved by the FDA for the treatment of B-NHL and more recently for rheumatoid arthritis. Our initial findings demonstrated that rituximab treatment of B-NHL cell lines sensitize the cells to apoptosis by various chemotherapeutic drugs. The molecular mechanism of chemosensitization was examined and demonstrated rituximab-mediated inhibition of several survival anti-apoptotic pathways leading to inhibition of the anti-apoptotic gene products Bcl-2/Bcl-xL. These studies have been recently reviewed [61]. We have also explored the role of rituximab in sensitizing tumor cells to TNF-α family members (e.g. Fas ligand and TRAIL) as potential novel therapeutics to overcome resistance. Our findings revealed, for the first time, that rituximab sensitizes drugs/Fas ligand/TRAIL-resistant B-NHL cells to Fas ligand and TRAIL-induced apoptosis. Immunosensitization was the result of rituximab-mediated inhibition of NF-κB activity and inhibition of the transcription repressor YY1 as its inhibition by rituximab upregulated Fas and DR5 expression and sensitized cells to death-receptor apoptosis [57, 58, 60]. The above findings are mainly representative of other antibodies that may also mediate sensitization of resistant tumor cells.

6.5.4 Pharmacologic Inhibitors

Tumor cell resistance to cytotoxic immunotherapy results from multiple mechanisms as described above, namely, the development of resistance to apoptosis-inducing stimuli. The hyperactivation of survival pathways and overexpression of anti-apoptotic gene products regulate, in large part, immunoresistance [62]. Hence, interference with these anti-apoptotic survival pathways by pharmacological inhibitors targeting gene products in these pathways should inhibit synthesis and transcription of anti-apoptotic gene products and, hence, could sensitize the cells to various apoptotic stimuli. Several studies have been performed with a large number of pharmacologic and proteasome inhibitors which resulted in the sensitization to death ligand-induced apoptosis. Representative findings will illustrate their applications as immunosensitizing agents. For instance, the NF-κB inhibitors Bay11-7085 and DHMEQ have been shown to sensitize tumor cells to Fas ligand-induced apoptosis through inhibition of the transcription repressor YY1, that is under the regulation of NF-kB [57]. Likewise, inhibitors of the p38 MAPK pathways, such as PP2, SB203580, also resulted in the inhibition of NF-kB and YY1 and sensitized tumor cells to Fas ligand-induced apoptosis [58]. Similar findings achieved for Fas ligand were also obtained for sensitizing tumor cells to TRAIL-induced apoptosis through upregulation of DR5 [36]. In addition, proteasome inhibitors such as Bortezomib and NPI 0052 were also found to sensitize tumor cells to both Fas ligand and TRAIL-induced apoptosis [63–66]. A recent study has addressed the role of the tumor microenvironment in the regulation of tumor cells sensitivity to TRAIL. Resistance to TRAIL might have been due to hypoxia [known to induce the upregulation of the hypoxy-inducing factor- 1α(HIF- 1α)] and this inhibits TRAIL-induced apoptosis [67, 68].

6.6 Influence of the Tumor Microenvironment on the Development of Tumor Cell Resistance to Cytotoxic Therapy

Several potential mechanisms have been outlined above that regulate tumor cell resistance to cytotoxic immunotherapy. In addition, several approaches have been outlined above to reverse resistance. Also, in addition to tumor cell development of resistance outside the environment, the tumor microenvironment can influence and drive into the selection of tumor cell resistance to cytotoxic immunotherapy. Therefore, the tumor microenvironment might be considered, as well, in its ability to reverse the resistance and the selection of appropriate sensitizing agents that need to be used. Below, several factors have been reported to contribute to tumor cell immune resistance. Noteworthy, a recent study by [69] demonstrated that mesothelioma cell lines grown *in vitro* as monolayers were sensitive to TRAIL in combination with cyclohexamide. However, patient's mesothelionma grown as spheroids *in vitro* were resistant to the same treatment. These spheroids contained, in addition to tumor cells, malignant cells and tumor extracellular matrix. This exemplifies the influence of the microenvironment on the response to TRAIL. In a recent review by [70], they describe the role of several factors in the tumor microenvironment that influence tumor cell immunosensitivity to TRAIL. Such factors include fibroblasts as source of growth factor, angiogenesis, and extravasation of cytotoxic lymphocytes. Fibroblasts secrete many growth factors (e.g. IGF, MMPs, EGF, etc.) that play a role in cancer [71]. Stroma cells secrete osteoprotegerin (OPG), a soluble decoy receptor for TRAIL, and may block TRAIL activity [72]. Tumor fibroblasts co-cultured can degrade type-1 collagen via MMP secretion [73]. The tumor microenvironment is abundant of pro-inflammatory growth factors and cytokines such as IFN-γ and IL-8. For example IL-8 can inhibit tumor cell sensitivity to TRAIL [74]. IL-8 reduces DR4 significantly and prevents caspase activation. Matrix metalloproteinases constitute a multi-gene family of over 25 secreted and cell surface enzymes that are mainly produced by non-malignant stroma cells and are involved in tumor growth invasion and metastasis [75]. MMP inhibitors or RNA interference resulted in caspase activation and DNA fragmentation [76]. Tissue inhibitors of MMPs have been shown to sensitize tumor cells to TRAIL apoptosis [77]. These findings identify several targets for intervention.

6.7 Therapeutic Implications

It is important to consider not only the sensitivity of the tumor cells, but also the potential of factors inhered in the complex microenvironment that impacts the sensitivity of the tumor as a whole. Combination therapies that target the tumor cells (sensitizing agents) as well as the microenvironment may be used. For instance, one may consider the use of MMPs inhibitors, and find ways for better delivery of the drug through vasodilation of the tumor vasculator by liposomal drug delivery and nanoparticles [78]. It has been reported that human leukemic cells from patients can develop resistance to perforin-mediated cytotoxicity due to impaired binding of perforin on the surface of tumor cells [21]. Otten et al. [79] have studied perforin and Fas ligand resistance in lymphoid and myeloid types of leukemia. Perforin resistance was demonstrated in some leukemic cells from some patients with AML and more pronounced resistance in leukemic cells from patients with CML. *In vitro*, leukemic cells were more susceptible to killing by perforin than by anti-Fas. The mechanism of resistance to anti-Fas was reported to involve lack of expression of Fas, secretion of soluble Fas to block Fas ligand and low Bax/Bcl-2 ratio [80–82].

6.8 Concluding Remarks

Immunotherapy in the treatment of cancer has made major strides during the last decade. For instance, antibody-mediated immunotherapy has been applied since 1997 when the first FDA approved monoclonal antibody for cancer treatment, rituximab, was used to treat patients with B-NHL. Many other monoclonal antibodies directed against cell surface tumor associated antigens were developed and many of those are currently being used for the treatment of several tumor types with significant success. In addition to antibody-mediated immunotherapy, several strategies have been developed to generate antitumor cell-mediated cytotoxic cells with great success. However, the clinical response in the presence of a significant antitumor cytotoxic response still remains very poor. Several underlying mechanisms have been proposed for the poor clinical response including tumor cell modification and secretion of suppressor factors as well as the tumor microenvironment that participates in preventing killer cells to either interact and/or kill corresponding tumor cells. An important mechanism that has come to the attention of investigators deal with tumor cell development of mechanisms to resist killing by both cytotoxic cells and antibody-mediated cytotoxicity. Such mechanisms of immune resistance may account for the poor correlation between the presence of antitumor cytotoxic antibody or effector cells and the poor clinical response. Tumor cell development of resistance to killing is an important mechanism for tumor escape from immune surveillance. Several representative mechanisms have been described in this review which illustrate how tumor cells can resist killing by cytotoxic immunotherapeutic approaches. A better understanding of the molecular, biochemical, and genetic mechanisms that regulate immune resistance should provide a better understanding as well as develop novel approaches for therapeutic interventions to reverse resistance. Studies have demonstrated that interference with resistant factors in the tumor cells by a variety of sensitizing agents can reverse immune resistance. Such studies suggest that reversal of immune resistance may occur by combination treatment of immunosensitizing agents and immunotherapy. Thus, it may be possible in the presence of antitumor cytotoxic immunotherapy that the combination treatment in patients with sensitizing agents should alter the tumor resistant phenotype of the tumor cells and the patients can respond successfully to immunotherapy. In addition, the tumor microenvironment should not be neglected, as other factors in the microenvironment may still prevent cell killing due to blocking factors or factors that act on the cytotoxic lymphocytes. Here again one has to devise novel intervention approaches to counteract these factors. Hence, we may face novel combinations of therapeutic approaches that are complex and multi-targeted, namely sensitizing agents on the tumor cells, agents that potentiate the interaction of cytotoxic cells with tumor cells in the microenvironment and agents that potentiate and activate the cytotoxic cells.

Acknowledgements I acknowledge the Jonsson Comprehensive Cancer Center for their valuable contribution. In addition, I acknowledge the various individuals that have contributed in the studies described: Ali Jazirehi, Ph.D.; Mario I. Vega, Ph.D.; Sara Huerta-Yepez, Ph.D.; Stavroula Baritaki, Ph.D.; Kazuo Umezawa, Ph.D.; and Eriko Suzuki. The author also acknowledges the assistance of Maggie Yang, Alina Katsman, and Amy Wu for the preparation of this manuscript.

References

1. Patel, N.H. and Rothenberg, M.L., 1994, Multidrug resistance in cancer chemotherapy, *Invest New Drugs* **12**:1.
2. Kurnick, J.T. and Kradin, R.L., 1991, Adoptive immunotherapy with recombinant interleukin 2, LAK and TIL, *Allergol. Immunopathol. (Madr)* **19**:209.

3. Sogn, J.A., 1998, Tumor immunology: the glass is half full, *Immunity* **9**:757.
4. Mule, J.J., Shu, S., Schwarz, S.L., and Rosenberg, S.A., 1984, Adoptive immunotherapy of established pulmonary metastases with LAK clls and recombinant interleukin-2, *Science* **225**:1487.
5. Rosenberg, S.A., Lotze, M.T., Muul, L.M., Chang, A.E., Avis, F.P., Leitman, S., Linehan, W.M., Robertson, C.N., Lee, R.E., Rubin, J.T., et al., 1987, A progress report on the treatment of 157 patients with advanced cancer using lymphokine-activated killer cells and interleukin-2 or high-dose interleukin-2 alone, *N. Engl. J. Med.* **316**:889.
6. Tan, Y., Xu, M., Wang, W., Zhang, F., Li, D., Xu, X., Gu, J., and Hoffman, R.M., 1996, IL-2 gene therapy of advanced lung cancer patients, *Anticancer Res.* **16**:1993.
7. Cormier, J.N., Salgaller, M.L., Prevette, T., Barracchini, K.C., Rivoltini, L., Restifo, N.P., Rosenberg, S.A., and Marincola, F.M., 1997, Enhancement of cellular immunity in melanoma patients immunized with a peptide from MART-1/Melan A, *Cancer J. Sci. Am.* **3**:37.
8. Reed, J.C., 1999, Mechanisms of apoptosis avoidance in cancer, *Curr. Opin. Oncol.* **11**:68.
9. Martin, S.J. and Green, D.R., 1994, Apoptosis as a goal of cancer therapy, *Curr. Opin. Oncol.* **6**:616.
10. Thompson, C.B., 1995, Apoptosis in the pathogenesis and treatment of disease, *Science* **267**:1456.
11. Shresta, S., Pham, C.T., Thomas, D.A., Graubert, T.A., and Ley, T.J., 1998, How do cytotoxic lymphocytes kill their targets? *Curr. Opin. Immunol.* **10**:581.
12. Ashkenazi, A. and Dixit, V.M., 1998, Death receptors: signaling and modulation, *Science* **281**:1305.
13. Thomas, W., and Hersey P., 1998, TNF-related apoptosis-inducing ligand (TRAIL) induces apoptosis in Fas ligand-resistant melanoma cells and mediates CD4 T cell killing of target cells, *J. Immunol.* **161**:2195.
14. Frost, P.J., Butterfield, L.H., Dissette, V.B., Economou, J.S., and Bonavida, B., 2001, Immunosensitization of Melanoma Tumor Cells to Non-MHC Fas-Mediated Killing by MART-1-Specific CTL Cultures, *J. Immunol.* **166**:3564.
15. Hengartner, M.O., 2000, The biochemistry of apoptosis, *Nature* **407**:770.
16. Luo, X., Budihardjo, I., Zou, H., Slaughter, C., and Wang, X., 1998, Bid, a Bcl2 interacting protein, mediateds cytochrome c release from mitochondria in response to activation of cell surface death receptors, *Cell* **94**:481.
17. Los, M., Herr, I., Friesen, C., Fulda, S., Schulze-Osthoff, K., and Debatin, K.M., 1997, Cross-resistance of CD95- and drug-induced apoptosis as a consequence of deficient activation of caspases (ICE/Ced-3 proteases), *Blood* **90**:3118.
18. Wang, G.Q., Gastman, B.R., Wieckowski, E.U., Goldstein, L.A., Rabinovitz, A., Yin, X.M., and Rabinowich, H., 2000, Apoptosis-resistant mitochondria in T cells selected for resistance to FAS signaling, *J. Biol. Chem.* **276**:3610.
19. Safrit, J.T. and Bonavida, B., 1992, Hierarchy of in vitro sensitivity and resistance of tumor cells to cytotoxic effector cells, cytokines, drugs and toxins, *Cancer Immunol. Immunother.* **34**:321.
20. Landowski, T.H., Gleason-Guzman, M.C., and Dalton, W.S., 1997, Selection for drug resistance results in resistance to Fas-mediated apoptosis, *Blood* **89**:1854.
21. Lehmann, C., Zeis, M., Schmitz, N., and Uharek, L., 2000, Impaired binding of perforin on the surface of tumor cells is a cause of target cell resistance against cytotoxic effector cells, *Blood* **96**:594.
22. Chouaib, S., Thiery, J., Gati, A., Guerra, N., El Behi, M., Dorothee, G., Mami-Chouaib, F., Bellet, D., and Caignard, A., 2002, Tumor escape from killing: role of killer inhibitory receptors and acquisition of tumor resistance to cell death, *Tissue Antigens* **60**:273.
23. Medema, J.P., Schuurhuis, D.H., Rea, D., van Tongeren, J., de Jong, J., Bres, S.A., Laban, S., Toes, R.E., Toebes, M., Schumacher, T.N., Bladergroen, B.A., Ossendorp, F., Kummer, J.A., Melief, C.J., and Offringa, R., 2001, Expression of the serpin serine protease inhibitor 6 protects dendritic cells from cytotoxic T lymphocyte-induced apoptosis: differential modulation by T helper type 1 and type 2 cells, *J. Exp. Med.* **194**:657.

24. Bots, M., VAN Bostelen, L., Rademaker, M.T., Offringa, R., and Medema, J.P., 2006, Serpins prevent granzyme-induced death in a species-specific manner, *Immunol. Cell Biol.* **84**:79.
25. Ruberti, G., Cascino, I., Papoff, G., and Eramo, A., 1996, Fas splicing variants and thier effect on apoptosis, *Adv. Exp. Me. Biol.* **406**:125.
26. Eberstadt, M., Huang, B., Olejniczak, E.T., and Fesik, S.W., 1997, The lymphoproliferation mutation in Fas locally unfolds the Fas death domain, *Nat. Struct. Biol.* **4**:983.
27. Martinez-Lorenzo, M.J., Gamen, S., Etxeberria, J., Lasierra, P., Larrad, L., Pineiro, A., Anel, A., Naval, J., and Alava, M.A., 1998, Resistance to apoptosis correlates with a highly proliferative phenotype and loss of Fas and CPP32 (caspase-3) expression in human leukemia cells, *Int. J. Cancer* **75**:473.
28. Gajate, C., and Mollinedo, F., 2005, Cytoskeleton-mediated death receptor and ligand concentration in lipid rafts forms apoptosis-promoting clusters in cancer chemotherapy, *J. Biol. Chem.* **280**:11641.
29. Abouzahr, S., Bismuth, G., Gaudin, C., Caroll, O, Van Endert, P., Jalil, A., Dausset, J., Vergnon, I., Richon, C., Kauffmann,A., Galon, J., Raposo,G., Mami-Chouaib, F., and Chouaib, S., 2006, Identification of target actin content and polymerization status as a mechanism of tumor resistance after cytolytic T lymphocyte pressure, *PNAS* **103**:1428.
30. Vousden, K.H., and Lu, X., 2002, Live or let die: the cell's response to p53, *Nat. Rev. Cancer* **2**:594.
31. Brown, J.M., and Wouters, B.G., 1999, Apoptosis, p53, and tumor cell sensitivity to anticancer agents, *Cancer Res.* **59**:1391.
32. Thiery, J., Abouzahr, S., Dorothee, G., Jalil, A., Richon, C., Vergnon, I., Mami-Chouaib, F., and Chouai, S., 2005, p53 potentiation of tumor cell susceptibility to CTL involves Fas and mitochondrial pathways, *J. Immunol.* **174**:871.
33. Mizutani, Y., Yoshida, O., and Bonavida, B., 1998, Sensitization of human bladder cancer cells to FAS-mediated cytotoxicity by cix-diamminedichloroplatinum (II), *J. Urol.* **309**:160.
34. Mori, S., Murakami-Mori, K., Nakamura, S., Ashkenazi, A., and Bonavida, B., 1999, Sensitization of AIDS-Kaposi's sarcoma cells to Apo-2 ligand-induced apoptosis by actinomycin D, *J. Immunol.* **162**:5616.
35. Wen, J., Ramadevi, N., Nguyen, D., Perkins, C., Worthington, E., and Bhalla, K., 2000, Antileukemic drugs increase death receptor 5 levels and enhance Apo-2L induced apoptosis of human acute leukemia cells, *Blood* **96**:3900.
36. Baritaki, V., Huerta-Yepez, S., Sakai, T., Spandidos, D., and Bonavida, B., 2007, Chemotherapeutic drugs sensitize cancer cells to TRAIL-mediated apoptosis: upregulation of DR5 and inhibition of YY1, *Mol. Cancer Thers.* (In Press).
37. Jazirehi, A.R., Ng, C.P., Schiller, G., and Bonavida, B., 2001, Adrenomycin sensitizes the adrenomycin-resistant 8226/Dox 40 human multiple myeloma cells to TRAIL/Apo2L-mediated apoptosis, *Clin. Cancer Res.* **7**:3874.
38. Fulda, S., Meyer, E., and Debatin, K.M., 2000, Metabolic inhibitors sensitize for CD95 (APO-1/Fas)-induced apoptosis by down-regulating Fas-associated death domain-like interleukin 1-converting enzyme inhibitory protein expression, *Cancer Res* **60**:3947.
39. Solary, E., Plenchette, S., Sordet, O., Rebe, C., Ducoroy, P., Filomenko, R., Bruey, J.M., Droin, N., and Corcos, L., 2001, Modulation of apoptotic pathways triggered by cytotoxic agents, *Therapie* (*Review*) **56**(5):511.
40. Uslu, R., Borsellino, N., Frost, P., Garban, H., Ng, C.P., Mizutani, Y., Belldegrun, A., and Bonavida, B., 1997, Chemosensitization of human prostate carcinoma cell lines to anti-fas-mediated cytotoxicity and apoptosis, *Clin. Cancer Res.* **3**:963.
41. Petak, I., Tillman, D.M., and Houghton, J.A., 2007, p53 dependence of Fas induction and acute apoptosis in response to 5-Fluorouracil-Leucovorin in human colon carcinoma cell lines, *Clin. Cancer Res.* **6**:4432.
42. Huerta-Yepez, S., Vega, M., Jazirehi, A.R., Garban, H., Hongo, F., Cheng, G., and Bonavida, B., 2004, Nitric oxide sensitizes prostate carcinoma cell lines to TRAIL-mediated apoptosis via inactivation of NF-kB and inhibition of Bcl-xL expression, *Oncogene* **23**:4993.
43. Mueller, et al., 1998.

44. Garban, H., and Bonavida, B., 2001, Nitric oxide inhibits the transcription receptor Yin-Yang 1 binding activity at the silencer region of the Fas Promoter: a pivotal role for nitric oxide in the upregulation of Fas gene expression in human tumor cells, *J. Immunol.* **167**:75.
45. Peter, M.E., Kischkel, F.C., Scheuerpflug, C.G., Medema, J.P., Debatin, K.M., and Krammer, P.H., 1997, Resistance of cultured peripheral T cells towards activation-induced cell death involves a lack of recruitment of FLICE (MACH/caspase 8) to the CD95 death-inducing signaling complex, *Eur. J. Immunol.* **27**:1207.
46. Micheau, O., Hammann, A., Solary, E., and Dimanche-Boitrel, M.T., 1999, STAT-1-independent upregulation of FADD and procaspase-3 and -8 in cancer cells treated with cytotoxic drugs, *Biochem. Biophys. Res. Commun.* **256**:603.
47. Ng, C.P., Frost, P., Garban, H., and Bonavida, B., 1998, Mechanism of chemo-immunosensitization of prostate carcinoma target cells to cytotoxic lymphocytes-mediated apoptosis, *Proc. AACR* **39**:356.
48. Zou, H., Li, Y., Liu, X., and Wang, X., 1999, An APAF-1 cytochrome c multimeric complex is a functional apoptosome that activates procaspase-9, *J. Biol. Chem.* **274**:11549.
49. Kinoshita, H., Yoshikawa, H., Shiiki, K., Hamada, Y., Nakajima, Y., and Tasaka, K., 2000, Cisplatin (CDDP) sensitizes human osteosarcoma cell to Fas/CD95-mediated apoptosis by down-regulating FLIP-L expression, *Int. J. Cancer* **88**:986.
50. Droin, N., Dubrez, L., Eymin, B., Renvoizé, C., Bréard, J., Dimanche-Boitrel, M.T., and Solary, E., 1998, Upregulation of *CASP* genes in human tumor cells undergoing etoposide-induced apoptosis, *Oncogene* **16**:2885.
51. Green, D.R. and Reed, J.C., 1998, Mitochondria and apoptosis, *Science* **281**:1309.
52. Zhang, L., Yu, J., Park, B.H., Kinzler K.W., and Vogelstein, B., 2000, Role of *BAX* in the apoptotic response to anticancer agents, *Science* **290**:989.
53. Yang, Y.L. and Li, X.M., 2000, The IAP family: endogenous caspase inhibitors with multiple biological activities, *Cell Res.* **10**:169.
54. Deveraux, Q.L., Roy, N., Stennicke, H.R., Van Arsdale, T., Zhou, Q., Srinivasula, S.M., Alnemri, E.S., Salvesen, G.S., and Reed, J.C., 1998, IAPs block apoptotic events induced by caspase-8 and cytochrome c by direct inhibition of distinct caspases, *EMBO J.* **17**:2215.
55. Zisman, A., Ng, C.P., Pantuck, A.J., Bonavida, B., and Belldegrun, A.S., 2001, Actinomycin D and gemcitabine synergistically sensitize androgen-independent prostate cancer cells to Apo2L/TRAIL-mediated apoptosis, *J. Immunother.* **24**:459.
56. Suzuki, A., Tsutomi, Y., Akahane, K., Araki, T., and Miura, M., 1998, Resistance to Fas-mediated apoptosis: activation of caspase 3 is regulated by cell cycle regulator p21 WAF1 and IAP gene family ILP, *Oncogene* **17**:931.
57. Vega, M.I., Jazirehi, A.R., Huerta-Yepez, S., and Bonavida, B., 2005, Rituximab-induced inhibition of YY1 and Bcl-x_L expression in Ramos non-Hodgkin's lymphoma cell line via inhibition of NF-κB activity: role of YY1 and Bcl-x_L in Fas resistance and chemoresistance, respectively, *J. Immunol.* **175**:2174.
58. Vega, M.I., Huerta-Yepez, S., Jazirehi, A.R., Garban, H., and Bonavida, B., 2005, Rituximab (chimeric anti-CD20) sensitizes B-NHL cell lines to Fas-induced apoptosis, *Oncogene* **24**:8114.
59. Hongo, F., Huerta-Yepez, S., Vega, M., Garban, H., Jazirehi, A., Mizutani, Y., Miki, T., and Bonavida, B., 2005, Inhibition of the transcription factor Yin Yang 1 (YY1) activity by S-nitrosation, *Biochem. Biophys. Res. Commun.* **336**:692.
60. Bonavida, B., Khineche S., Huerta-Yepez S., and Garban H., 2006, Therapeutic potential of nitric oxide in cancer, *Drug Resist Updat.* **9**:157.
61. Jazirehi, A., and Bonavida, B., 2005, Cellular and molecular signal transduction pathways modulated by Rituximab (Rituxan, anti-CD20 mAb) in Non-Hodgkin's lymphoma: implications in chemo-sensitization, *Oncogene* **24**:2142.
62. Ng, C-P., and Bonavida, B., 2002, A new challenge to immunotherapy by tumors that are resistant to apoptosis: two complementary signals to overcome cross-resistance, *Adv. Cancer Res.* **85**:145.
63. Jazirehi, A., Vega., MI., and Bonavida, B., 2007, Development of rituximab-resistant lymphoma clones with altered cell signaling and cross-resistance to chemotherapy, *Cancer Res.* **67**:1270.

64. Vega, et al., 2006, in preparation.
65. Suzuki, et al., 1999, in preparation.
66. Katsman, A., Umezawa, K., and Bonavida, B., 2007, Reversal of resistance to cytotoxic cancer therapies: DHMEQ as a chemo-sensitizing and immuno-sensitizing agent, *Drug Resist Updat*. Epub [In Press].
67. Wouters, B.G., and Brown, J.M., 1997, Cells at intermediate oxygen levels can be more important than the "hypoxic fraction" in determining tumor response to fractionated radiotherapy, *Radiat. Res*. **147**:541.
68. Kwon, S.J., Song, J.J., and Lee, Y.J., 2005, Signal pathway of hypoxia-inducible factor-1alpha phosphorylation and its interaction with von Hippel-Lindau tumor suppressor protein during ischemia in MiaPaCa-2 pancreatic cancer cells, *Clin. Cancer Res*. **11**:7607–13.
69. Kim, K.U., Wilson, S.M., Abayasiriwardana, K.S., Collins, R., Fjellbirkeland, L., Xu, Z., Jablons, D.M., Nishimura, S.L., and Broaddus, V.C., 2005, A novel in vitro model of human mesothelioma for studying tumor biology and apoptotic resistance, *Am. J. Respir. Cell Mol. Biol*. **33**:541.
70. Mace, T.A., Yamane, N., Cheng, J., Hylander, B.L., and Repasky, E.A., 2006, The potential of the tumor microenvironment to influence Apo2L/TRAIL induced apoptosis, *Immunol. Invest*. **35**:279.
71. Bhowmick, N.A., Neilson, E.G., and Moses, H.L., 2004, Stromal fibroblasts in cancer initiation and progression, *Nature* **432**:332.
72. Miyashita, T., Kawakami, A., Nakasima, T., Yamasaki, D., Tamai, M., Tanaka, F., Kamachi, M., Ida, H., Migi ta, K., Origuchi, T., Nakao, K., and Eguchi, K., 2004, Osteoprotegerin (OPG) acts as an endogenous decoy receptor in tumour necrosis factor-related apoptosis-inducing ligand (TRAIL)-mediated apoptosis of fibroblast-like synovial cells, *Clin. Exp. Immunol*. **137**:430.
73. Rosenthal, E.L., Zhang, W., Talbert, M., Raisch, K.P., and Peters, and G.E., 2005, Extracellular matrix metalloprotease inducer-expressing head and neck squamous cell carcinoma cells promote fibroblast-mediated type I collagen degradation in vitro, *Mol. Cancer Res*. **3**:195.
74. Abdollahi, T., Robertson, N.M., Abdollahi, A., and Litwack, G., 2003, Identification of interleukin 8 as an inhibitor of tumor necrosis factor-related apoptosis-inducing ligand-induced apoptosis in the ovarian carcinoma cell line OVCAR3, *Cancer Res*. **63**:4521.
75. Albertsson, P., Kim, M.H., Jonges, L.E., Kitson, R.P., Kuppen, P.J., Johansson, B.R., Nannmark, U., and Goldfarb, R.H., 2000, Matrix metalloproteinases of human NK cells, *In Vivo* **14**:269–76.
76. Limb, G.A., Daniels, J.T., Cambrey, A.D., Secker, G.A., Shortt, A.J., Lawrence, J.M., and Khaw, P.T., 2005, Current prospects for adult stem cell-based therapies in ocular repair and regeneration, *Curr. Eye Res*. **31**:381.
77. Nyormoi, O., Mills, L., and Bar-Eli, M., 2003, An MMP-2/MMP-9 inhibitor, 5a, enhances apoptosis induced by ligands of the TNF receptor superfamily in cancer cells. *Cell Death Differ*. **10**:558.
78. Sengupta, S., Eavarone, D., Capila, I., Zhao, G., Watson, N., Kiziltepe, T., and Sasisekharan, R., 2005, Temporal targeting of tumour cells and neovasculature with a nanoscale delivery system, *Nature* **436**:568.
79. Otten, H.G., van Ginkel, W.G.J., Hagenbeek, A., and Petersen, E.J., 2004, Prevalence and clinical significance of resistance to perforin- and FAS-mediated cell death in leukemia, *Leukemia* **18**:1401.
80. Griffith, T.S., Brunner, T., Fletcher, S.M., Green, D.R., and Ferguson, T.A., 1995, Fas ligand-induced apoptosis as a mechanism of immune privilege, *Science* **270**:1189.
81. Snell, V., Clodi, K., Zhao, S., Goodwin, R., Thomas, E.K., Morris, S.W., Kadin, M.E., Cabanillas, F., Andreeff, M., and Younes, A., 1997, Activity of TNF-related apoptosis-inducing ligand (TRAIL) in haematological malignancies, *Br. J. Haematol*. **99**:618.
82. Del Poeta, G., Venditti, A., Del Principe, M.I., Maurillo, L., Buccisano, F., Tamburini, A., Cox, M.C., Franchi, A., Bruno, A., Mazzone, C., Panetta, P., Suppo, G., Masi, M., and Amadori, S., 2003, Amount of spontaneous apoptosis detected by Bax/Bcl-2 ratio predicts outcome in acute myeloid leukemia (AML), *Blood* **101**:2125.

83. Mayes, P.A., Campbell, L., Ricci, M.S., Plastaras, J.P., Dicker D.T., and el-Deiry, W.S., 2005, Modulation of TRAIL-induced tumor cell apoptosis in a hypoxic environment, *Cancer Biol. Ther*. **4**:1068.
84. Mueller, A., Odze, R., Jenkins, T.D., Shahsesfaei, A., Nakagawa, H., Inomoto, T., and Rustgi, A.K., 1997, A transgenic mouse model with cyclin D1 overexpression results in cell cycle, epidermal growth factor receptor, and p53 abnormalities, *Cancer Res*. **57**:5542.
85. Sheikh, M.S., Burns, T.F., Huang, Y., Wu, G.S., Amundson, S., Brooks, K.S., Fornace, A.J. Jr., and el-Deiry, W.S., 1998, p53-dependent and –independent regulation of the death receptor KILLER/DR5 gene expression in response to genotoxic stress and tumor necrosis factor alpha, *Cancer Res*. **58**:1593.

Chapter 7
The Tumor Microenvironment as a Model for Tissue-Specific Rejection

Silvia Selleri[1,2], Sara Deola[1], Cristiano Rumio[2], and Francesco M. Marincola*[1]

Abstract Since the discovery in cancer patients of tumor infiltrating and circulating lymphocytes that can recognize and kill autologous cancer cells, research have been perplexed by the paradoxical coexistence in the same organism of effector immune responses and their targets. This observation suggests that while the afferent arm of the immune response can properly exert its cognitive functions, the efferent arm displays insufficient effector activity.

Two main categories of explanations may be hypothesized: either the immune system of cancer patients is systemically hampered by cancer-specific immune tolerance or a general status of immune suppression, or the cross talk between tumor and immune cells is modulated by adaptive changes of tumor cells that may escape recognition by masking or loosing the target antigens, by providing insufficient co-stimulation for T cell activation or producing immune-modulatory factors.

In this chapter, we will present and discuss the present understanding of the relationship between immune and cancer cells in the context of the tumor microenvironment. Far from presenting a comprehensive explanation, our goal is to offer an update of the current status and foster interest in the pursue of studies directed at the ex vivo analysis of human samples that may spark the identification of novel hypotheses in the frame of human reality.

Keywords Tumor microenvironment, inflammation, immune rejection

7.1 Introduction

The modern view of immunology suggests that the immune system can recognize potential antigens independent of self non-self discrimination if they are presented in association with sufficient co-stimulatory signals as those that occur in special

[1] Immunogenetics Section, Department of Transfusion Medicine, Clinical Center, National Institutes of Health, Bethesda, MD, 20892, USA

[2] Universita' degli Studi di Milano, Department of Human Morphology, via Mangiagalli 31, 20133 Milan Italy

E. Yefenof (ed.), *Innate and Adaptive Immunity in the Tumor Microenvironment.*

pathophysiological conditions such as pathogen infection, autoimmune reactions and presence of allogeneic tissues [1]. Thus, it may be suggested that allograft rejection may not only reflect the presence of non-self antigens on the surface of heterologous cells but other conditions may contribute to immune-mediated rejection. Those conditions may be similar to those necessary for immune-mediated tissue-specific destruction in other circumstances in which self non-self discrimination is not pertinent such as acute flares of auto-immunity when self molecules are target of immune effector cells. Astride between the inevitable allo-recognition of transplanted organs in the absence of immune suppression and the occasional self-recognition when autoimmunity occurs, is the variable response of immune cells to pathogen invasion. In this case, although pathogens represent non-self entities, the response to the immune system may vary broadly from rapid clearance of the offending organism during an acute infectious event to a chronic reaction that perpetuates in some cases for the lifetime of the affected organism as in the case of chronic hepatitis C virus (HCV) infection [2].

The immune response to cancer sits in between the chronic course of unresolving infections and the persistence of mild autoimmune diseases. In particular, some cancer such as melanoma display a natural tendency to elicit recognition of autologous non-mutated tumor antigens [3] and prime systemic immune responses [4] that, however, are most often insufficient to eradicate the disease. Thus, like in the case of HCV infection, recognition of the antigen is not sufficient and a lingering chronic inflammatory status persists indefinitely wedding indefinitely the pathogen with its host. An overview of various types of immune pathologies, suggests that acute inflammation is necessary for destruction of the pathogenic process whether with beneficial (clearance of pathogen, rejection of cancer) or detrimental (allograft rejection, autoimmunity) results [5]. Thus, inflammatory processes are ultimately required to activate immune effector cells within the target tissue. We have recently proposed that, independently of the triggering events associated with different immune pathologies, such acute activation of immune effectors follows a common final pathway to which we referred to as the immunological constant of rejection [5]. If insufficient pro-inflammatory signals are produced by the pathogenic process, a lingering immune response persists that self perpetuates without eliminating the pathogenic stimulus. This is the most frequent occurrence in cancer. As we will discuss later, the literature has increasingly produced extensive information about immunological or broader biological reasons to explain the frequent incompetence of the immune response toward cancer. Here, we will discuss briefly such mechanism but in the end we will try consider those cases in which tumors are rejected in autologous settings in humans to try to understand how immune manipulation seems to overcome regulatory mechanisms physiologically present in all individuals or induced by the cancer-bearing status.

Before addressing the potentials of antitumor immune responses, it is worth addressing the conflicting perception of the role of the immune response (also referred to as inflammation) in cancer. In fact, the role played by the immune-system in modulating cancer growth and/or its rejection remains perplexing and a matter of active debate. Several investigators propose that immune responses that generate a chronic inflammatory status foster cancer growth [5–11]. For instance, chronic viral and bacterial infections may contribute to more than 1 million cases of infection-related malignancies per year [12, 13]. Two among many interesting examples of virally-induced carcinogenesis are Epstein-Barr virus (EBV) latent infections and chronic HCV. While the former may have direct oncogenic properties intrinsic to the complex biology of this DNA virus, HCV induced hepato-cellular carcinoma is most likely and predominantly induced by the continuous destruction and remodeling of liver tissue induced by chronic HCV infection [14, 15]. Latent EBV infection can induce two types of neoplastic processes: in immunosuppressed individuals, EBV induces lymphoproliferative disorders that are effectively eliminated by withdrawal of immune suppression [16]. In addition, complete regression of post-transplant lymphoproliferative disorders can be mediated by adoptive transfer of Human Leukocyte Antigen (HLA)-matched EBV-specific cytotoxic T cells (CTL) [17–19] and prophylactic administration of CTL can prevent their insurgence [20]. This observation suggests that in conditions, in which the latent EBV infection is not associated with a known chronic inflammatory process affecting a specific organ, viruses can promote oncogenesis directly but the neoplastic process is kept in check by a competent immune system through adaptive responses. It is presently unknown what biological characteristic of EBV-infected cells can sustain a competent immune response in non-immune suppressed hosts. This information is of extreme interest because lack of lymphoproliferative disease in immune competent hosts represents a most outstanding human model of anti-cancer immune surveillance.

In contrast, EBV-associated tumors such as Burkitt's lymphoma, Hodgkin's disease and nasopharyngeal cancer occur in immune competent individuals and display a restricted expression of EBV proteins [21] which may explain the reduced effectiveness of adoptively transferred EBV specific CTL [21–23]. Thus, the same virus, acting as a potent oncogene, may induce various types of cancers whose insurgence is predicated upon the immune status of the host. In the latter case, however, it can be postulated that a combination of oncogenic properties of the virus and a chronic inflammatory process induced by the latent infection or other factors affecting the upper airways may be together responsible for the insurgence of the neoplasia. In particular, since there is no evidence that patients with nasopharyngeal carcinoma have dramatically perturbed immune function, it seems that in such cases, the virally-transformed cells may have acquired novel immune-regulatory properties that facilitate their escape from immune recognition as suggested by experimental animal models [24]. Alternatively, these EBV-associated cancers may lack the appropriate co-stimulatory properties displayed by EBV-infected B cells.

To the extreme, HCV-induced hepatocellular carcinoma is probably associated with the remodeling and inflammation that characterizes the progression of this cancer from a liver damaged by chronic inflammation and cirrhosis. In fact, a similar disease arises in livers damaged by other chronic conditions such as alcoholic hepatitis or in association with severe hemochromatosis [25]. Hepatocellular carcinoma stands, therefore, as a salient example of a neoplastic process fostered by chronic inflammation independently of its primary cause. This information clearly suggests that chronic inflammatory processes and the immune cells that mediate them favor cancer growth.

While it is clear that the lingering immune responses associated with chronic inflammation foster the growth of several human cancers [5, 11], it is also apparent that in some favorable circumstances activation of the same immune responses can provoke cancer regression in humans providing evidence that tumors can be recognized and rejected by a properly activated immune system [5]. Treatment of metastatic melanoma is a prototype model of the potential immune responsiveness of human cancers [26–28]. One of the peculiarities of this cancer is that it is naturally immunogenic eliciting autologous immune responses against endogenous, non-mutated self-proteins [29, 30]. In this case, a particularly immunogenic micro-environment may be conducive to the induction of a cancer-specific autoimmune response directed against non-mutated self molecules. Possibly, other solid tumors that lack these immune properties are less likely to respond to immune manipulation [31]. Yet, in either case, in the absence of immune stimulation, tumors incrementally grow irrespectively of their potential immune responsiveness suggesting that additional immune stimulation is necessary to complement the naturally occurring immune responses. It is also not clear whether adaptive immune responses (the immune responses that are acquired throughout life after exposure to a given antigen) of the kind frequently observed in the context of melanoma are causative in promoting cancer rejection in humans or represent an epiphenomenon resulting from the intrinsic immunogenicity of melanoma; renal cell cancer is not as efficient in inducing the kind of antigen-specific immune responses observable in melanoma; yet, this cancer is as likely to regress in response to systemic interleukin-2 administration [32, 33]. In addition, recent experimental evidence suggests that the innate immune response that includes effector mechanisms present in each individual impendent of previous antigen exposure may play a prominent role in cancer rejection [34].

A major contribution to the understanding of the complex relationship between melanoma (as a prototype immune responsive cancer model) and its host was the identification and characterization of tumor antigen-specific CD8+ and CD4+ T cells within melanoma metastases [35]. If these tumor antigen-specific immune cells are consistently produced by the host and they can reach the tumor site, why is the rejection of cancer such a rare event [36]? Obviously the simple presence of antigen-specific T cells in not sufficient to induce rejection as illustrated by experimental models in which immune responses foster rather than hamper cancer growth [7, 13, 37] and evidence in humans that chronic infection of bacterial [38] or viral origin [39–42], or other pro-inflammatory stimuli promote cancer growth [43]. Two categories of reasons may explain why effector immune mechanisms cannot control tumor growth: a systemic insufficiency of the immune response (due to immune tolerance and immune suppression), or an insufficiency within the tumor that may be induced by unfavorable conditions within the tumor microenvironment and may include tumor escape mechanisms [44, 45] or simply lack of sufficient co-stimulatory properties of cancer tissues [46, 5, 31]. In this review, we will focus on the latter hypothesis referring the reader to other manuscripts discussing the role of systemic immune deficiencies in tumor bearing hosts [47, 48]. In particular, we will discuss the possible role that various components of the immune system may play within the tumor micro-environment and how their cross talk with cancer or other by-stander cells may modulate their function.

7.1.1 Innate Immunity and Inflammation

Chronic inflammation fosters tumor growth [7, 10, 13]; many components typical of an inflammatory status are commonly observed at the tumor site including immune cells (leukocytes or macrophages) and soluble factors (cytokines and chemokines). It has long been known that the presence of tumor associated macrophages is correlated with poor prognosis. Several molecular pathways (i.e. CD68) associated with leukocytes and macrophage infiltration and activation have been described that provide some mechanistic insight about this role of macrophages [49]. Tumor necrosis factor, interleukin-1 and the chemokine CCL-2 (a macrophage attractant) also have been shown to favor tumor growth and metastasis. However, the simple observation of a cellular population within the tumor microenvironment without information about its functional status may be misleading. For instance, macrophages display a particular level of plasticity and heterogenicity [50–53] and according to the composition of the external environment can assume M1 or M2 phenotype. Classically activated M1 macrophages are induced by IFN-γ alone or in concert with microbial stimuli or cytokines. M1 macrophages have potent pro-inflammatory properties and can elicit recruitment and activation of effective adaptive immune responses within the tumor microenvironment [45]. A tumor microenvironment enriched with IL-4 and IL-13 may induce an alternative form of macrophage referred to as M2 [51]. M2 macrophages promote chronic inflammation, tissue repair and angiogenesis and lack immune stimulatory properties. M2 macrophages can also be induced by the presence of immune complexes, IL-10, glucocorticoid or secosteroid hormones [52]. M2 macrophages are also able to induce a slow and persistent chronic inflammation [10] that promotes tumor proliferation and progression [54], stromal deposition and remodelling, stimulates angiogenesis [55, 56] lymphangiogenesis, and disorients or inhibits adaptive immunity [57]. At the tumor site macrophages tend to assume the M2 phenotype, where they promote tumor progression and remodelling [58] and can have immune-regulatory function [52]. However, their phenotype can be changed according to the surrounding immune environment and we have argued that the predominant mechanism of tumor rejection during systemic therapy with interleukin-2 or local immune stimulation with Toll-like receptor (TLR)-7 agonists is a direct activation of M2 type of macrophages into fully potent antigen presenting cells capable of producing chemoattractant for T and B cells and sustaining their activation and proliferation at the tumor site [45, 59–61].

7.1.2 Dendritic Cells

Dendritic cells (DC) represent another face of macrophage physiology and possibly are the end product of their activation after antigen loading through the phagocytic properties of non activated M2 type cells. Physiologically, DCs are believed to play a central role as modulators of immune function. In their immature state, DC can present self antigens to T lymphocytes inducing tolerance; as they mature in the presence of strong pro-inflammatory stimuli such as TLR stimulation [62] as it may occur during pathogen infection, they effectively present foreign antigens through human leukocyte antigen (HLA) class I and class II molecules to activate effector and helper T cell mechanisms. DCs are able to activate effector T, B, NK and NKT cells [63], but also T regulatory/suppressor (Treg) cells [64]. Occasionally, tumors are infiltrated by mature DCs that contrary to macrophages seem to participate in cancer rejection by activating adoptive immune responses. However, it has been reported that often mature DCs are segregated at the periphery of tumor masses, while the prevalent phenotype within the tumor environment consists of immature DC. This observation suggests that cancer cells can re-direct the differentiation of DCs in order to escape immune rejection. For example, by producing VEGF cancer cells interfere with DCs maturation [65]. Melanoma cells can secrete IL-10 and breast cancer IL-6, that in turn lead to tolerogenic DCs promoting monocyte differentiation toward macrophages rather than DCs [66, 67]. Moreover, breast cancer cells can secrete MUC-1, a glycoprotein that provokes incorrect antigen processing and presentation [68]. It should be pointed out that although these mechanisms may be responsible for an indolent immune responses within the tumor microenvironment that do not cause irreversible differentiation of antigen presenting cells that can be reactivated through immune manipulation such as the systemic administration of recombinant interleukin-2 [59, 60].

Recently a new DC subset was described in mice, called IKDC (4849, 4850). These engulf characteristics of NK cells, plasmacytoid DC (similar to both DC and B cells) and normal DC. Like NK cells, active IKDCs can kill cells lacking MHC molecules; like plasmacytoid DC they can secrete IFN-γ; and like conventional DC they can process and present antigens in order to activate T cells. IKDC has been proposed as important mediators of tumor surveillance: in a mouse model they are able to induce rejection of tumors poorly recognized by NK cells, by secreting high levels of IFN-γ and mediate TRAIL-dependent lysis of tumor cells [69]. The existence of the subpopulation of DC remains to be confirmed in humans. While it is likely that DCs may play various roles in the tumor microenvironment in natural conditions or in response to immunotherapy, controversy exists about their value as antigen presenting adjuvants during vaccine trials as discrepant outcomes have been reported [70–73]. However, such discussion is beyond the purposes of this review since it refers to a specific role played by DC's in inducing systemic immune responses through the afferent loop of immune responses, while the focus of this article is the role of various immune cells and their cross talk within the target organ: the tumor microenvironment [31].

7.1.3 NK Cells

Natural killer (NK) cells are involved in the innate immune-response against tumors and virally-infected cells. NK cells kill target cells both through perforin/granzyme secretion and through Fas ligand and TRAIL induced apoptosis. NK cells also produce cytokines that modulate T cell function [74]. Tumor cells typically lose or downregulate HLA expression in order to escape T cell recognition [44]. This characteristic makes them susceptible to NK-mediate lysis. NK cells can recognize HLA class I molecules through the inhibitory Killer immunoglobulin-like receptors (KIR): cells that express an abnormally low number of HLA class I molecules on their surface cannot activate the suppressor mechanisms modulated by the KIR and, consequently, NK cells can exert cytotoxic and other effector functions against autologous cells (the "missing self" hypothesis). NK cells activity is, however, modulated by other stimulatory receptors like some KIR missing the immune modulatory intracellular domain [75], 2B4 and NKG2D. Ligands identified for these receptors include CD48 for 2B4 [76], MHC class I chain related (MIC) molecules [77] and UL16-binding proteins (ULBP1, ULBP2, ULBP3 and ULBP4) [78, 79]. CD48 is a membrane-bound protein ubiquitously expressed on hematopoietic cells [80], whereas MIC and ULBP have a restricted tissue distribution and are preferentially induced by viral and bacterial infections as well as malignant transformation of cells [81]. Nevertheless, clinically the loss of HLA is normally associated with poor prognosis, indicating that NK cells are not able to substitute efficiently T cell response at the tumor site in natural conditions and in the absence of therapeutically-induced immune stimulation. This could be explained by the lack of co-stimulatory signals on tumor cell surface [82]. These findings in conjunction with those of the inhibitory ligand/receptor system support the notion that NK cell activity is regulated by a balance between inhibitory and activating signals and shifting of this balance is largely dependent on the ligand availability on target cells while the role of microenvironment stimuli has been less well characterized. Recently, we studied the early transcriptional events induced by the TLR-7 agonist imiquimod applied topically to basal cell carcinomas. This treatment eliminates these cancerous lesions in about 90% of the times. Among the most significantly activated transcripts, we identified several genes associated with powerful activation of NK cell as well as CD8 T cell function [61]. These findings were corroborated by the documentation by immunohistochemistry of an increased number of CD56 and CD8 expressing cells in regressing lesions during treatment compared with the placebo control-treated lesions. Similar findings have recently been reported by Torres et al. [83] following the effects of topical treatment of actinic keratosis with imiquimod. Thus, although in natural conditions NK cells may not be sufficient in number or sufficiently activated to induce tumor regress, given appropriate activatory stimuli they appear to actively participate in tumor destruction.

7.1.4 T Cells

The identification of tumor infiltrating lymphocytes (TIL) that can be expanded from melanoma metastases in the presence of the T cell growth factor interleukin-2 and that can recognize autologous and HLA-matched tumor cells but normal cells strongly supports the notion that the immune system can naturally mount cellular adaptive immune responses against cancer [3]. This characterization, however, triggers the paradoxical and yet un-explained observation of a balanced co-existence of tumor antigen-specific T cells with their target is a progressively growing tumor mass [45]. The identification and characterization of tumor-associated antigens, subsequently allowed a higher resolution to the study of the interactions between T cell and tumor cells adding an accurately-defined molecular dimension to the study of tumor host interactions [84]. In addition, when these antigens were used for active specific immunization a consistent enhancement of circulating tumor antigen-specific T cells cold be documented that could recognize specifically autologous and/or appropriately HLA matched tumor cells [85–88]. In addition, ex vivo comparisons of fine needle aspirates obtained from melanoma metastases before and after immunization strongly supported the notion that tumor-antigen specific T cells do localize at tumor site [89] and recognize tumor cells [90], yet localization and recognition are not sufficient for tumor regression.

The identification of TA recognized by CD8+ cytolytic T lymphocytes (CTL) enabled the development of TA-specific assays such as fluorescent tetrameric HLA/epitope complexes (tHLA) to monitor TA-specific T cell in complex populations [91]. Ex vivo characterization of tHLA-binding circulating vaccine-induced CD8+ T cells portrayed a quiescent phenotype that can recognize HLA-matched tumor targets by producing interferon-γ but they cannot kill the same targets unless they were activated *in vitro* by exposure to antigen recall and exogenous interleukin-2 [46]. This phenotype may explain why the dramatically enhanced number of tumor recognizing CD8+ T cells induced by immunization does not correspond to tumor regression suggesting that a second activatory stimulus should be applied to CD8+ T cells at the receiving within the tumor microenvironment which is naturally inert [31, 36].

Similarly to vaccine-induced tumor-specific CD8+ T cells, TIL isolated from biopsies and expanded *in vitro* in the presence of IL-2 maintain specificity and ability to produce IFN-γ after TA stimulation and regain the ability to kill target tumor cells [45]. However, like immunization-induced T cells that reach the tumor site, it is possible that TIL do not exert sufficient effector function at the tumor site. Interestingly, Cesana et al. [92] observed that patients with metastatic melanoma or renal cell carcinoma are burdened by a higher number of T regulatory (suppressor, Treg) cells compared with normal non tumor-bearing individuals. Their number increases during treatment with interleukin-2 and remains elevated in patients with disease progression, while their levels return to those observed in non tumor-bearing individuals within 4 weeks in patients experiences objective clinical response. The presence of Treg cells in the circulation might be paralleled by their enhanced presence at the tumor site which may result in a direct down-modulation of T cell effector functions. Experimental models suggest that Treg cells can infiltrate tumors [93, 94–97] where they are potent inhibitors of T cell function [98]. Treg cells could be recruited at the tumor site both by tumor-associated macrophages trough TGF-β release and by cancer cells expressing COX-2 [99]. Accumulation of tumor Treg cells predicts poor survival in individuals with ovarian carcinoma [94], breast and pancreatic cancer [100]. Alternatively, Treg cells may indirectly affect the tumor microenvironment by altering the function of T cell at the systemic level. Interestingly, polymorphisms in genes involved in immune response [101] may mediate different responses to IL-2 in individual patients [101] and, in the case of Treg cells a polymorphism in the chemokine receptor-5 (CCR5) may be responsible for the different response to interleukin-2 therapy in cancer patients [92].

It has also been suggested that TIL undergo exhaustion because of the continuous chronic stimulation (the relationship between CTL and cancer cells is near 1:10) by cancer cells that occurs in an inappropriate co-stimulatory environment that does not foster their expansion *in situ*. Secretion by tumor or tumor-infiltrating normal cells of soluble factors such as VEGF, IL-10 of TGF-β could also hamper T cell activation and proliferation [102]. In a murine model of lymphocytic choriomeningitis virus (LCMV), Barber et al. [103] observed that PD-1 (programmed death 1; also known as Pdcd1) was selectively upregulated in exhausted T cells, and that *in vivo* administration of antibodies that blocked the interaction of this inhibitory receptor with its ligand, PD-L1 (also known as B7-H1), restored T-cell responses. This could be very important if we consider the status of chronic inflammation induced by tumors in human, and open an encouraging perspective in the study of ligands expressed on target cells that can regulate CTL activity and in the possible application in cancer immunotherapy.

The role of helper T (Th) cells has been recently re-evaluated in modulating cancer rejection. Two predominant Th cell subtypes exist, Th1 and Th2. Th1 cells, characterized by secretion of IFN-γ and TNF-α, are primarily responsible for activating and regulating the development and persistence of CD8+ T cells and sustain their effector function. In addition, Th1 cells activate antigen-presenting cells (APC) and promote the production of antibodies that can enhance the uptake of antigens shed from infected or tumor cells by APC. Th2 cells favor a predominantly humoral response and limit the extent of cellular effector mechanisms. Occasionally, Th1 cells can kill tumor cells by direct recognition of MHC class II expressing cancer cells and exerting cytotoxic functions through perforin/granzyme secretion [104]. Moreover, Th1 cells can induce apoptosis of cancer cells through Fas/Fas ligand interaction [105] or activating death signal through the TNF-related apoptosis-inducing ligand (TRAIL) [106].

The role played by T cells in controlling of cancer growth has been well documented in experimental models [107]. However, their relevance in human tumors has been only recently documented; Pages et al. [108] observed a correlation between the presence of memory T cells and longer disease-free and overall survival of patients with colorectal cancer. Indeed, some propose to add histopathological information about T cell and other immune infiltrates as a staging parameter in colon cancer [109–111,112].

Our experience derived from direct human observation in real-time of human cancers in natural conditions or under treatment with immune stimulatory agents [5, 59, 61, 113] that lack of tumor regression in most cases represent the normal physiological condition that prevents the onset of autoimmune disease in most normal individuals; in addition, studies performed in human are likely to overestimated the significance of regulatory mechanisms within the tumor microenvironment. The study of tumors that are growing in patients represents a very peculiar time point in which the host and the cancer have stricken a balance that progressively favors tumor growth; other balances would simply not be seen because they would result in clearance of tumor cells. It would be probably more appropriate to add a temporal dimension to the study of the tumor microenvironment particularly in conditions in which the balance between the host immune system its cancer may be altered by immune intervention [114]. In that case, a very dynamic reality appears that in many instances recapitulates the molecular events observable during other immune phenomena characterized by acute inflammation. Our studies strongly emphasize that tolerance of tumors by immune mechanisms can be reversed if appropriate stimulation is provided within the tumor microenvironment [31].

7.1.5 Tumor Cells

It should be recognized that tumor cells sit at the center of the immune biology of cancer and they are likely to be directly or indirectly involved in orchestrating the nature and characteristics of the environment in which they are growing. Tumor cells can cunningly produce molecules that improve their growth, survival and migration potential and at the same time may suppress unwanted immune-responses. As proposed by others [115], tumors can be characterized by seven hallmarks: self sufficiency in growth signals, insensitivity to anti-growth signals, evasion of apoptosis, limitless replicative potential, sustained angiogenesis and tissue invasion and metastasis, and avoidance of immune-surveillance. All these processes involve molecules that affect the immune response, like IL-4 and 10 (inducing Th-2 differentiation), IL-6 that suppresses inflammation, TGF-β that inhibits T cells proliferation and antigen presentation and NK cytotoxicity and activate Treg cells. Obviously, things are not as simple as represented in fabricated experimental models. For instance, we have noted that IL-10 expression at the transcriptional and protein level is a positive predictor of immune responsiveness [116, 117]. This observation parallels pre-clinical experimental models in which the systemic or local effects of IL-10 are against tumor growth [60]. We hypothesized that while IL-10 may hamper immune responsiveness in the steady state biology of naturally growing tumors, they may precondition APC [60] and or NK cells [117] to become powerful immune stimulators and effectors during immune therapy. Although tumor cells may alter immune responses by presenting a mechanical barrier composed of infiltrating stroma [118] that may preventing efficient T cell priming and expansion [119], they may also express factors like the tumor-necrosis factor superfamily member LIGHT that can that induce massive naive T lymphocytes infiltration associated with upregulation of both chemokine production and expression of adhesion molecules [120]. Overall, the direct study of the human tumor microenvironment as a predictor of immune responsiveness has not been sufficiently explored although tools are available for such studies [114]. We observed studying a limited number of melanoma metastases that those likely to respond to immunotherapy consisting of the systemic administration of interleukin-2 display immunological signatures different from those unlikely to respond to therapy even before the treatment is administered [113]. In particular, metastases that respond to therapy are characterized by a chronic inflammatory state that does not appear to be sufficient to induce spontaneous tumor regression but facilitates the activation of an acute inflammatory state under systemic immune stimulation [45]. This notion is also supported by the finding that melanomas with better prognosis display transcriptional signatures suggesting of enhanced immunological activity [121]. Similarly, renal cell carcinomas likely to respond to immunotherapy with systemic interleukin-2 display before treatment high level of expression of carbonic anhydrase IX [122] at the transcriptional and protein level which is, in turn, associated with signatures of immune activation [123].

In summary, we believe that tumor cells are the major orchestrators of immune responsiveness. Melanoma and renal cell carcinomas, which are characterized by enhanced immune responsiveness, display a particular immune environment reflected by tissue-restricted signatures suggesting of enhanced immune activation [124]. The biology of these immune responsive tumors cannot induce spontaneous tumor regression in most cases; however, it fosters micro environmental conditions conducive to the acute inflammatory process necessary for the successful activation of immune effector mechanisms [5] if appropriate stimulation is provided and future clinical studies should investigate systemic and local strategies aimed at the activation of the efferent arm of the immune response within the target organ [31].

References

1. Matzinger, P. Danger model of immunity. Scand J Immunol, *54*: 2–3, 2001.
2. Rehermann, B. and Nascimbeni, M. Immunology of hepatitis B virus and hepatitis C virus infection. Nat Rev Immunol, *5*: 215–229, 2005.
3. Wolfel, T., Klehmann, E., Muller, C., Schutt, K. H., Meyer zum Buschenfelde, K. H., and Knuth, A. Lysis of human melanoma cells by autologous cytolytic T cell clones. Identification of human histocompatibility leukocyte antigen A2 as a restriction element for three different antigens. J Exp Med, *170*: 797–810, 1989.
4. Marincola, F. M., Rivoltini, L., Salgaller, M. L., Player, M., and Rosenberg, S. A. Differential anti-MART-1/MelanA CTL activity in peripheral blood of HLA-A2 melanoma patients in comparison to healthy donors: evidence for in vivo priming by tumor cells. J Immunother, *19*: 266–277, 1996.
5. Mantovani, A., Romero, P., Palucka, A. K., and Marincola, F. M. Tumor immunity: effector response to tumor and the influence of the microenvironment. Lancet, in press, 2006.
6. Bergers, G., Brekken, R., McMahon, G., Vu, T. H., Itoh, T., Tamaki, K., Tanzawa, K., Thorpe, P., Itohara, S., Werb, Z., and Hanahan, D. Matrix metalloproteinase-9 triggers the angiogenic switch during carcinogenesis. Nat Cell Biol, *2*: 737–744, 2000.
7. Coussens, L. M. and Werb, Z. Inflammation and cancer. Nature, *420*: 860–867, 2002.
8. Hanahan, D., Lanzavecchia, A., and Mihich, E. Fourteenth Annual Pezcoller Symposium: the novel dichotomy of immune interactions with tumors. Cancer Res, *63*: 3005–3008, 2003.
9. De Visser, K. E., Korets, L. V., and Coussens, L. M. De novo carcinogenesis promoted by chronic inflammation is B lymphocyte dependent. Cancer Cell, *7*: 411–423, 2005.

10. Balkwill, F., Charles, K. A., and Mantovani, A. Smoldering and polarized inflammation in the initiation and promotion of malignant disease. Cancer Cell, *7*: 211–217, 2005.
11. Mantovani, A. Cancer: inflammation by remote control. Nature, *435*: 752–753, 2005.
12. Kuper, H., Adami, H. O., and Trichopoulos, D. Infections as a major preventable cause of human cancer. J Intern Med, *248*: 171–183, 2000.
13. Balkwill, F. and Mantovani, A. Inflammation and cancer: back to Virchow? Lancet, *357*: 539–545, 2001.
14. Young, L. S. and Murray, P. G. Epstein-Barr virus and oncogenesis: from latent genes to tumours. Oncogene, *22*: 5108–5121, 2003.
15. Block, T. M., Mehta, A. S., Fimmel, C. J., and Jordan, R. Molecular viral oncology of hepatocellular carcinoma. Oncogene, *22*: 5093–5107, 2003.
16. Green, M. Management of Epstein-Barr virus-induced post-tranplant lymphoproliferative disease in recipients of solid organ transplanation. Am J Transplant, *1*: 103–108, 2001.
17. Heslop, H. E. and Rooney, C. M. Adoptive cellular immunotherapy for EBV lymphoproliferative disease. Immunol Rev, *157*: 217–222, 1997.
18. Khanna, R., Bell, S., Sherritt, M., Galbraith, A., Burrows, S. R., Rafter, L., Clarke, B., Slaughter, R., Falk, M. C., Douglass, J., Williams, T., Elliott, S. L., and Moss, D. J. Activation and adoptive transfer of Epsten-Barr virus-specific cytotoxic T cells in solid organ transplant patients with posttransplant lymphoproliferative disease. Proc Natl Acad Sci USA, *96*: 10391–10396, 1999.
19. Haque, T., Taylor, C., Wilkie, G. M., Murad, P., Amlot, P. L., Beath, S., McKiernan, P. J., and Crawford, D. H. Complete regression of posttransplant lymphoproliferative disease using partially HLA-matched Epstein Barr virus-specific cytotoxic T cells. Transplantation, *72*: 1399–1402, 2001.
20. Rooney, C. M., Roskrow, M. A., Smith, C. A., Brenner, M. K., and Heslop, H. E. Immunotherapy of Epstein-Barr virus-associated cancer. J Natl Cancer Institute Monogr, *23*: 89–93, 1998.
21. Gottschalk, S., Heslop, H. E., and Rooney, C. M. Treatment of Epstein-Barr virus-associated malignancies with specific T cells. Adv Cancer Res, *84*: 175–201, 2002.
22. Chua, D., Huang, J., Zheng, B., Lau, S. Y., Luk, W., Kwong, D. L., Sham, J. S., Moss, D., Yuen, K. Y., Im, S. W., and Ng, M. H. Adoptive transfer of autologous Epstein-Barr virus-specific cytotoxic T cells for nasopharyngeal carcinoma. Int J Cancer, *94*: 73–80, 2001.
23. Straathof, K. C., Bollard, C. M., Popat, U., Huls, M. H., Lopez, T., Morriss, M. C., Gresik, M. V., Gee, A. P., Russell, H. V., Brenner, M. K., Rooney, C. M., and Heslop, H. E. Treatment of nasopharyngeal carcinoma with Epstein-Barr virus – specific T lymphocytes. Blood, *105*: 1898–1904, 2005.
24. Dunn, G. P., Old, L. J., and Schreiber, R. D. The three Es of cancer immunoediting. Annu Rev Immunol, *22*: 329–360, 2004.
25. Motola-Kuba, D., Zamora-Valdes, D., Uribe, M., and Mendez-Sanchez, N. Hepatocellular carcinoma. An overview. Ann Hepatol, *5*: 16–24, 2006.
26. Minev, B. R. Melanoma vaccines. Semin Oncol, *29*: 479–493, 2002.
27. Parmiani, G., Castelli, C., Dalerba, P., Mortarini, R., Rivoltini, L., Marincola, F. M., and Anichini, A. Cancer immunotherapy with peptide-based vaccines: what have we achieved? Where are we going? J Natl Cancer Inst, *94*: 805–818, 2002.
28. Parmiani, G., Castelli, C., Rivoltini, L., Casati, C., Tully, G. A., Novellino, L., Patuzzo, A., Tosi, D., Anichini, A., and Santinami, M. Immunotherapy of melanoma. Sem Cancer Biol, *13*: 391–400, 2003.
29. Boon, T. and Van Der, B. P. Human tumor antigens recognized by T lymphocytes. J Exp Med, *183*: 725–729, 1996.
30. Old, L. J. and Chen, Y. T. New Paths in Human Cancer Serology. J Exp Med, *187*: 1163–1167, 1998.
31. Monsurro', V., Wang, E., Panelli, M. C., Nagorsen, D., Jin, P., Smith, K., Ngalame, Y., Even, J., and Marincola, F. M. Active-specific immunization against cancer: is the problem at the receiving end? Sem Cancer Biol, *13*: 473–480, 2003.

32. Abrams, J. S., Rayner, A. A., Wiernik, P. H., Parkinson, D. R., Eisenberger, M., Aronson, F. R., Gucalp, R., Atkins, M. B., and Hawkins, M. J. High-dose recombinant interleukin-2 alone: a regimen with limited activity in the treatment of advanced renal cell carcinoma. J Natl Cancer Inst, *82*: 1202–1206, 1990.
33. Atkins, M. B., Regan, M., and McDermott, D. Update on the role of interleukin 2 and other cytokines in the treatment of patients with stage IV renal carcinoma. Clin Cancer Res, *10*: 6342S–6346S, 2004.
34. Hicks, A. M., Riedlinger, G., Willingham, M. C., Alexander-Miller, M. A., von Kap-Herr, C., Pettenati, M. J., Sanders, A. M., Weir, H. M., Du, E., Kim, J., Simpson, A. J. G., Old, L. J., and Cui, Z. Transferable anticancer innate immunity in spontaneous regression/complete resistance mice. Proc Natl Acad Sci USA, *103*: 7753–7758, 2006.
35. Boon, T., Cerottini, J.-C., Van den Eynde, B., van der Bruggen, P., and Van Pel, A. Tumor antigens recognized by T lymphocytes. Annu Rev Immunol, *12*: 337–365, 1994.
36. Marincola, F. M. A balanced review of the status of T cell-based therapy against cancer. J Transl Med, *3*: 16, 2005.
37. Lodish, H. F. Should cell biologists study human disease? ASCB Newsletter, *27*: 2–4, 2004.
38. Mager, D. L. Bacteria and cancer: cause, coincidence or cure? J Transl Med, *4*: 14, 2006.
39. Boland, C. R., Luciani, M. G., Gasche, C., and Goel, A. Infection, inflammation, and gastrointestinal cancer. Gut, *54*: 1321–1331, 2005.
40. Chan, H. L. and Sung, J. J. Hepatocellular carcinoma and hepatitis B virus. Semin Liver Dis, *26*: 153–161, 2006.
41. Levrero, M. Viral hepatitis and liver cancer: the case of hepatitis C. Oncogene, *25*: 3834–3847, 2006.
42. Bertout, J. and Thomas-Tikhonenko, A. Infection & neoplastic growth 101: the required reading for microbial pathogens aspiring to cause cancer. Cancer Treat Res, *130*: 167–197, 2006.
43. Peloponese, J. M., Yeung, M. L., and Jeang, K. T. Modulation of nuclear factor-kappaB by human T cell leukemia virus type 1 Tax protein: implications for oncogenesis and inflammation. Immunol Res, *34*: 1–12, 2006.
44. Marincola, F. M., Jaffe, E. M., Hicklin, D. J., and Ferrone, S. Escape of human solid tumors from T cell recognition: molecular mechanisms and functional significance. Adv Immunol, *74*: 181–273, 2000.
45. Marincola, F. M., Wang, E., Herlyn, M., Seliger, B., and Ferrone, S. Tumors as elusive targets of T cell-directed immunotherapy. Trends Immunol, *24*: 334–341, 2003.
46. Monsurro', V., Wang, E., Yamano, Y., Migueles, S. A., Panelli, M. C., Smith, K., Nagorsen, D., Connors, M., Jacobson, S., and Marincola, F. M. Quiescent phenotype of tumor-specific CD8+ T cells following immunization. Blood, *104*: 1970–1978, 2004.
47. Malmberg, K. J. Effective immunotherapy against cancer: a question of overcoming immune suppression and immune escape? Cancer Immunol Immunother, *53*: 879–892, 2004.
48. Overwijk, W. W. Breaking tolerance in cancer immunotherapy: time to ACT. Curr Opin Immunol, *17*: 187–194, 2005.
49. Paik, S., Shak, S., Tang, G., Kim, C., Baker, J., Cronin, M., Baehner, F. L., Walker, M. G., Watson, D., Park, T., Hiller, W., Fisher, E. R., Wickerham, D. L., Bryant, J., and Wolmark, N. A multigene assay to predict recurrence of tamoxifen-treated, node-negative breast cancer. N Engl J Med, *351*: 2817–2826, 2004.
50. Mantovani, A., Sozzani, S., Locati, M., Allavena, P., and Sica, A. Macrophage polarization: tumor-associated macrophage as a paradigm for polarized M2 mononuclear phagocytes. Trends Immunol, *23*: 549–555, 2002.
51. Gordon, S. Alternative activation of macrophages. Nature Reviews, *3*: 23–35, 2003.
52. Mantovani, A., Sica, A., Sozzani, S., Allavena, P., Vecchi, A., and Locati, M. The chemokine system in diverse forms of macrophage activation and polarization. Trends Immunol, *25*: 677–686, 2004.
53. Mantovani, A., Sica, A., and Locati, M. Macrophage polarization comes of age. Immunity, *23*: 344–346, 2005.

54. Wyckoff, J., Wang, W., Lin, E. Y., Wang, Y., Pixley, F., Stanley, E. R., Graf, T., Pollard, J. W., Segall, J., and Condeelis, J. A paracrine loop between tumor cells and macrophages is required for tumor cell migration in mammary tumors. Cancer Res, *64*: 7022–7029, 2004.
55. De Palma, M., Venneri, M. A., Galli, R., Sergi, L. S., Politi, L. S., Sampaolesi, M., and Naldini, L. Tie2 identifies a hematopoietic lineage of proangiogenic monocytes required for tumor vessel formation and a mesenchymal population of pericyte progenitors. Cancer Cell, *8*: 211–226, 2005.
56. Albini, A., Tosetti, F., Benelli, R., and Noonan, D. M. Tumor inflammatory angiogenesis and its chemoprevention. Cancer Res, *65*: 10637–10641, 2005.
57. Nakao, S., Kuwano, T., Tsutsumi-Miyahara, C., Ueda, S., Kimura, Y. N., Hamano, S., Sonoda, K. H., Saijo, Y., Nukiwa, T., Strieter, R. M., Ishibashi, T., Kuwano, M., and Ono, M. Infiltration of COX-2-expressing macrophages is a prerequisite for IL-1beta-induced neovascularization and tumor growth. J Clin Invest, *115*: 2979–2991, 2005.
58. Wynn, T. A. Fibrotic disease and the T(H)1/T(H)2 paradigm. Nat Rev Immunol, *4*: 583–594, 2004.
59. Panelli, M. C., Wang, E., Phan, G., Puhlman, M., Miller, L., Ohnmacht, G. A., Klein, H., and Marincola, F. M. Genetic profiling of peripheral mononuclear cells and melanoma metastases in response to systemic interleukin-2 administration. Genome Biol, *3*: RESEARCH0035, 2002.
60. Mocellin, S., Panelli, M. C., Wang, E., Nagorsen, D., and Marincola, F. M. The dual role of IL-10. Trends Immunol, *24*: 36–43, 2002.
61. Panelli, M. C., Stashower, M., Slade, H. B., Smith, K., Norwood, C., Abati, A., Fetsch, P. A., Filie, A., Walters, S. A., Astry, C., Arico, E., Zhao, Y., Selleri, S., Wang, E., and Marincola, F. M. Sequential gene profiling of basal cell carcinomas treated with Imiquimod in a placebo-controlled study defines the requirements for tissue rejection. Genome Biol, *8*: R8, 2006.
62. Ramakrishna, V., Vasilakos, J. P., Tario, J. D., Jr., Berger, M. A., Wallace, P. K., and Keler, T. Toll-like receptor activation enhances cell-mediated immunity induced by an antibody vaccine targeting human dendritic cells. J Transl Med, *5*: 5, 2007.
63. Banchereau, J. and Steinman, R. M. Dendritic cells and the control of immunity. Nature, *392*: 245–252, 1998.
64. Yamazaki, S., Iyoda, T., Tarbell, K., Olson, K., Velinzon, K., Inaba, K., and Steinman, R. M. Direct expansion of functional CD25+ CD4+ regulatory T cells by antigen-processing dendritic cells. J Exp Med, *198*: 235–247, 2003.
65. Gabrilovich, D. I., Chen, H. L., Girgis, K. R., Cunningham, H. T., Meny, G. M., Nadaf, S., Kavanaugh, D., and Carbone, D. P. Production of vascular endothelial growth factor by human tumors inhibits the functional maturation of dendritic cells [published erratum appears in Nat Med 2(11) (Nov): 1267, 1996]. Nat Med, *2*: 1096–1103, 1996.
66. Roncarolo, M. G., Levings, M. K., and Traversari, C. Differentiation of T regulatory cells by immature dendritic cells. J Exp Med, *193*: F5–F9, 2001.
67. Chomarat, P., Banchereau, J., Davoust, J., and Palucka, A. K. IL-6 switches the differentiation of monocytes from dendritic cells to macrophages. Nat Immunol, *1*: 510–514, 2000.
68. Vlad, A. M., Kettel, J. C., Alajez, N. M., Carlos, C. A., and Finn, O. J. MUC1 immunobiology: from discovery to clinical applications. Adv Immunol, *82*: 249–293, 2004.
69. Taieb, J., Chaput, N., Menard, C., Apetoh, L., Ullrich, E., Bonmort, M., Pequignot, M., Casares, N., Terme, M., Flament, C., Opolon, P., Lecluse, Y., Metivier, D., Tomasello, E., Vivier, E., Ghiringhelli, F., Martin, F., Klatzmann, D., Poynard, T., Tursz, T., Raposo, G., Yagita, H., Ryffel, B., Kroemer, G., and Zitvogel, L. A novel dendritic cell subset involved in tumor immunosurveillance. Nat Med, *12*: 214–219, 2006.
70. Dubsky, P., Ueno, H., Piqueras, B., Connolly, J., Banchereau, J., and Palucka, A. K. Human dendritic cell subsets for vaccination. J Clin Immunol, *25*: 551–572, 2005.
71. Tanaka, F., Yamaguchi, H., Haraguchi, N., Mashino, K., Ohta, M., Inoue, H., and Mori, M. Efficient induction of specific cytotoxic T lymphocytes to tumor rejection peptide using functional matured 2 day-cultured dendritic cells derived from human monocytes. Int J Oncol, *29*: 1263–1268, 2006.
72. Houtenbos, I., Westers, T. M., Ossenkoppele, G. J., and van de Loosdrecht, A. A. Feasibility of clinical dendritic cell vaccination in acute myeloid leukemia. Immunobiology, *211*: 677–685, 2006.

73. Burgdorf, S. K., Fischer, A., Claesson, M. H., Kirkin, A. F., Dzhandzhugazyan, K. N., and Rosenberg, J. Vaccination with melanoma lysate-pulsed dendritic cells, of patients with advanced colorectal carcinoma: report from a phase I study. J Exp Clin Cancer Res, *25*: 201–206, 2006.
74. Hallett, W. H. and Murphy, W. J. Natural killer cells: biology and clinical use in cancer therapy. Cell Mol Immunol, *1*: 12–21, 2004.
75. Parham, P. Immunogenetics of killer-cell immunoglobulin-like receptors. Tissue Antigens, *62*: 194–200, 2003.
76. Boles, K. S., Stepp, S. E., Bennett, M., Kumar, V., and Mathew, P. A. 2B4 (CD244) and CS1: novel members of the CD2 subset of the immunoglobulin superfamily molecules expressed on natural killer cells and other leukocytes. Immunol Rev, *181*: 234–249, 2001.
77. Bauer, S., Groh, V., Wu, J., Steinle, A., Phillips, J. H., Lanier, L. L., and Spies, T. Activation of NK cells and T cells by NKG2D, a receptor for stress-inducible MICA. Science, *285*: 727–729, 1999.
78. Kubin, M., Cassiano, L., Chalupny, J., Chin, W., Cosman, D., Fanslow, W., Mullberg, J., Rousseau, A. M., Ulrich, D., and Armitage, R. ULBP1, 2, 3: novel MHC class I-related molecules that bind to human cytomegalovirus glycoprotein UL16, activate NK cells. Eur J Immunol, *31*: 1428–1437, 2001.
79. Chalupny, N. J., Sutherland, C. L., Lawrence, W. A., Rein-Weston, A., and Cosman, D. ULBP4 is a novel ligand for human NKG2D. Biochem Biophys Res Commun, *305*: 129–135, 2003.
80. Boles, K. S., Stepp, S. E., Bennett, M., Kumar, V., and Mathew, P. A. 2B4 (CD244) and CS1: novel members of the CD2 subset of the immunoglobulin superfamily molecules expressed on natural killer cells and other leukocytes. Immunol Rev, *181*: 234–249, 2001.
81. Gonzalez, S., Groh, V., and Spies, T. Immunobiology of human NKG2D and its ligands. Curr Top.Microbiol Immunol, *298*: 121–138, 2006.
82. Chang, C. C. and Ferrone, S. NK cell activating ligands on human malignant cells: molecular and functional defects and potential clinical relevance. Semin Cancer Biol, *16*: 383–392, 2006.
83. Torres, A., Storey, L., Anders, M., Miller, R. L., Bulbulian, B. J., Jin, J., Raghavan, S., Lee, J., Slade, H. B., and Birmachu, W. Immune-mediated changes in actinic Keratosis following topical treatment with Imiquimod 5% cream. J Transl Med, *5*: 7, 2007.
84. Boon, T., Coulie, P. G., and Van den Eynde, B. Tumor antigens recognized by T cells. Immunol Today, *18*: 267–268, 1997.
85. Boon, T., Gajewski, T. F., and Coulie, P. G. From defined human tumor antigens to effective immunization? Immunol Today, *16*: 334–336, 1995.
86. Rosenberg, S. A., Yang, J. C., Schwartzentruber, D., Hwu, P., Marincola, F. M., Topalian, S. L., Restifo, N. P., Dufour, E., Schwartzberg, L., Spiess, P., Wunderlich, J., Parkhurst, M. R., Kawakami, Y., Seipp, C., Einhorn, J. H., and White, D. Immunologic and therapeutic evaluation of a synthetic tumor associated peptide vaccine for the treatment of patients with metastatic melanoma. Nat Med, *4*: 321–327, 1998.
87. Lee, K.-H., Wang, E., Nielsen, M.-B., Wunderlich, J., Migueles.S., Connors, M., Steinberg, S. M., Rosenberg, S. A., and Marincola, F. M. Increased vaccine-specific T cell frequency after peptide-based vaccination correlates with increased susceptibility to in vitro stimulation but does not lead to tumor regression. J Immunol, *163*: 6292–6300, 1999.
88. Slingluff, C. L., Jr. and Speiser, D. E. Progress and controversies in developing cancer vaccines.. J Transl Med, *3*: 18, 2005.
89. Panelli, M. C., Riker, A., Kammula, U. S., Lee, K.-H., Wang, E., Rosenberg, S. A., and Marincola, F. M. Expansion of Tumor/T cell pairs from Fine Needle Aspirates (FNA) of Melanoma Metastases. J Immunol, *164*: 495–504, 2000.
90. Kammula, U. S., Lee, K.-H., Riker, A., Wang, E., Ohnmacht, G. A., Rosenberg, S. A., and Marincola, F. M. Functional analysis of antigen-specific T lymphocytes by serial measurement of gene expression in peripheral blood mononuclear cells and tumor specimens. J Immunol, *163*: 6867–6879, 1999.
91. Altman, J. D., Moss, P. H., Goulder, P. R., Barouch, D. H., McHeyzer-Williams, M. G., Bell, J. I., McMichael, A. J., and Davis, M. M. Phenotypic analysis of antigen-specific T lymphocytes [published erratum appears in Science 1998 Jun 19;280(5371):1821]. Science, *274*: 94–96, 1996.

92. Cesana, G. C., DeRaffele, G., Cohen, S., Moroziewicz, D., Mitcham, J., Stoutenburg, J., Cheung, K., Hesdorffer, C., Kim-Schulze, S., and Kaufman, H. L. Characterization of CD4+ CD25+ regulatory T cells in patients treated with high-dose interleukin-2 for metastatic melanoma or renal cell carcinoma. J Clin Oncol, *24*: 1169–1177, 2006.
93. Woo, E. Y., Yeh, H., Chu, C. S., Schlienger, K., Carroll, R. G., Riley, J. L., Kaiser, L. R., and June, C. H. Cutting edge: Regulatory T cells from lung cancer patients directly inhibit autologous T cell proliferation. J Immunol, *168*: 4272–4276, 2002.
94. Curiel, T. J., Coukos, G., Zou, L., Alvarez, X., Cheng, P., Mottram, P., Evdemon-Hogan, M., Conejo-Garcia, J. R., Zhang, L., Burow, M., Zhu, Y., Wei, S., Kryczek, I., Daniel, B., Gordon, A., Myers, L., Lackner, A., Disis, M. L., Knutson, K. L., Chen, L., and Zou, W. Specific recruitment of regulatory T cells in ovarian carcinoma fosters immune privilege and predicts reduced survival. Nat Med, *10*: 942–949, 2004.
95. Badoual, C., Hans, S., Rodriguez, J., Peyrard, S., Klein, C., Agueznay, N. H., Mosseri, V., Laccourreye, O., Bruneval, P., Fridman, W. H., Brasnu, D. F., and Tartour, E. Prognostic value of tumor-infiltrating CD4+ T-cell subpopulations in head and neck cancers. Clin Cancer Res, *12*: 465–472, 2006.
96. Albers, A. E., Ferris, R. L., Kim, G. G., Chikamatsu, K., Deleo, A. B., and Whiteside, T. L. Immune responses to p53 in patients with cancer: enrichment in tetramer+ p53 peptide-specific T cells and regulatory T cells at tumor sites. Cancer Immunol Immunother, *54*: 1072–1081, 2005.
97. Badoual, C., Hans, S., Rodriguez, J., Peyrard, S., Klein, C., Agueznay, N. H., Mosseri, V., Laccourreye, O., Bruneval, P., Fridman, W. H., Brasnu, D. F., and Tartour, E. Prognostic value of tumor-infiltrating CD4+ T-cell subpopulations in head and neck cancers. Clin Cancer Res, *12*: 465–472, 2006.
98. Yamaguchi, T. and Sakaguchi, S. Regulatory T cells in immune surveillance and treatment of cancer. Semin Cancer Biol, 2005.
99. Ghiringhelli, F., Puig, P. E., Roux, S., Parcellier, A., Schmitt, E., Solary, E., Kroemer, G., Martin, F., Chauffert, B., and Zitvogel, L. Tumor cells convert immature myeloid dendritic cells into TGF-beta-secreting cells inducing CD4+ CD25+ regulatory T cell proliferation. J Exp Med, *202*: 919–929, 2005.
100. Liyanage, U. K., Moore, T. T., Joo, H. G., Tanaka, Y., Herrmann, V., Doherty, G., Drebin, J. A., Strasberg, S. M., Eberlein, T. J., Goedegebuure, P. S., and Linehan, D. C. Prevalence of regulatory T cells is increased in peripheral blood and tumor microenvironment of patients with pancreas or breast adenocarcinoma. J Immunol, *169*: 2756–2761, 2002.
101. Jin, P. and Wang, E. Polymorphism in clinical immunology. From HLA typing to immunogenetic profiling. J Transl Med, *1*: 8, 2003.
102. Zou, W. Immunosuppressive networks in the tumour environment and their therapeutic relevance. Nat Rev Cancer, *5*: 263–274, 2005.
103. Barber, D. L., Wherry, E. J., Masopust, D., Zhu, B., Allison, J. P., Sharpe, A. H., Freeman, G. J., and Ahmed, R. Restoring function in exhausted CD8 T cells during chronic viral infection. Nature, *439*: 682–687, 2006.
104. Echchakir, H., Bagot, M., Dorothee, G., Martinvalet, D., Le Gouvello, S., Boumsell, L., Chouaib, S., Bensussan, A., and Mami-Chouaib, F. Cutaneous T cell lymphoma reactive CD4+ cytotoxic T lymphocyte clones display a Th1 cytokine profile and use a fas-independent pathway for specific tumor cell lysis. J Invest Dermatol, *115*: 74–80, 2000.
105. Schattner, E. J., Mascarenhas, J., Bishop, J., Yoo, D. H., Chadburn, A., Crow, M. K., and Friedman, S. M. CD4+ T-cell induction of Fas-mediated apoptosis in Burkitt's lymphoma B cells. Blood, *88*: 1375–1382, 1996.
106. Thomas, W. D. and Hersey, P. TNF-related apoptosis-inducing ligand (TRAIL) induces apoptosis in Fas ligand-resistant melanoma cells and mediates CD4 T cell killing of target cells. J Immunol, *161*: 2195–2200, 1998.
107. Shankaran, V., Ikeda, H., Bruce, A. T., White, J. M., Swanson, P. E., Old, L. J., and Schreiber, R. D. IFN-g and lymphocytes prevent primary tumour development and shape tumour immunogenicity. Nature, *410*: 1107–1111, 2001.

108. Pages, F., Berger, A., Camus, M., Sanchez-Cabo, F., Costes, A., Molidor, R., Mlecnik, B., Kirilovsky, A., Nilsson, M., Damotte, D., Meatchi, T., Bruneval, P., Cugnenc, P. H., Trajanoski, Z., Fridman, W. H., and Galon, J. Effector memory T cells, early metastasis, and survival in colorectal cancer. N Engl J Med, *353*: 2654–2666, 2005.
109. Galon, J., Costes, A., Sanchez-Cabo, F., Kirilovsky, A., Mlecnik, B., Lagorce-Pages, C., Tosolini, M., Camus, M., Berger, A., Wind, P., Zinzindohoue, F., Bruneval, P., Cugnenc, P. H., Trajanoski, Z., Fridman, W. H., and Pages, F. Type, density, and location of immune cells within human colorectal tumors predict clinical outcome. Science, *313*: 1960–1964, 2006.
110. Cianchi, F., Messerini, L., Palomba, A., Boddi, V., Perigli, G., Pucciani, F., Bechi, P., and Cortesini, C. Character of the invasive margin in colorectal cancer: does it improve prognostic information of Dukes staging? Dis Colon Rectum, *40*: 1170–1175, 1997.
111. Bryne, M., Boysen, M., Alfsen, C. G., Abeler, V. M., Sudbo, J., Nesland, J. M., Kristensen, G. B., Piffko, J., and Bankfalvi, A. The invasive front of carcinomas. The most important area for tumour prognosis? Anticancer Res, *18*: 4757–4764, 1998.
112. Klintrup, K., Makinen, J. M., Kauppila, S., Vare, P. O., Melkko, J., Tuominen, H., Tuppurainen, K., Makela, J., Karttunen, T. J., and Makinen, M. J. Inflammation and prognosis in colorectal cancer. Eur J Cancer, *41*: 2645–2654, 2005.
113. Wang, E., Miller, L. D., Ohnmacht, G. A., Mocellin, S., Petersen, D., Zhao, Y., Simon, R., Powell, J. I., Asaki, E., Alexander, H. R., Duray, P. H., Herlyn, M., Restifo, N. P., Liu, E. T., Rosenberg, S. A., and Marincola, F. M. Prospective molecular profiling of subcutaneous melanoma metastases suggests classifiers of immune responsiveness. Cancer Res, *62*: 3581–3586, 2002.
114. Wang, E. and Marincola, F. M. A natural history of melanoma: serial gene expression analysis. Immunol Today, *21*: 619–623, 2000.
115. Zitvogel, L., Tesniere, A., and Kroemer, G. Cancer despite immunosurveillance: immunoselection and immunosubversion. Nat Rev Immunol, *6*: 715–727, 2006.
116. Mocellin, S., Wang, E., and Marincola, F. M. Cytokine and immune response in the tumor microenvironment. J Immunother, *24*: 392–407, 2001.
117. Mocellin, S., Panelli, M., Wang, E., Rossi, C. R., Pilati, P., Nitti, D., Lise, M., and Marincola, F. M. IL-10 stimulatory effects on human NK cells explored by gene profile analysis. Genes Immun, *5*: 621–630, 2004.
118. Singh, S., Ross, S. R., Acena, M., Rowley, D. A., and Schreiber, H. Stroma is critical for preventing or permitting immunological destruction of antigenic cancer cells. J Exp Med, *175*: 139–146, 1992.
119. Ochsenbein, A. F., Klenerman, P., Karrer, U., Ludewig, B., Pericin, M., Hengartner, H., and Zinkernagel, R. M. Immune suerveillance against a solid tumor fails because of immunological ignorance. Proc Natl Acad Sci USA, *96*: 2233–2238, 1999.
120. Yu, P., Lee, Y., Liu, W., Chin, R. K., Wang, J., Wang, Y., Schietinger, A., Philip, M., Schreiber, H., and Fu, Y. X. Priming of naive T cells inside tumors leads to eradication of established tumors. Nat Immunol, *5*: 141–149, 2004.
121. Mandruzzato, S., Callegaro, A., Turcatel, G., Francescato, S., Montesco, M. C., Chiarion-Sileni, V., Mocellin, S., Rossi, C. R., Bicciato, S., Wang, E., Marincola, F. M., and Zanovello, P. A gene expression signature associated with survival in metastatic melanoma. J Transl Med, *4*: 50, 2006.
122. Atkins, M. B., Regan, M., McDermott, D., Mier, J., Stanbridge, E., Youmans, A., Febbo, P., Upton, M., Lechpammer, M., and Signoretti, S. Carbonic anhydrase IX expression predicts outcome in interleukin-2 therapy of renal cancer. Clin Cancer Res, *11*: 3714–3721, 2005.
123. Panelli, M. C., Wang, E., and Marincola, F. M. The pathway to biomarker discovery: carbonic anhydrase IX and the prediction of immune responsiveness. Clin Cancer Res, *11*: 3601–3603, 2005.
124. Wang, E., Panelli, M. C., Zavaglia, K., Mandruzzato, S., Hu, N., Taylor, P. R., Seliger, B., Zanovello, P., Freedman, R. S., and Marincola, F. M. Melanoma-restricted genes. J Transl Med, *2*: 34, 2004.

Chapter 8
Functional Cytotoxicity of T Cells in the Tumor Microenvironment

Michal Lotem, Arthur Machlenkin, Shoshana Frankenburg, and Tamar Peretz

Abstract Cytotoxic T cells (CTLs) take a central part in tumor cell destruction. Initially, naïve CD8+ T cells are primed by antigen presenting cells in lymphoid tissue or, alternatively, by antigen processing tumor cells at the tumor microenvironment. The contact region of CTLs with antigen bearing cells, the immune synapse, consists of clusters of T cell receptors and adhesion molecules with reorientation of the cytoskeleton to direct secretion of cytotoxic molecules to a specific location. Membrane fragments carrying surface peptide-MHC complexes are captured by tumor responsive CTLs. The incorporation of tumor membrane exposes CTLs to fratricide killing, or turns them to secondary APCs. Activated CTLs release cytotoxins including perforin and Granzyme B, which activate several pathways leading to apoptosis and target cell death. Membrane bound proteins of the tumor necrosis factor receptor family induce cell damage, and do not necessitate perforin activation. Interferon-γ is an important cytokine secreted by activated CTLs essential for tumor regression through its paracrine effect on other immune cells and generation of inflammation.

With the advent of gene transfer to T lymphocytes, the major goal in the future will be to increase the tumor destructive properties of CTLs and amplify immune response.

Keywords Cytotoxic T Cells; trogocytosis; functional cytotoxicity

8.1 Introduction

The ultimate purpose of all immunotherapeutic strategies is tumor cell destruction. The road to this goal involves the generation of immune effector cells that will carry out the task. CD8+ cytotoxic T cells (CTL) are the critical effector cells of adaptive immunity that actually destroy tumors. CTLs are equipped with specializing cellular mechanisms and molecules that cast cell damage. Models in mice and in humans show that tumors are susceptible to T cell killing and that T cells effectively destroy their targets [1, 2]. Following encounter with tumors or tumor vaccines, large numbers of tumor antigen-specific CTLs can clearly be generated [3]. But yet, expanded CTL clones only rarely mediate tumor regression [4, 5].

Sharett Institute of Oncology, Hadassah Hebrew University Hospital, Jerusalem, Israel

E. Yefenof (ed.), *Innate and Adaptive Immunity in the Tumor Microenvironment.*

Many mechanisms were explored that inhibit tumor rejection. Strategies employed by the tumor induce suppression through immunoinhibitory cytokines, impairment of antigen presentation [6, 7] and activation of negative costimulatory signals [8]. CTLA4 competes with CD28 on co-stimulatory ligands and block T cell activation [9]. Tolerizing effects can be generated by immature dendritic cells and alternative antigen presenting cells (APCs) such as tumor-associated macrophages [10, 11]. Regulatory effector cells including CD4+ CD25+ T cells and CD4+ natural killer T cells inhibit immune response [12, 13]. In addition to these active, negative regulatory mechanisms, improper localization of antitumor T cells and their death following activation lead to abrogation of antitumor immunity [14].

The larger volume of literature focuses on factors that hinder anti-cancer immunity while the potential efficacy of the immune response is deduced from non-tumoral models, many of them based on viral diseases, where success is easier to achieve. The current review will focus on mechanisms which mediate tumor destruction or hold the potential to exert this effect in the arena of the tumor microenvironment.

8.2 Early Events: Antigen Presentation, CTL Activation and Migration

Tumor cell components are digested by immature dendritic cells (DC) at the tumor site, followed by migration of DC to lymphoid tissues while processing the digested debris. Antigen presentation in conjunction with MHC class I and class II, and initial priming of T cells happens in the lymphoid tissue [15–17]. During their transfer from tumor to lymph nodes (LN), dendritic cells mature and express costimulatory molecules that support the transition of antigen-inexperienced naïve T cells into activated cells. Among these, ligation of CD28 by co-stimulatory molecules B7.1 and B7.2 is required for clonal expansion and proliferation of naïve T cells [18]. Once activated, other ligands, including CD40L and 4–1BB, are over-expressed, and pair with their counterparts on APC to sustain the development of a CTL response [19]. In the absence of co stimulation, TCR binding to its peptide will fail to activate the cell and lead to a state of anergy [20].

Tumor cells create an unfavorable milieu for DC activation. Production of prostaglandinE2 by tumors induces over expression of indoleamine 2,3-dioxygenase (IDO) protein in DC and significant expression of cell-surface and soluble CD25 protein [21]. As a consequence, T-cell proliferation and clonal expansion are significantly inhibited [22]. Supplying the host with DC from an external source can initiate an antitumor response, but the tolerizing environment created by the tumor will still exert a suppressive effects on T cells [23].

Presentation of tumor antigen is not the mandate of antigen presenting cells only. Cancer cells can present antigens independently. Intact processing machinery and antigen presentation in the MHC class I context is often preserved [24]. There is no need for APC to cross-present digested tumor to evoke CD8+ T cell responses. Direct contact with the tumor suffices [25]. Loss of MHC expression by tumor cells occurs, and impacts the course of disease [26], but is not the rule. If the barrier created by the tumor stroma is overcome, naïve T cells can directly be primed at the tumor site [27, 28]. Tumor stromal cells are capable of taking up tumor components and cross presenting them on MHC class II molecules, resulting in tumor destruction via damage to its stroma [29]. T cells themselves may acquire antigen presentation capacities through acquisition of peptide-MHC complexes (pMHC) from APCs and tumor, as will be described below. Intriguingly, such T-APCs can stimulate Ag-specific CD8+ central memory T cells, and elicit antitumor immunity [30].

Prior to recruitment into the tumor, naïve CTLs migrate from peripheral blood to lymph nodes, where they constantly scan APCs for surface peptides. Only one naïve T cell in 10^4 to 10^6 is likely to be specific for a particular antigen [31]. Upon encounter with APCs presenting their cognate epitope, naïve CTL stop their scanning path for several days. This period is necessary for the lymphocytes to proliferate and transform to armed effectors.

Activated CTLs leave the lymphoid tissue and home to the tumor. The organ of antigen acquisition by APC influences the final destination of T cell trafficking. For example, through specific expression of ligands and chemokine receptors, T cells imprinted in peripheral lymph nodes migrate to the skin [32], and T cells encountering antigen presented by intestinal DC migrate to the intestine [33]. An inflammatory milieu may enhance T cell homing through vasodilation and upregulation of adhesion molecules and selectins [32, 34].

Once a contact is made, higher levels of CD2 and LFA-1 will enhance CTL capacity to firmly adhere to the tumor cell, and develop strong engagement – the immune synapse – that potentially result in tumor death.

8.3 The Immune Synapse

The term "immune synapse" (IS) was adopted from neuron physiology to represent a reciprocal attachment-activation-secretion system of intercellular communication occurring at specialized sites, between an immune effector cell and its target counterpart [35, 36]. The concept of immune synapse and its original definition are based on APC-lymphocyte interactions. In the broader context, IS may be generated by any specialized cell:cell interaction. CD4 and CD8 T cells, NK cells and B cells create immune synapses with antigen presenting cells, tumor targets, and with each other [37–42].

DC:T cell interaction and the subsequent transition of naïve CD8+ T cells into effector CTLs are but two steps towards the final goal which is target cell damage. It is surprising that little data has been provided thus far on the course of CTLs in peripheral tissues and their physical interaction with tumor cells, considering that this is their primary function. Due to paucity of information, the description of the IS is derived from models established with CTL:APC interactions. However, it is likely that these interactions follow a similar pattern when they occur between CTLs and tumor target, as will be described below.

In essence, IS refers to the redistribution of TCRs, their pMHC and co-receptors, with a reorientation of the cytoskeleton that polarizes and focuses release of effector molecules at the site of contact [37, 39, 40]. The immunological synapse serves several functions: (1) it increases T cell sensitivity to small amounts of antigen presented on the cell surface; (2) a stable attachment is created between the reacting cells; (3) effector molecules are directed and secreted to a specific site where their function is required; (4) the IS generates its own regulation and down-tuning.

Kupfer et al. [41] delineated the conformation of supramolecular activation clusters (SMAC) between APC and B cells, and later with T cells, in which CD4, LFA-1, IL-4, talin, and protein kinase C participate [41, 43]. A central domain of TCR cluster was defined as cSMAC, and a peripheral adhesive ring, enriched for the integrin LFA-1, as pSMAC [40, 44]. In addition to the TCR, cSMAC contains accessory molecules such as CD2 and CD28. These molecular clustering sites bring together ligands and receptors, thus increasing the likelihood of T cell activation. Stable binding at the IS creates tight, close contact between the interacting cells and is contributed by chemokines expressed on APCs. Chemokines are a group of highly charged small molecules easily adsorbed onto the cell surface. The increased adhesiveness of chemokine-expressing APCs facilitates binding of T cell receptors to their ligands most dominantly ICAM-1 [45]. This may be a first antigen independent step, that leads to an increased probability of TCR-pMHC encounter, and to enhanced T cell responsiveness to APC-generated peptides [46, 47]. Tumor cells may employ a similar mechanism. It was demonstrate that CD8+ T cells are directly activated by tumor cells depending on ICAM-1/LFA-1 interaction, as a CD80/CD86-CD28 independent costimulation [27]. Conversely, TCR triggering by pMHC results in denser display of LFA-1 and consequently stronger adhesion of T cell to APC [48]. As a result, both agonistic and "nule" peptides (peptides that bind to the MHC but do not activate TCR) binding to MHC lead to molecular clustering at the IS [49].

The reverse effect of chemokines, chemo-repulsion, was also shown, demonstrated in mice for CXCL12 expressed on tumor and its receptor CXCR4 on T cells [50].

The prime event that leads to the formation of IS is TCR recognition of its relevant peptide [44]. Using direct visualization, Davis et al. [51] found that a single pMHC complex was sufficient to arrest a CTL near an APC, to stimulate minimal calcium flux, and to induce transient immunological synapse formation. Stable immunological synapse formation required engagement with ten pMHC complexes [52]. In fact, possessing this extreme receptiveness to minute peptide load makes T cells a sensory organ [51]. Recently, it was proposed that the ability of T cells to augment TCR sensitivity to a given antigenic peptide is regulated by micro RNA (miRNA)181a, a 22 nt RNA piece that can repress multiple phosphatases and lead to reduction of the T cell receptor signaling threshold [53].

8.4 Functional CTL Capture Membrane Fragments from Tumors

Armed T cells acquire membrane fragments from APCs with which they establish an IS. Initially, it was shown that T cells acquire pMHC from APCs [54]. This process occurs in an MHC-restricted antigen-specific manner and is an early and rapid event taking place at the IS. Within minutes of T cell-APC interaction, pMHC are taken up and internalized through endocytosis [55]. This process, expressed at the cellular level by the incorporation of membrane fragments from target into effector cell outer cytoplasmic membrane, was demonstrated using fluorescently labeled pMHCs [55, 56]. Furthermore, non-specific tagging of APC membranes using biotinylation or fluorochrome dye was sufficient to label cognate T cells (Machlenkin et al., in press). TCR signaling by its antigen ligand was found to be the prime requirement for membrane capture [57]. Trogocytosis, a term coined by Joly and Hudrisier [58], describes transfer of membrane fragments from target to reactive immune cells. Based on this phenomenon, the TRAP (T cell recognition by antigen presenting cells) method was established and proved reliable to quantify antigen-specific CTL frequencies [59]. As TRAP implies, membrane capture was documented with APCs as the target cell from which T cells acquire membranes [57]. The role of tumor cells as a source of pMHC was not explored in this context, although it is clear that pMHC complexes can be obtained from tumor cells and presented unchanged on immune cells, a phenomenon named "cross dressing" and demonstrated with dendritic cells and tumors [60].

In a recent study we postulated that activated human T cells will generate an IS-like interaction with tumor cells similarly to that described with dendritic cells [61] and that as a consequence membrane fragments will be acquired by T cells. Using fluorochrome-tagged melanoma cells or peptide-loaded T2 cells, we showed that antigen-specific CD8+ T cells incorporate membrane fragments from tumor cells and fuse them with their outer cytoplasmic membrane (Fig. 8.1). T cells that captured labeled membranes were detectable by flow cytometry and sorted by FACS

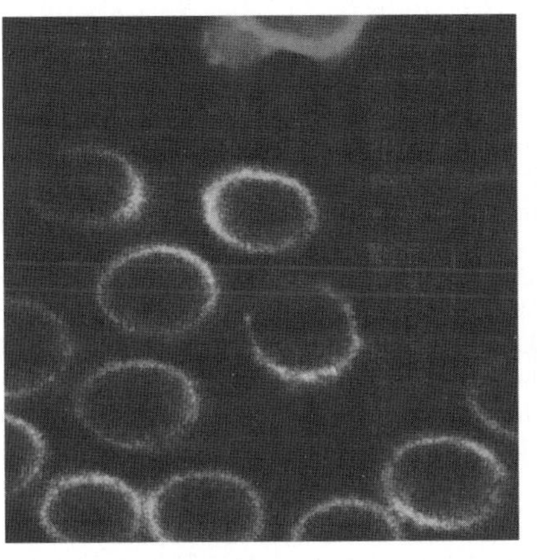

A

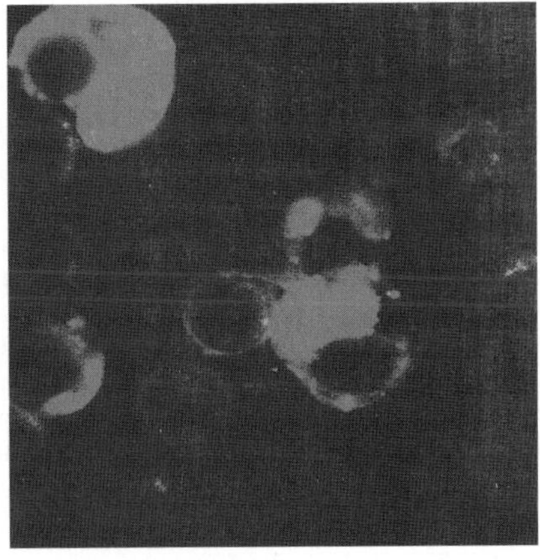

B

Fig. 8.1 (A) Confocal microscopy of a membrane capturing MART1$_{27-35}$-reactive T cells. Right: T cell showing patches of tumor membrane (red, DiI) fused with T cell membrane (green, CD8-FITC); (B) T cell wrapped with target cell membrane suggesting adhesion and rolling of effector over target

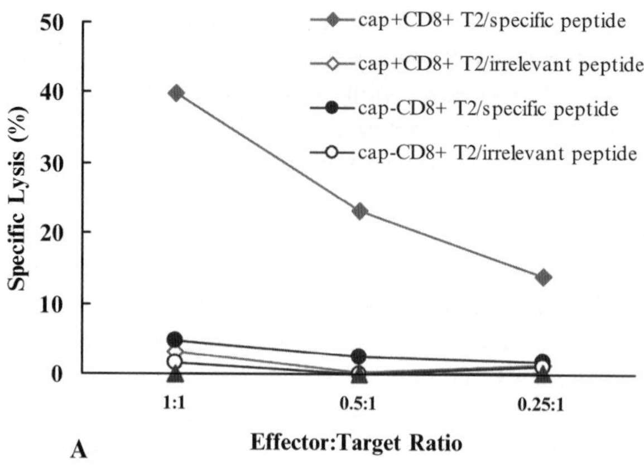

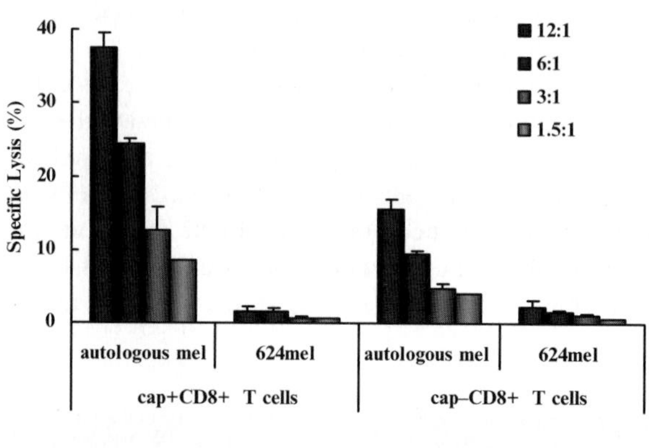

Fig. 8.2 Membrane capturing CD8+ T cells show increased cytotoxicity against peptide pulsed targets and autologous melanoma cells. Cytotoxic activity of membrane capturing (cap+) or non-capturing (cap–) T cells against peptide-pulsed targets and melanoma cell lines. (A) gp100$_{154–162}$-reactive A2-restricted CD8+ T cells were co-incubated with DiI-labeled 624mel melanoma cells (A2+ /gp100+) and sorted by FACS into cap+ and cap– populations of CD8+ T cells. CTL assays were done with [^{35}S]-labeled specific (gp100$_{154–162}$)- or irrelevant (HIV-derived) peptide-pulsed T2 cells as targets. (B) Melanoma patient derived CD8+ T cells were co-incubated with biotinylated autologous melanoma cells, followed by separation into cap+ and cap– fractions by streptavidin-coated magnetic beads. CTL assays were done with [^{35}S]-labeled autologous and HLA-mismatched 624mel melanomas, as specific and non-specific targets

or magnetic beads. Only a fraction of the T cells of a given clone were membrane capture capable. Cytotoxic ability was exclusively conferred to membrane capturing CD8+ T cells, whereas non-capturing CD8+ lymphocytes possessed reduced killing capacity (Fig. 8.2 A, B). Ex vivo expansion of membrane capturing T cells did not eliminate their functional cytotoxicity, implying that capturing capacity is an inherent trait of these cells. The therapeutic relevance of membrane capturing T cells was verified in a murine model of human melanoma. Tumor progression of melanoma transplanted into nude mice was significantly inhibited by membrane capturing T cells but not by non-capturing T cells of the same antigen specificity.

So far, the phenomenon of membrane capture was exploited as a new assay for antigen-specific T cell quantification and isolation. The use of tumor cells to identify antigen-specific functionally cytotoxic T cells is of paramount importance, because CTLs are selected not only by virtue of their antigen specificity, but by their capacity to damage tumor cells bearing naturally expressed antigens.

The biologic significance of membrane capture by tumor-reactive T cells is not clear. By acquiring pMHCs and other membrane-bound determinants from a target, T cells cross-dress themselves to become secondary APC (T-APC). The outcome can be manifold. T-APC may enhance antigen presentation and prolong T cell activation: membrane capturing CD4+ T cells, lacking co-stimulation, were as efficient as CD40+ B-APC for triggering epitope-specific CD8+ T cells [62]. Captured pMHC-TCR complexes undergo recycling and repeatedly generate TCR signaling and activation. Alternatively, Ag-presenting T cells expose themselves to fratricide killing by T cells of the same TCR specificity [54]. This process requires a functional perforin pathway [63], and could represent a form of activation-induced cell death.

Membrane capture by CTL from tumor cells suggests that an immune synapse of a kind occurs in the tumor microenvironment as part of the antitumor response. Live cell video microscopy of T cell:tumor interactions will be essential to illustrate the course of events that take place in the tumor microenvironment.

8.5 Killing Mechanisms of CTLs

The endpoint of target cell killing is traditionally measured by the release of radioisotope following disintegration of the cell membrane. However, the progression of events leading to target cell destruction by CTLs is complex, and only partly related to physical attack on integrity. A repertoire of effector molecules produced in the activated CD8 T cell is involved in this process:

8.5.1 Cytotoxins

Cytotoxins are a group of molecules contained within intracellular cytotoxic granules that can damage any cell membrane close to their site of release [64]. Cytotoxins include specific proteases – granzymes, and the pore forming protein perforin [64, 65]. Granzyme B is a 32 kDa serine protease that enters the target cell by endocytosis, either by binding the mannose-6-phosphate receptor [66], or by constitutive fluid phase uptake. Once released inside the target cell from the endocytic vescicle, Granzyme B activates several pathways that altogether lead to apoptosis. Among the molecules involved is the pro-apoptotic Bid, a member of the Bcl-2 family that triggers mitochondrial outer membrane permeabilization and release of cytochrome C. In turn, caspase 9 and then caspase 3 are then activated [67]. Damage to mitochondrial outer membrane is the critical point in the apoptosis pathway, since, following this event, cells are committed to die [68].

Perforin is the single most critical component of target cell killing. Although pore formation may directly cause cell death, it may also create ports of entry for other granule components that will synergistically induce apoptosis [69]. Sublytic perforin levels are necessary for Granzyme B to break the endosomal vesicle in which it is pinocytosed, be released into the target cell and exert its effect [70].

Granzymes A, C, K, and M are other proteolytic granule components that induce cell death independently of Granzyme B, but in synergy with perforin. Other components of the lytic granules, including Granzyme H and Granulysin, have been characterized, but the mechanisms by which they mediate cell death have not been elucidated.

8.5.2 Cytokines

There are two distinct secretory pathways of CTLs: one is directed into the synaptic gap; the other spills over into the surrounding microenvironment and sustains inflammation [71]. The latter paracrine system is mediated by cytokines. They are a large group of molecules acting on diverse targets on the secreting cell and on other effector cells. Their function is mediated by intracellular signaling, usually through a pathway that includes molecules belonging to the Janus family of cytoplasmic tyrosine kinases and their targets, the signal transducing activators of transcription (STATs). The main cytokine secreted by activated CTLs is interferon-γ (IFNγ). The efficacy of CD8+ T cells to reject tumors *in vivo* correlated better with their ability to secrete IFN-γ than with their *in vitro* cytotoxic activity [72]. IFN-γ secretion was shown to be essential for tumor regression through its paracrine effect on other participating cells [29].

8.5.3 Membrane-associated Proteins

More recently, it was shown that cells from perforin deficient mice could still reject tumors, and that in double knock out mice for IFN-γ and peforin, control of tumor can be hampered by administration of anti-TNF antibody [73, 74]. These data imply that other mechanisms, that do not necessitate perforin activity, can be involved in CTL-induced cell damage. And indeed, Fas ligand that is present on the cytoplasmic membrane of activated CD8 T cells can bind to FAS on target cells and lead to apoptotic cell death. Fas (APO-1/CD95) is a member of the TNF-R family involved in mediating death signals [75]. It is a type I transmembrane receptor widely expressed in human tissues. Upon binding of FAS to its specific ligand in a trimeric configuration, a series of signals lead to formation of active death-inducing signaling complex (DISC). Caspase-8-mediated activation of Caspase-3, following activation mediated by caspase-8, kills cells through cleavage and destruction of multiple cellular target structures [76]. Fas ligand is another TNF-related transmembrane molecule; under physiological conditions, its expression is confined to a limited spectrum of cell types, in particular to activated T cells [77]. Although the direction of signal transduction is usually from ligand to receptor, reverse signaling may occur between partner activated T cells, and modulate the immune reaction. FasL-mediated signaling was recently shown to act as CD28-independent T cells costimulation that synergizes with TCR-specific stimulation to enhance IFN-γ- production [78]. Following antigen contact and under the influence of IL-2, CTLs upregulate FasL, increasing target cell killing while resisting Fas-induced apoptosis.

8.6 Future Research Directions to Improve Functional Cytotoxicity of Tumor-specific CTLs

Until the near past, much effort was invested in generating ample amounts of tumor-specific T cells. The quantitative barrier was considered a major obstacle to achieve tumor regression. The feasibility of ex vivo T cell expansion allowed the clinical use of selective tumor-reactive T cells out of TILs and PBMCs [79]. The progress made in TCR gene transfer into T cells is now enabling the selection of high affinity TCRs in order to produce large numbers of antigen-specific T cells [80]. This technology will eventually bring to an end the dependency on preexisting effectors necessary to establish T cell lines for treatment. Clearly, a major goal of research over the next years will be to increase the tumor destructive properties of CTLs, making them better killers and less vulnerable to the consequences of activation. Gene transfer opens exciting possibilities for T cell manipulation beyond the TCR. This tool, and the emerging new monoclonal antibodies that act to restore T cell functions in cancer patients, may change the outlook for tumor immunotherapy.

Critical elements affecting tumor-T cell interaction are potential targets for intervention. Because lytic functions are dependent on TCR signaling [81], a primary goal would be to enhance TCR-pMHC interaction directly or through the establishment of a stable IS. So far, the strategy employed was to strengthen peptide-MHC complex using improved antigenic epitopes with high affinity binding to MHC molecules. TCR signaling may be directly augmented, independent of the binding affinity to its antigenic peptide. For example, the use of miR 181a may augment TCR sensitivity to threshold levels of ligand, and enhance T cell cytotoxic activity [53]. Stabilizing the immune synapse could yield stronger TCR signaling, for example by increasing LFA1 expression on CTL. Using a different strategy, one could endow T cells with improved resistance to apoptotic signals through transfer of anti-apoptotic genes, such as Bcl-2 [82]. Most encouraging, clinical trials recorded objective cancer regressions achieved by anti-CTLA-4, a blocking antibody that antagonizes suppressive signals and unleashes CTLs [83]. Once again the potency of cytotoxic T cells for the destruction of malignant tumors was shown.

References

1. Boon T, Coulie PG, Van den Eynde BJ, van der Bruggen P (2006) Human T cell responses against melanoma. Annu Rev Immunol 24: 175–208
2. Overwijk WW, Theoret MR, Finkelstein SE, Surman DR, de Jong LA, Vyth-Dreese FA, Dellemijn TA, Antony PA, Spiess PJ, Palmer DC, Heimann DM, Klebanoff CA, Yu Z, Hwang LN, Feigenbaum L, Kruisbeek AM, Rosenberg SA, Restifo NP (2003) Tumor regression and autoimmunity after reversal of a functionally tolerant state of self-reactive CD8+ T cells. J Exp Med 198: 569–580
3. Lee PP, Yee C, Savage PA, Fong L, Brockstedt D, Weber JS, Johnson D, Swetter S, Thompson J, Greenberg PD, Roederer M, Davis MM (1999) Characterization of circulating T cells specific for tumor-associated antigens in melanoma patients. Nat Med 5: 677–685
4. Nelson DJ, Mukherjee S, Bundell C, Fisher S, van Hagen D, Robinson B (2001) Tumor progression despite efficient tumor antigen cross-presentation and effective "arming" of tumor antigen-specific CTL. J Immunol 166: 5557–5566
5. Ochsenbein AF, Klenerman P, Karrer U, Ludewig B, Pericin M, Hengartner H, Zinkernagel RM (1999) Immune surveillance against a solid tumor fails because of immunological ignorance. Proc Natl Acad Sci USA 96: 2233–2238
6. Khong HT, Restifo NP (2002) Natural selection of tumor variants in the generation of "tumor escape" phenotypes. Nat Immunol 3: 999–1005
7. Wang Z, Cao Y, Albino AP, Zeff RA, Houghton A, Ferrone S (1993) Lack of HLA class I antigen expression by melanoma cells SK-MEL-33 caused by a reading frameshift in beta 2-microglobulin messenger RNA. J Clin Invest 91: 684–692
8. Blank C, Gajewski TF, Mackensen A (2005) Interaction of PD-L1 on tumor cells with PD-1 on tumor-specific T cells as a mechanism of immune evasion: implications for tumor immunotherapy. Cancer Immunol Immunother 54: 307–314
9. Perez VL, Van Parijs L, Biuckians A, Zheng XX, Strom TB, Abbas AK (1997) Induction of peripheral T cell tolerance in vivo requires CTLA-4 engagement. Immunity 6: 411–417
10. Sotomayor EM, Borrello I, Rattis FM, Cuenca AG, Abrams J, Staveley-O'Carroll K, Levitsky HI (2001) Cross-presentation of tumor antigens by bone marrow-derived antigen-presenting cells is the dominant mechanism in the induction of T-cell tolerance during B-cell lymphoma progression. Blood 98: 1070–1077
11. Steinman RM, Hawiger D, Nussenzweig MC (2003) Tolerogenic dendritic cells. Annu Rev Immunol 21: 685–711
12. Antony PA, Piccirillo CA, Akpinarli A, Finkelstein SE, Speiss PJ, Surman DR, Palmer DC, Chan CC, Klebanoff CA, Overwijk WW, Rosenberg SA, Restifo NP (2005) CD8+ T cell immunity against a tumor/self-antigen is augmented by CD4+ T helper cells and hindered by naturally occurring T regulatory cells. J Immunol 174: 2591–2601
13. Terabe M, Matsui S, Noben-Trauth N, Chen H, Watson C, Donaldson DD, Carbone DP, Paul WE, Berzofsky JA (2000) NKT cell-mediated repression of tumor immunosurveillance by IL-13 and the IL-4R-STAT6 pathway. Nat Immunol 1: 515–520
14. Schmitz I, Krueger A, Baumann S, Schulze-Bergkamen H, Krammer PH, Kirchhoff S (2003) An IL-2-dependent switch between CD95 signaling pathways sensitizes primary human T cells toward CD95-mediated activation-induced cell death. J Immunol 171: 2930–2936
15. Chiodoni C, Paglia P, Stoppacciaro A, Rodolfo M, Parenza M, Colombo MP (1999) Dendritic cells infiltrating tumors cotransduced with granulocyte/macrophage colony-stimulating factor (GM-CSF) and CD40 ligand genes take up and present endogenous tumor-associated antigens, and prime naive mice for a cytotoxic T lymphocyte response. J Exp Med 190: 125–133
16. Marzo AL, Lake RA, Lo D, Sherman L, McWilliam A, Nelson D, Robinson BW, Scott B (1999) Tumor antigens are constitutively presented in the draining lymph nodes. J Immunol 162: 5838–5845

17. Preynat-Seauve O, Schuler P, Contassot E, Beermann F, Huard B, French LE (2006) Tumor-infiltrating dendritic cells are potent antigen-presenting cells able to activate T cells and mediate tumor rejection. J Immunol 176: 61–67
18. Bonnevier JL, Mueller DL (2002) Cutting edge: B7/CD28 interactions regulate cell cycle progression independent of the strength of TCR signaling. J Immunol 169: 6659–6663
19. Uno T, Takeda K, Kojima Y, Yoshizawa H, Akiba H, Mittler RS, Gejyo F, Okumura K, Yagita H, Smyth MJ (2006) Eradication of established tumors in mice by a combination antibody-based therapy. Nat Med 12: 693–698
20. Van Gool SW, Vermeiren J, Rafiq K, Lorr K, de Boer M, Ceuppens JL (1999) Blocking CD40 - CD154 and CD80/CD86 - CD28 interactions during primary allogeneic stimulation results in T cell anergy and high IL-10 production. Eur J Immunol 29: 2367–2375
21. von Bergwelt-Baildon MS, Popov A, Saric T, Chemnitz J, Classen S, Stoffel MS, Fiore F, Roth U, Beyer M, Debey S, Wickenhauser C, Hanisch FG, Schultze JL (2006) CD25 and indoleamine 2,3-dioxygenase are up-regulated by prostaglandin E2 and expressed by tumor-associated dendritic cells in vivo: additional mechanisms of T-cell inhibition. Blood 108: 228–237
22. Mellor AL, Baban B, Chandler P, Marshall B, Jhaver K, Hansen A, Koni PA, Iwashima M, Munn DH (2003) Cutting edge: induced indoleamine 2,3 dioxygenase expression in dendritic cell subsets suppresses T cell clonal expansion. J Immunol 171: 1652–1655
23. Anderson MJ, Shafer-Weaver K, Greenberg NM, Hurwitz AA (2007) Tolerization of tumor-specific T cells despite efficient initial priming in a primary murine model of prostate cancer. J Immunol 178: 1268–1276
24. Kageshita T, Ishihara T, Campoli M, Ferrone S (2005) Selective monomorphic and polymorphic HLA class I antigenic determinant loss in surgically removed melanoma lesions. Tissue Antigens 65: 419–428
25. Ochsenbein AF, Sierro S, Odermatt B, Pericin M, Karrer U, Hermans J, Hemmi S, Hengartner H, Zinkernagel RM (2001) Roles of tumour localization, second signals and cross priming in cytotoxic T-cell induction. Nature 411: 1058–1064
26. Anichini A, Mortarini R, Nonaka D, Molla A, Vegetti C, Montaldi E, Wang X, Ferrone S (2006) Association of antigen-processing machinery and HLA antigen phenotype of melanoma cells with survival in American Joint Committee on Cancer stage III and IV melanoma patients. Cancer Res 66: 6405–6411
27. Jenkinson SR, Williams NA, Morgan DJ (2005) The role of intercellular adhesion molecule-1/LFA-1 interactions in the generation of tumor-specific CD8+ T cell responses. J Immunol 174: 3401–3407
28. Yu P, Lee Y, Liu W, Chin RK, Wang J, Wang Y, Schietinger A, Philip M, Schreiber H, Fu YX (2004) Priming of naive T cells inside tumors leads to eradication of established tumors. Nat Immunol 5: 141–149
29. Schuler T, Blankenstein T (2003) Cutting edge: CD8+ effector T cells reject tumors by direct antigen recognition but indirect action on host cells. J Immunol 170: 4427–4431
30. Xia D, Hao S, Xiang J (2006) CD8+ cytotoxic T-APC stimulate central memory CD8+ T cell responses via acquired peptide-MHC class I complexes and CD80 costimulation, and IL-2 secretion. J Immunol 177: 2976–2984
31. Blattman JN, Antia R, Sourdive DJ, Wang X, Kaech SM, Murali-Krishna K, Altman JD, Ahmed R (2002) Estimating the precursor frequency of naive antigen-specific CD8 T cells. J Exp Med 195: 657–664
32. Grover A, Kim GJ, Lizee G, Tschoi M, Wang G, Wunderlich JR, Rosenberg SA, Hwang ST, Hwu P (2006) Intralymphatic dendritic cell vaccination induces tumor antigen-specific, skin-homing T lymphocytes. Clin Cancer Res 12: 5801–5808
33. Mora JR, Cheng G, Picarella D, Briskin M, Buchanan N, von Andrian UH (2005) Reciprocal and dynamic control of CD8 T cell homing by dendritic cells from skin- and gut-associated lymphoid tissues. J Exp Med 201: 303–316
34. Dudda JC, Simon JC, Martin S (2004) Dendritic cell immunization route determines CD8+ T cell trafficking to inflamed skin: role for tissue microenvironment and dendritic cells in establishment of T cell-homing subsets. J Immunol 172: 857–863

35. Delon J, Germain RN (2000) Information transfer at the immunological synapse. Curr Biol 10: R923–R933
36. Kupfer A, Swain SL, Singer SJ (1987) The specific direct interaction of helper T cells and antigen-presenting B cells. II. Reorientation of the microtubule organizing center and reorganization of the membrane-associated cytoskeleton inside the bound helper T cells. J Exp Med 165: 1565–1580
37. Anton van der Merwe P, Davis SJ, Shaw AS, Dustin ML (2000) Cytoskeletal polarization and redistribution of cell-surface molecules during T cell antigen recognition. Semin Immunol 12: 5–21
38. Chen X, Trivedi PP, Ge B, Krzewski K, Strominger JL (2007) Many NK cell receptors activate ERK2 and JNK1 to trigger microtubule organizing center and granule polarization and cytotoxicity. Proc Natl Acad Sci USA 104: 6329–6334
39. Dustin ML (2005) A dynamic view of the immunological synapse. Semin Immunol 17: 400–410
40. Grakoui A, Bromley SK, Sumen C, Davis MM, Shaw AS, Allen PM, Dustin ML (1999) The immunological synapse: a molecular machine controlling T cell activation. Science 285: 221–227
41. Kupfer A, Mosmann TR, Kupfer H (1991) Polarized expression of cytokines in cell conjugates of helper T cells and splenic B cells. Proc Natl Acad Sci USA 88: 775–779
42. Monks CR, Kupfer H, Tamir I, Barlow A, Kupfer A (1997) Selective modulation of protein kinase C-theta during T-cell activation. Nature 385: 83–86
43. Monks CR, Freiberg BA, Kupfer H, Sciaky N, Kupfer A (1998) Three-dimensional segregation of supramolecular activation clusters in T cells. Nature 395: 82–86
44. Lee KH, Holdorf AD, Dustin ML, Chan AC, Allen PM, Shaw AS (2002) T cell receptor signaling precedes immunological synapse formation. Science 295: 1539–1542
45. Parameswaran N, Suresh R, Bal V, Rath S, George A (2005) Lack of ICAM-1 on APCs during T cell priming leads to poor generation of central memory cells. J Immunol 175: 2201–2211
46. Friedman RS, Jacobelli J, Krummel MF (2006) Surface-bound chemokines capture and prime T cells for synapse formation. Nat Immunol 7: 1101–1108
47. Dustin ML, Springer TA (1989) T-cell receptor cross-linking transiently stimulates adhesiveness through LFA-1. Nature 341: 619–624
48. Mueller KL, Daniels MA, Felthauser A, Kao C, Jameson SC, Shimizu Y (2004) Cutting edge: LFA-1 integrin-dependent T cell adhesion is regulated by both ag specificity and sensitivity. J Immunol 173: 2222–2226
49. Wulfing C, Sumen C, Sjaastad MD, Wu LC, Dustin ML, Davis MM (2002) Costimulation and endogenous MHC ligands contribute to T cell recognition. Nat Immunol 3: 42–47
50. Vianello F, Papeta N, Chen T, Kraft P, White N, Hart WK, Kircher MF, Swart E, Rhee S, Palu G, Irimia D, Toner M, Weissleder R, Poznansky MC (2006) Murine B16 melanomas expressing high levels of the chemokine stromal-derived factor-1/CXCL12 induce tumor-specific T cell chemorepulsion and escape from immune control. J Immunol 176: 2902–2914
51. Davis MM, Krogsgaard M, Huse M, Huppa J, Lillemeier BF, Li QJ (2007) T Cells as a Self-Referential, Sensory Organ. Annu Rev Immunol 25: 681–695
52. Purbhoo MA, Irvine DJ, Huppa JB, Davis MM (2004) T cell killing does not require the formation of a stable mature immunological synapse. Nat Immunol 5: 524–530
53. Li QJ, Chau J, Ebert PJ, Sylvester G, Min H, Liu G, Braich R, Manoharan M, Soutschek J, Skare P, Klein LO, Davis MM, Chen CZ (2007) miR-181a is an intrinsic modulator of T cell sensitivity and selection. Cell 129: 147–161
54. Huang TT, Zong Y, Dalwadi H, Chung C, Miceli MC, Spicher K, Birnbaumer L, Braun J, Aranda R (2003) TCR-mediated hyper-responsiveness of autoimmune Galphai2(-/-) mice is an intrinsic naive CD4(+) T cell disorder selective for the Galphai2 subunit. Int Immunol 15: 1359–1367
55. Hudrisier D, Riond J, Mazarguil H, Gairin JE, Joly E (2001) Cutting edge: CTLs rapidly capture membrane fragments from target cells in a TCR signaling-dependent manner. J Immunol 166: 3645–3649

56. Tomaru U, Yamano Y, Nagai M, Maric D, Kaumaya PT, Biddison W, Jacobson S (2003) Detection of virus-specific T cells and CD8+ T-cell epitopes by acquisition of peptide-HLA-GFP complexes: analysis of T-cell phenotype and function in chronic viral infections. Nat Med 9: 469–476
57. Hudrisier D, Aucher A, Puaux AL, Bordier C, Joly E (2007) Capture of target cell membrane components via trogocytosis is triggered by a selected set of surface molecules on T or B cells. J Immunol 178: 3637–3647
58. Joly E, Hudrisier D (2003) What is trogocytosis and what is its purpose? Nat Immunol 4: 815
59. Beadling C, Slifka MK (2006) Quantifying viable virus-specific T cells without a priori knowledge of fine epitope specificity. Nat Med 12: 1208–1212
60. Harshyne LA, Watkins SC, Gambotto A, Barratt-Boyes SM (2001) Dendritic cells acquire antigens from live cells for cross-presentation to CTL. J Immunol 166: 3717–3723
61. Machlenkin A, Uzana R, Frankenburg S, Pitcovsky Y, Peretz T, Lotem M (2007) Membrane capture by tumor reactive T-cells: an innovative concept of selection and isolation of tumor reactive functionally cytotoxic lymphocytes (Abstract # 225) Keystone Symposium: The New Potent anti-cancer Immunotherapies. Banff, Canada
62. Adamopoulou E, Diekmann J, Tolosa E, Kuntz G, Einsele H, Rammensee HG, Topp MS (2007) Human CD4+ T Cells Displaying Viral Epitopes Elicit a Functional Virus-Specific Memory CD8+ T Cell Response. J Immunol 178: 5465–5472
63. Su MW, Pyarajan S, Chang JH, Yu CL, Jin YJ, Stierhof YD, Walden P, Burakoff SJ (2004) Fratricide of CD8+ cytotoxic T lymphocytes is dependent on cellular activation and perforin-mediated killing. Eur J Immunol 34: 2459–2470
64. Peters PJ, Borst J, Oorschot V, Fukuda M, Krahenbuhl O, Tschopp J, Slot JW, Geuze HJ (1991) Cytotoxic T lymphocyte granules are secretory lysosomes, containing both perforin and granzymes. J Exp Med 173: 1099–1109
65. Smyth MJ, Trapani JA (1995) Granzymes: exogenous proteinases that induce target cell apoptosis. Immunol Today 16: 202–206
66. Motyka B, Korbutt G, Pinkoski MJ, Heibein JA, Caputo A, Hobman M, Barry M, Shostak I, Sawchuk T, Holmes CF, Gauldie J, Bleackley RC (2000) Mannose 6-phosphate/insulin-like growth factor II receptor is a death receptor for granzyme B during cytotoxic T cell-induced apoptosis. Cell 103: 491–500
67. Waterhouse NJ, Sedelies KA, Browne KA, Wowk ME, Newbold A, Sutton VR, Clarke CJ, Oliaro J, Lindemann RK, Bird PI, Johnstone RW, Trapani JA (2005) A central role for Bid in granzyme B-induced apoptosis. J Biol Chem 280: 4476–4482
68. Waterhouse NJ, Clarke CJ, Sedelies KA, Teng MW, Trapani JA (2004) Cytotoxic lymphocytes; instigators of dramatic target cell death. Biochem Pharmacol 68: 1033–1040
69. Pinkoski MJ, Hobman M, Heibein JA, Tomaselli K, Li F, Seth P, Froelich CJ, Bleackley RC (1998) Entry and trafficking of granzyme B in target cells during granzyme B-perforin-mediated apoptosis. Blood 92: 1044–1054
70. Froelich CJ, Orth K, Turbov J, Seth P, Gottlieb R, Babior B, Shah GM, Bleackley RC, Dixit VM, Hanna W (1996) New paradigm for lymphocyte granule-mediated cytotoxicity. Target cells bind and internalize granzyme B, but an endosomolytic agent is necessary for cytosolic delivery and subsequent apoptosis. J Biol Chem 271: 29073–29079
71. Huse M, Lillemeier BF, Kuhns MS, Chen DS, Davis MM (2006) T cells use two directionally distinct pathways for cytokine secretion. Nat Immunol 7: 247–255
72. Barth RJ, Jr, Mule JJ, Spiess PJ, Rosenberg SA (1991) Interferon gamma and tumor necrosis factor have a role in tumor regressions mediated by murine CD8+ tumor-infiltrating lymphocytes. J Exp Med 173: 647–658
73. Poehlein CH, Hu HM, Yamada J, Assmann I, Alvord WG, Urba WJ, Fox BA (2003) TNF plays an essential role in tumor regression after adoptive transfer of perforin/IFN-gamma double knockout effector T cells. J Immunol 170: 2004–2013
74. Seki N, Brooks AD, Carter CR, Back TC, Parsoneault EM, Smyth MJ, Wiltrout RH, Sayers TJ (2002) Tumor-specific CTL kill murine renal cancer cells using both perforin and Fas

ligand-mediated lysis in vitro, but cause tumor regression in vivo in the absence of perforin. J Immunol 168: 3484–3492

75. Itoh N, Nagata S (1993) A novel protein domain required for apoptosis. Mutational analysis of human Fas antigen. J Biol Chem 268: 10932–10937
76. Peter ME, Krammer PH (1998) Mechanisms of CD95 (APO-1/Fas)-mediated apoptosis. Curr Opin Immunol 10: 545–551
77. Greil R, Egle A, Villunger A (1998) On the role and significance of Fas (Apo-1/CD95) ligand (FasL) expression in immune privileged tissues and cancer cells using multiple myeloma as a model. Leuk Lymphoma 31: 477–490
78. Sun M, Ames KT, Suzuki I, Fink PJ (2006) The cytoplasmic domain of Fas ligand costimulates TCR signals. J Immunol 177: 1481–1491
79. Dudley ME, Wunderlich JR, Yang JC, Sherry RM, Topalian SL, Restifo NP, Royal RE, Kammula U, White DE, Mavroukakis SA, Rogers LJ, Gracia GJ, Jones SA, Mangiameli DP, Pelletier MM, Gea-Banacloche J, Robinson MR, Berman DM, Filie AC, Abati A, Rosenberg SA (2005) Adoptive cell transfer therapy following non-myeloablative but lymphodepleting chemotherapy for the treatment of patients with refractory metastatic melanoma. J Clin Oncol 23: 2346–2357
80. Morgan RA, Dudley ME, Wunderlich JR, Hughes MS, Yang JC, Sherry RM, Royal RE, Topalian SL, Kammula US, Restifo NP, Zheng Z, Nahvi A, de Vries CR, Rogers-Freezer LJ, Mavroukakis SA, Rosenberg SA (2006) Cancer regression in patients after transfer of genetically engineered lymphocytes. Science 314: 126–129
81. Koneru M, Schaer D, Monu N, Ayala A, Frey AB (2005) Defective proximal TCR signaling inhibits CD8+ tumor-infiltrating lymphocyte lytic function. J Immunol 174: 1830–4180
82. Charo J, Finkelstein SE, Grewal N, Restifo NP, Robbins PF, Rosenberg SA (2005) Bcl-2 overexpression enhances tumor-specific T-cell survival. Cancer Res 65: 2001–2008
83. Phan GQ, Yang JC, Sherry RM, Hwu P, Topalian SL, Schwartzentruber DJ, Restifo NP, Haworth LR, Seipp CA, Freezer LJ, Morton KE, Mavroukakis SA, Duray PH, Steinberg SM, Allison JP, Davis TA, Rosenberg SA (2003) Cancer regression and autoimmunity induced by cytotoxic T lymphocyte-associated antigen 4 blockade in patients with metastatic melanoma. Proc Natl Acad Sci USA 100: 8372–8377

Chapter 9
Natural Killer Cells at the Tumors Microenvironment

Roi Gazit and Ofer Mandelboim

Abstract NK cells have the innate ability to recognize and eliminate transformed cells. However, tumors still form and progress in seemingly immune-competent individuals. The tumor microenvironment controls NK recruitment and maintenance *in situ*. Current understanding of NK basic biology explain in part their limitations, that might turn the anti-cancerous devotion into a tumor progression outcome. Recent progress in understanding of the evasion from NK recognition and counterattack against the killer cells are discussed, along with unresolved questions.

Keywords Natural Killer, Activating NK receptors, NK inhibition, NK suppression, tumor-ligands, tumor microenvironment.

9.1 Introduction

Natural killer cells were initially recognized as able to kill tumor cells without the need for a prior stimulation [1, 2]. Now, after more than three decades, the significance of the antitumor activity of NK cells *in vivo* is still not completely understood. Accumulating data suggest that the tumor microenvironment may play a major role in the prevention and the modulation of NK cell activities. Therefore the identification and understanding of tumor microenvironment components that affects NK activity *in vivo* will enable the development of new therapeutic approaches, aiming at increasing NK cytotoxicity against various tumors.

9.2 Basic Properties of Natural Killers

NK cells cytotoxicity is executed by the massive degranulation of perforin and granzymes, eliciting a fast apoptosis of the opponent target cell [3, 4]. Perforin proteins polymerize to form permeable pores in the membrane of targeted cells. Through

---

Lautenberg Center for General and Tumor Immunology, The Hebrew University–Hadassah Medical School, Jerusalem, Israel

these pores granzymes enter and activates a robust caspase-pathway of apoptosis which results in a rapid cell death [5]. Naïve NK cells contain granules that store perforin and granzymes, prior to stimulation, giving them the Large Granular Lymphocytes (LGL) appearance [3]. This is why NK cells are immediate potent killers. However, in most cases tumors are not efficiently killed by circulating unstimulated NK cells, and the efficient cytotoxicity is evident only upon IL-2, IL-12, IL-15 or IFNα stimulation [6, 5]. NK cells can also use other means of destruction such as the FasL, TNFα and TNF related apoptosis inducing ligand (TRAIL), that induce apoptosis of cancerous cells, albeit with a slower kinetics [3, 7]. These alternative killing pathways play a secondary role as evident from the poor NK cytotoxicity in perforin knockout mice [8], and in human perforin-deficiency cases [9].

Beside their efficient killing ability, NK cells also secret cytokines and chemokines, notably IFNγ [3, 5]. The secretion of growth factors influences immunity in general, and NK cells were demonstrated, for example, to play an essential role in the polarization of immune response towards the T_H1type by providing the major source of IFNγ in the lymph node [10]. In addition, NK cells by themselves may act as Antigen Presenting Cells (APCs) [11], and they demonstrate a complex direct interactions with professional APCs such as Dendritic cells (DCs) [12] and macrophages [13]. This APCs network might be useful for example on the transfer of MHC loaded with peptides from one cell type to another (NK, DCs or macrophages), for fine-tuning of the quality and magnitude of T cell response.

9.3 The Presence of NK Cells Inside Tumors

One of the major difficulties in the detection of NK cells in tumors is the lack of a single NK specific marker. The classical NK marker in human (CD56) is expressed also on NKT cells, and is absent in mice. To identify NK cells in the mouse the DX5 or NK1.1 mAb are used and these two receptors are also not specific to NK cells only, or are expressed only in certain mouse strains [4]. The only NK specific marker known to date is the NKp46 in human, or NCR1 in mouse. Indeed, NK cells were recently identified in Renal-Cell-Carcinoma patients by using antibodies against the NKp46 receptor [14]. Developing additional antibodies to the human NKp46 and the mice NCR1 will facilitate easier detection of NK cells in the future. Several studies reported on the presence and functions of NK in tumors [15]. For example, intratumor NK cells were found in squamous cell lung cancer, gastric and colorectal carcinomas [16–18] and these studies suggested that the presence of NK cells inside the tumor is an indicative for a good prognosis. NK cells derived from malignant biopsies of ductal invasive breast cancer had enhanced IFNγ and IL-4 production [19]. In addition, NK cells were demonstrated to specifically infiltrate tumors metastasis tissue after IL-2 treatment [20]. However, in this case, the *in situ* detection of NK cells was performed by using CD56 only (which also detect NKT cells). In mouse model of B16 melanoma metastases in the lung, NK cells were identified with anti-asialo GM1, and increased NK numbers were demonstrated after IL-2 treatment [21].

While NK cells are considered to provide a first line of innate defense against any transformed cells, the question of whether they are present at the initial site of the pre-cancerous stages is still unknown. In addition, the emerging concept of epithelial-mesenchymal transition (EMT) [15] was not investigated with regard to NK surveillance. In this concept, while primary carcinoma cells might be less accessible behind the basement membrane, they also possess a less dangerous treat. Once cells go through the EMT and become invasive and consequently more dangerous, they are also more exposed to NK cell attack [15]. It will be interesting to characterize whether NK cells will better-recognize such EMT-phenotype cancer cells, which represent a more dangerous stage of tumor progression. In addition, invasive edge of carcinomas and the outermost cells of tumors which are accessible to NK-recognition are preferentially at EMT condition [15]. Finally, tumor cells were demonstrated to revert from this EMT phenotype after establishment of metastases due to their new stromal environment [15]. This phenotypic change enables further growth, and according to the above hypothesis might also decrease the *in vivo* sensitivity to NK attack. Thus the tumor microenvironment influences the tumor cell phenotype and might also determine tumor susceptibility to NK cell killing.

9.4 Trafficking to the Tumor Site

NK trafficking, like other lymphocytes, is controlled by chemokine receptors and by chemokines gradients. We have demonstrated that the chemokine receptor repertoire is different between the CD56^{dim}CD16^{+} (more cytotoxic) and CD56^{bright}CD16^{-} (more cytokine-secreting) subsets [22]. Once in the tumor microenvironment, increased proliferation of the CD56^{bright}CD16^{-} compared with the CD56^{dim}CD16^{+} subset might occur by interactions with both mDCs and pDCs [23, 24] . Indeed, while the CD56^{bright}CD16^{-} population consists less than 10% of the NK in the blood, inside tumors infiltrates this subset was found to be the predominant NK population [25]. The complexity of examining NK cell phenotype inside tumors is further illustrated by the observations suggested that there are at least 48 different phenotypic NK subsets in human blood [26]. Hence, it is most likely that detailed characterization of tumor infiltrating NK cells for the expression of additional markers will reveal novel NK subsets distribution.

In mice, NK are commonly found in many organs, including bone-marrow, spleen, thymus, lung, lymph nodes and liver [4, 27, 28]. NK cells were also found in the human skin [29]. However, NK cells are not equally distributed in all organs. Thus, when tumors develop, NK might be recruited to the tumor site from the blood stream by mechanisms such as induced fractalkine expression and its recognition by the CX3CR1 receptor which is found on most NK cells [30, 31]. Importantly, it was suggested that the major difficulty in adoptive transfer of NK cells in tumor patients is to bring the effector killer cells to the tumor site [20]. The tumor microenvironment probably plays a major role in NK cell recruitment or preclusion.

After exiting the blood vessels, NK cells needs to pass through the extracellular matrix in order to directly contact their targets. Multiple matrix metalloproteinase

(MMPs) [20], urokinase plasminogen activator (uPA) and its receptor are expressed by NK cells [32], and can be augmented upon stimulation. Intriguingly, the journey through extra cellular matrix was suggested not to wear out NK cells but rather to further activate them for killing [33–35]. Such activation might be enhanced *in vivo* due to the release of cytokines, including IL-2, from collagens by the proteases activity in the extracellular matrix [36]. Histological examinations revealed that most tumor infiltrating NK cells reside in the stromal area, and are not in direct contact with the tumor cells [37]. This suggest for another protective mechanism, by which the extracellular matrix and stroma around the tumor absorb the infiltrating NK cells and reduce their trafficking toward the tumor cells [37]. The characterization of NK cells behavior at the tumor territory would lead to better tumor targeting. A direct imaging of NK cell movements *in vivo* was recently demonstrated and NK cells showed to have a slower-speed in the lymph node (2.75 ∝m/min) compared to T cell (9.6 ∝m/min) [38]. It would be interesting to perform such an intravital imaging of NK cells in the tumor microenvironment.

9.5 Innate Recognition of Tumor Cells

In the "missing self" hypothesis Karre and colleagues suggested that normal healthy self cells are protected from NK cell attack due to the recognition of a self-antigen [39]. This hypothesis was confirmed by the discovery of the inhibitory NK receptors which recognize self class-I MHC molecules [40]. Thus, it was suggested that once tumor cells reduces the expression of the class-I MHC proteins to evade recognition by CTLs, they also become more susceptible to NK lysis [40, 39]. Today, we know that downregulation of MHC class I proteins per se is not enough to induce NK cytotoxicity and that activating receptors control NK cytotoxicity [41, 5]. In addition, other NK inhibitory receptors recognize non-MHC ligands (see below).

The major human activating NK receptors include the CD16, NKG2D, NKp30, NKp44, NKp46 and NKp80 [5]. Only few tumor ligands were identified for these receptors. CD16 is a low-affinity FcRγIII receptor that binds antibodies and unknown tumor ligands [42]. NKG2D recognize the stress-induced proteins ULBPs, MICA and MICB in human or Rae-1s, MULT and H60 in mice [43] (further discussed below), and NKp80 was recently reported to recognize the myeloidic antigen AICL [44].

Activation of the CD16 elicits a strong cytotoxic response even on naïve NK cells [45]. Several research reports demonstrated the presence of antitumor antibodies in the blood of patients [15]. It is still unknown if these antibodies can provide enough stimulation for the CD16 (or other Fc-receptors). The Antibody Dependent Cellular Cytotoxicity (ADCC) activity, which is mediated by NK cells activation through CD16, can have a significant antitumor activity, and is already used clinically [46].

Only viral ligands were identified for the three NCRs: NKp30, NKp44 and NKp46 receptors [47–49]. The tumor-ligands for the NCRs are unknown; however, it was demonstrated that these receptors specifically recognize various tumors [50–53] and that HSPG are involved in tumor recognition [54]. In addition to the direct *in vitro* data regarding the killing of tumor cells by these receptors, analysis of NK cells derived from acute myeloid leukemia (AML) revealed lower expression of the NKp30 and NKp46 receptors, and therefore reduced activity [55]. Since the unknown tumor ligands for NKp30, NKp44 and NKp46 are expressed in many different tumors [50, 56], it is likely to assume that they are needed to support tumor growth, or could not be easily removed by the tumor cells.

Additional NK receptors are also important for tumor eradication. For example, the DNAX-1 accessory molecule-1 (DNAM-1) was demonstrated to have a major role in NK cytotoxicity and cytokines production [57]. The KIR2DS4 activating receptor was demonstrated to have a tumor-ligand other than MHC class-I [58], and this antigen is yet to be identified. Another example is in the case of defective signaling of the 2B4 in XLP patients, causing inability of NK cells to eliminate EBV-infected cells [59] (further discussed below). Thus, understanding the complex balance of inhibitory, activating and co-stimulatory signals [40, 5] that control NK activity is needed to elucidate the multiple mechanisms by which tumors avoid NK cell attack.

TLR9 is expressed on NK cells, and is activated by non-methylated DNA [12]. Tumors have abnormal DNA-methylation, overall manifesting a general genome-wide hypomethylation, despite local hypermethylation islands. Necrotic and apoptotic tumor cells secret such non-methylated CpGs oligonucleotides and these could affect NK killing. Recently, the possible activation of NK cells through TLR9 was shown in human leukemia as stimulation of NK activity was observed via CpG oligonucleotides [60].

9.6 Modulation of Tumor Recognition by NK Cells

NKG2D is involved in the prevention of tumor spread and tumor formation in mice [43]. Its human ligands, ULBPs, MICA and MICB proteins are expressed by several tumors. Numerous studies reported on the frequent expression of MICA and MICB, both on primary lesions and on metastases of melanomas, leukemic cells, gliomas and various carcinomas [61]. The ULBPs were also expressed by the same tumors, albeit at a low frequencies [61]. NKG2D-ligands are specifically induced by genotoxic stress, which contributes significantly also for tumor formation [62]. Recently, the BCR/ABL human oncogene was shown to induce MICA expression in leukemia cells [63]. With this regard it is important to mention that chronic contact with NKG2D-ligand expressing target cells might be unsafe to the NK cells as it can reduce the expression of NKG2D and other receptors and consequently reduce cytotoxicity [64].

Certain tumors develop mechanisms to overcome NKG2D-mediated killing by the downregulation of NKG2D ligands [61]. A naturally defective allele of MICA (MICA*010) was found in tumor cells and is not expressed to the cell surface [65]. In addition, many tumors produce soluble MICA and MICB proteins and such proteins were detected in the sera of cancer patients [61]. These soluble NKG2D ligands could either block NKG2D activity or modulate its expression [61, 15]. The mechanisms of soluble NKG2D ligands formation are only poorly understood [61]. Inhibition of metalloproteinase activity resulted in the MICA accumulation on tumor cells by preventing soluble MICA formation [66]. Soluble MICs could serve as a good diagnostic tool for disease progression in prostate carcinomas [67]. More recently, soluble MICA levels were reported to be significantly elevated in a big cohort of cancer patients and healthy individuals [68]. Soluble MICB protein was also found in many patients, but at surprisingly low-correlation with the soluble MICA [69]. This suggested that there are different mechanisms employed for the modulation of the MICA and MICB [69]. Other soluble NKG2D ligands such as ULBP2 were found in the supernatant of cultured tumor cells [70]. One mechanism of soluble ULBPs formation was discovered in gastric cancer cell lines, in which soluble ULBPs were generated by phosphatidylinositol-specific phospholipase C release of the GPI anchor [71]. As for MICA and –B, the soluble ULBPs were demonstrated to reduce the expression of the NKG2D receptor and consequently the cytotoxicity of NK cells [71].

The frequent modulation of NKG2D ligands in tumors probably reflects their significance in the antitumor activity. The chronic binding of the NKG2D ligands to their receptor and the cognate downregulation of NKG2D, resembles the anergy condition of T-cell after sub-optimal stimulation [72, 73]. Understanding the precise molecular mechanism of NK repression might enable the development of new treatment aimed at increasing NK activity and killing of tumor cells.

Other activating receptors are also involved in tumor recognition [41]. AML patients demonstrate poor NK cytotoxicity, due to a severe reduced expression of NKp30, NKp46 and NKp44 [74]. The mechanisms accounting for the NCRs down-regulation are unknown, but they are TGFβ independent. Our *in vitro* experiments indicate that the expression of NKp30, NKp44 and NKp46 receptors is more stable on the surface of NK cells upon engagement, compared to NKG2D (Gross-Gonen T and Mandelboim O, manuscript in preparation). Thus, the NCRs might provide more rigid interactions with tumors. The critical role of the NCRs in tumor formation is demonstrated in mice lacking the NCR1 receptor. Upon 3MCA challenge the NCR1 knockout homozygous mice developed many more tumors [Gazit R and Mandelboim O, manuscript in preparation]. Thus, the NCRs play a key role in preventing tumor formation.

Other proteins which can impair NK activity are also secreted in the tumor microenvironment. For example, soluble HLA-A, B and C as well as the non-classical HLA-G molecule are secreted from tumor cells [75]. These soluble class-I MHC molecules aggregates to engage receptors on NK and T cells, resulting in the induction of apoptosis [75]. The ICAM-1 protein was also reported to appear as a soluble form in melanoma patients [76]. Soluble ICAM-1 can interact with the LFA-1 integrin (CD11a/CD18 heterodimer) [77], preventing NK adhesion to the tumor cells. With this regard, our study of LFA-1 on NK cells revealed an essential role for this receptor in killing of tumor cells-lines and primary melanoma cells [Sa'ar M and Mandelboim O, manuscript in preparation].

9.7 MHC Class-I Independent Inhibition of NK Cells

Inhibitory Killer Ig-like Receptors (KIR, in human), or the inhibitory Ly49 receptors (in mice), the CD94/NKG2A and LIR1 receptors recognizes both classical as well as non-classical MHC molecules. Downregulation of MHC molecules is frequently observed in tumors, presumably in order to avoid T cell recognition [40]. These pathways were intensively studied and reviewed before [78, 79]. However, in the recent years, additional inhibitory receptors and additional inhibitory pathways are discovered. The KLRG1 receptor contains ITIM motifs which transmit inhibitory signal upon binding to the cadherins-E, R or N [80]. This pathway is exploited by endothelial tumors to reduce NK cytotoxicity, and expression of these cadherins in the tumor area might inhibit all KLRG1$^+$ NK cells [80]. Aberrant expression of NCAM on certain tumor cell-lines reduced recognition and lysis by NK cells, probably due to reduced expression of ICAM-1 on the NCAM-transfected cells [81].

Characterization of NK cells from the lymph-nodes of melanoma patients revealed enhanced expression of the CD66a (CEACAM1) inhibitory receptor, and this inhibitory receptor reduced NK cytotoxicity [82]. A 10-year follow-up study established that CEACAM1 expression correlated with worse survival of melanoma patients [83]. We have also demonstrated that the tumor microenvironment is involved in this inhibition of NK cytotoxicity. NK cells isolated from lymph nods infiltrated with CEACAM1-postive tumors express CEACAM1, whereas NK cells isolated from lymph nods of CEACAM1-negative tumors did not express CEACAM1. The homophilic interaction between CEACAM1 of NK cells and tumor cells lead to inhibition NK cytotoxicity [82]. We also investigated the molecular interactions of CEACAM1 with other CEACAM family members, and revealed that in addition to its homophilic engagement, CEACAM1 also recognize the CEA protein, which is found on many tumors [84]. Such interactions between CEA and CEACAM1 lead to inhibition of NK cytotoxicity [84]. Our ongoing studies revealed that CEA is transferred from target cell to the surface of cognate NK cells in a highly efficient manner, leading to further inhibition of NK cytotoxicity [Stern N and Mandelboim O, 85]. Finally, recent characterization of TILs derived from melanomas showed that all of them express CEACAM1, and that CEACAM1 inhibit both cytotoxicity and IFNγ production of T cells [86].

9.8 Effector Functions and Tumor Development

The direct ability of NK cells and other immune effector cells to eliminate transformed cells suggests that we are continually protected from tumor growth by our efficient immune system [15]. As mentioned above, NK cells are efficient in killing class-I MHC negative targets, whereas class-I positive tumor cells are more likely to be killed by CTLs [39]. Indeed, NK-depletion or genetically ablation of

NK-effectors mechanisms conferred the mice more susceptible to tumor development [7]. NK cells were mostly depleted by using the anti-NK1.1 or the anti-asialo-GM1 antibodies, prior to challenge with tumor cells. These depletion experiments are limited due to the use of antibodies which are not specific to NK cells only [4]. In addition, such induced depletion is limited to a short time period. Currently there is no accurate model of NK cell deficiency except for one transgenic mice reported to lack NK cells only [87]. Using this mouse it was demonstrated that NK cells have an important role in IFNγ secretion and in the resistance to tumor spread. It was recently revealed that NK deficiency in this transgenic strain is the result of specific change in the Atf2 transcription factor [88]. Further support for the important role of NK cells in fighting tumor development came from knockout mice for perforin or for IFNγ-receptor mice which showed high incidence of tumor formation upon carcinogenic challenge [8, 15]. Finally, reduced clearance of certain tumor cell-line was observed in knockout mice of the NCR1 receptor [27].

Specific deficiency of human NK which do not include severe defects in other immune components is very rare. Unfortunately, most of the NK-deficient patient are short lived and the longer-lived patients develop cancers with age [89]. Several studies demonstrated that NK cells are essential for fighting EBV-transformed tumors [90]. Patients with defects in the SH2D1A gene showed NK inability to kill EBV infected cells due to conversion of the 2B4 receptor from its activating form to the inhibitory one [59]. These patients could not control EBV infection. NK cells were also reported to kill activated Hepatic Stellate Cells (HSC) [91]. Thus, NK cells manifest antitumor activity already at the precursor liver-fibrosis stage, which is pre-cancerous. Another compelling evidence for the antitumor activity of NK cells comes from immunocompromised organ-transplanted patients and HIV-infected individuals with elevated frequencies of various cancers [15]. HIV-infected individuals were demonstrated to have specific defects in NK functions [92, 93], and NK cells are known to play essential role in the resistance to at least some of the viruses that cause cancers among transplant patients (including EBV, HCV, etc.) [90]. Many tumors were demonstrated to have a global anti-NK effect *in vivo*, causing impairment in the IL-15 production which affects NK maturation and cytokine production [94].

9.9 TNF Family Ligands

In addition to the classical killing mechanism, NK cells uses several other proteins to induce tumor apoptosis mediated by caspase activation [95]. NK cells were shown to express the TNF, lymphotoxin (LT)-α, LT-ß, FasL, CD27L, CD30L, OX40L, 4–1BBL, and TRAIL [95]. However, several examples show these mechanisms are also used by the tumors to counterattack the immune cells. For example, FasL expression was significantly higher on NK cells of the TILs population derived from Hepatocellular carcinoma patients [96]. However, activated lymphocytes in general, and NK cells in particular also express the Fas receptor itself. Thus, the FasL expression on tumors might leads to the induction of apoptosis in the activated NK and T cells [97]. While Fas is expressed in most tissues, and is induced by IFNγ and TNFα, tumor cells are frequently found to be less sensitive to Fas-induced apoptosis. Reduced Fas expression was reported to result from p53 mutation [98], or from AKT signaling in hypoxic tumors [99].

Triggering of the TNFα pathway results in apoptotosis and death of various tumor cells [94]. However, in contrast to its intended antitumor effect, sustained exposure to TNFα was demonstrated to cause progression of tumors *in vivo*, through the activation of the NF-κB pathway [100]. Moreover, it was recently demonstrated that in melanoma cells treated with BRAF-inhibitors the TNFα specifically prevent apoptosis through NF-κB activation [101] . These examples of TNFα potential pro-tumor effect highlight the risk of this killing pathway.

9.10 The Influence of Physico-chemical Microenvironment on NK Cells

If we were to engineer a potent tumor-killer cell, we would design it to have an induced cytotoxicity at the peculiar tumor microenvironment conditions. Surprisingly however, NK cells demonstrate the opposite behavior and in contrast to the desired augmentation of cytotoxicity at tumor site, they are repressed by hypoxia, acidification and low glucose levels which are considered as a hallmark of tumor microenvironment [20]. The local conditions in tumor are long recognized to be significantly hypoxic at less than half the pO_2 of normal tissue, and down to anoxic [102]. Acidic pH is commonly referred to the tumor microenvironment, and lymphocytes cytotoxicity and proliferation abilities are suppressed by acidic environment [103]. Different measurements revealed that while lower pH is frequent, higher pH is also sometimes observed, and tumors might better be described to have less-stable pH ranging on wider spectrum than normal tissue, rather than only "acidic" [102].

Reactive oxygen metabolites in tumor microenvironment also suppress NK cytotoxicity, IL-2 responsiveness, and augment apoptosis [104]. Specific downregulation of the zeta chain was observed in infiltrating NK cells of colorectal carcinoma [105]. NK cytotoxicity against the K562 target cells and against liver cancer cells was significantly reduced upon prolonged incubation at 1% oxygen, compared to 21% oxygen of ambient air and low oxygen also diminished the induction of cytotoxicity by IFNγ [106]. It was recently demonstrated that human decidual NK cells promotes angiogenesis trough the secretion of PLGF and VEGF [107]. The pro-angiogenesis contribution of NK cells at early stages might aim to reduce the local danger of further transformation, while at later stages such activity might accidentally help to sustain tumor growth. In human clinical studies, the presence of infiltrating NK cells was reported to have a positive prognosis [16, 17], showing that the net effect of NK cells is antitumor, in spite of the suggested pro-tumor functions of other immune cells [108].

Tumors do not relay only on these environmental-conditions only and also secrete various factors that affect the function of local immune cells. For example, corticosteroids reduce the expression of NCRs [109]. Specific enzymatic oddities also impair NK activities, for example the expression of NKp46 and NKG2D was downregulated by the IDO product L-kynurenine [110]. The IL-10 and TGFβ were more extensively studied and are therefore specifically discussed below.

Taken together, it seems as if NK cells are not well designed to exhibit their abilities at the harsh conditions of progressed tumors. This fact suggest that the antitumor effect of NK cells should be studied at early stages of tumor formation and that the antitumor effects of NK cells should better be examined at the tumor conditions.

9.11 IL-10

NK cells, under certain conditions, produce IL-10 [111]. IL-10, and TGFβ, enhances the expression of the nonclassical class-I MHC molecule HLA-G [112] which in turn leads to the inhibition of NK activity [113]. IL-10 itself also inhibit the production IFNγ, TNFα by NK cells [114].Cancerous cells may secrete IL-10 and gain autocrine stimulation to enhance their proliferation [115]. On the other hand, IL-10 can also stimulate NK activity [116]. The *in vivo* administration of IL-10 increased the plasma levels of granzymes, indicating increased NK and CTL activity [117]. Enhancement of NK cytotoxicity against tumor cell-line *in vitro* was demonstrated to result from post-translational downregulation of class-I MHC presentation by IL-10 [118]. Preclinical studies demonstrated that IL-10 has antitumor effect *in vivo*, which is suggested to be due to NK cells mainly [119–121].

9.12 TGF-β

Transforming Growth Factor β (TGFβ) was demonstrated by many studies to play a major role in antitumor immunity [122]. NK cells lose IFNγ secretion, cytotoxicity and proliferation capabilities once exposed to TGFβ [123–125]. These effects were demonstrated *in vivo* by a gain-of-function experiments in which overexpression of TGFβ rendered a highly immunologic tumor to become resistant to immune rejection [126]. In addition, specific interruption of TGFβ signaling in NK and DCs was recently reported to affect NK homeostasis, and also had a major impact in regulation of NK activities such as IFNγ production [127].

The mechanisms by which TGFβ affects NK cell function are being revealed lately. TGFβ was shown to generally antagonize the activity of the pro-inflammatory cytokines through SMAD2, 3 and 4, inhibiting both IFNγ and T-bet expression in NK cells [128]. The expression of T-bet in NK cells is essential for their maturation, for proper expression of perforin, granzymeB and Runx1, and for the induction of enhanced IFNγ production [129]. In addition, TGFβ inhibit the expression of at least two major NK lysis receptors: NKp30 and NKG2D [130]. Beside the direct inhibitory effects of TGFβ on NK activity, this cytokine also influence NK cells indirectly in the tumor microenvironment, for example, by inducing FoxP3 expression in regulatory T cells that thereafter suppress NK cell activity locally [122]. Thus, TGFβ is a general inhibitor of NK functions and developing different strategies to overcome TGFβ inhibition are desired.

9.13 NK–DCs Interactions

The reciprocal interactions between NK cells and DCs gained much interest in recent years. It was demonstrated that DCs activate NK cells activity against tumors *in vivo*, and that DC–NK interaction increase both IFNγ production and NK cytotoxicity [131]. DCs also provide a direct stimulation for NK cells as they secret IL-12 which activates NK cell functions [132]. Additionally, IL-18 secretion by DCs was demonstrated to induce NK cells migration from the tissue to the lymph node (LN), where NK can further influence the adaptive immune response [133].

Direct killing of immature myloid-derived DCs by NK cells *in vitro* was demonstrated through the engagement of the NKp30 and DNAM-1 receptors [134–136]. Further studies reported that NK cells of AML patients, which express lower levels of NKp30are unable to eliminate immature DCs *in vitro* [55, 137]. As mentioned above, TGFβ induced NKp30 and NKG2D downregulation and such downregulation resulted in the inhibition of the killing of immature DCs [130]. The self killing of autologous DCs is in contrast to the "missing self" hyphothgesis which suggests that normal self cells are protected from NK cytotoxicity by the class-I MHC inhibitory receptors. Indeed, clonal analysis showed that the killing of iDCs is mainly executed by NK cells which lack the expression of inhibitory receptors recognizing HLA-A, B and C proteins and only express the NKG2A/CD94 receptor for HLA-E [138]. DC Maturation results in upregulation of MHC molecules (including HLA-E), and that is why mature DC are resistant to NK lysis [138]. It is suggested that the elimination of iDCs which have low levels of MHC molecules is an editing process, which selects for the most potent antigen-presenting cells [139]. Finally, NK cells that were treated with IL-12 had higher cytotoxicity against immature DCs and better induction of DCs maturation, while IL-4 exposed NK cells did not [140]. Thus suggesting that the cytokines milieu determines NK interactions with DCs.

Another outcome of NK–DCs interaction is the induced secretion of TNFα and other cytokines by NK cells which leads to DCs maturation [141]. Intriguingly, it was demonstrated *in vivo* that the DC-activation of NK cells is responsible for the Gleevec antitumor function in certain gastrointestinal stromal tumors [132]. Notably, a direct and specific DC-NK contacts were observed in lichenoid dermatitis biopsies from these patients (NK cells were recognized as $CD57^+$ and verified as $CD3^-$) [132].

Another impact of DCs interaction with NK is the induced proliferation of the cytokine-producer $CD56^{bright}CD16^-$ subset, after co-culturing *in vitro* with LPS [142].

The NK–DCs interactions are dependent on a prior NK activation, and were not efficient if fresh unstimulated NK cells were used. In the case of tumor microenvironment this is specifically relevant, as NK cells are expected to be found at activated state in the tumor site. Indeed, a recent study demonstrated that the NK-DCs cross-talk not only enable primary tumor rejection, but also induced a long-term CTLs response without need for $CD4^+$ helper function [143]. Such $CD4^+$ independent CTLs activity might be enabled by NK potential to perform APC functions [11].

The Plasmacytoid DCs (pDCs) are known to mainly function in the course of virus infection. However, pDCs were also demonstrated to efficiently activate NK cytotoxicity against tumor cells, and pDCs stimulation of NK cells was augmented by IL-2 [24]. In conclusion, reciprocal signals transduced between NK and DCs at the tumor microenvironment determine the local activation of both cells types, and also shape the later adaptive response.

9.14 Suppression of NK by Immune-regulatory Cells

Regulatory T cells (Treg) inhibition of NK functions was demonstrated *in vitro* and *in vivo* using anti-CD25 mAb against Tregs. In this model, CD25-depletion was also demonstrated to enhance NK activity *in vitro* [144, 145], suggesting that Tregs potentially inhibit NK cells. In mice with FoxP3 defect NK proliferation is enhanced [146], whereas adoptive Tregs transfer inhibit NK functions [147]. Similarly, human $CD4^+$ $CD25^+$ T cells were shown to impair NK cytotoxicity *in vitro* [148, 149]. Moreover, investigation of human gastrointestinal stromal tumor patients whom did not respond well to Gleevec demonstrated a higher Tregs numbers that were correlated with lower NK activity [150]. In agreement with these results, melanoma patients demonstrated increased NK functions if Tregs levels were low [151]. Tregs inhibition of NK activity was observed in a lung carcinoma model, where CD25-depletion enhanced the $NK1.1^+$ cytotoxicity and resistance to tumor metastasis [152]. The rejection of allogeneic Bone-marrow transplanted cells was enhanced upon $CD4^+$ $CD25^+$ depletion in the recipient mice [153], and transplanted $CD4^+$ $CD25^+$ cells could protect bone-marrow progenitors form NK elimination. Experiments with thrombospondin-1 (TSP-1) deficient mice revealed its essential role in the inhibition of NK cells by Tregs [154]. Myeloid suppressor cells ($CD11b^+$ $Gr\text{-}1^+$) were shown to expand during tumor growth, and to suppress NK functions *in vivo*, through cell-cell contact that affect Stat5 and inhibit perforin synthesis [155].

Indirect effects complicate the situation even further, for example the TGFβ production by iDCs is required for the optimal proliferation of Tregs [146]. Therefore, the elimination of iDCs by activated NK cells [139], would affect the physiological balance between activation and suppression by either omitting or allowing the initial signals. A comprehensive understanding of immune cells activities at the tumor site will enable the understanding of the complex relationships between these multiple cells types.

9.15 Killing of Metastatic Cancerous Cells by NK

In order to form metastases, tumor cells must leave the protective tumor microenvironment and travel to distinct sites, being exposed to destruction by NK and other immune cells. However, unique shield is provided to the cancerous cells in the blood stream by the form of platelets that form micro thrombi around them [15]. Experimentally, anticoagulant drugs treatment in mice inhibited metastases in an NK-dependent mechanism, and increased acute tumor cells clearance *in vivo* [156], suggesting that platelets aggregation and fibrin coating provide tumor protection of NK cell destruction in the blood. Clinical evidences have also pointed to coagulation role in tumor metastasis [157]. *In vitro*, fibrin coagulation around tumor cells, or around effector cells, reduced cytotoxicity by preventing cell-cell contact [158]. Thrombin was shown to contribute to platelet aggregation with tumor cells *in vitro* and consequently to enhance metastasis *in vivo* [159]. Thus, metastases which are the main cause of death in cancer patients are quite efficiently protected from NK cells attack.

9.16 Concluding Remarks

In this review we describe how NK cells communicate with tumor cells, especially at the tumor microenvironment which turns to have a major role in the modulation of NK functions. There are yet several vital issues to address; solving these questions will reveal some of the fundamentals of NK biology. Among these are: (1) Identification of the tumor-ligands for NK activating receptors, and in particular for the NCRs, to understand the basic recognition of transformed cells. (2) Elucidating NK surveillance of early-transformed cells at different tissues, and its disturbance in progressed tumors. (3) The characterization of trafficking of functional NK-subsets into the tumor site, and these subsets retention or further migration. (4) Development of intravital image techniques to observe NK functions inside tumors in order to understand the killers' activities in the crime scene. (5) Delineate NK relationships with other immune-cells at the tumor area. (6) Gain mechanistic comprehension of the local and the systemic NK suppression by tumors.

New insights can lead to new treatments that will harness the robust NK activity against tumors. Based on understanding how tumor-microenvironment disarms NK cells, we will be able to aim sustained NK activity against the tumor spread.

References

1. Kiessling, R., Klein, E., Pross, H., and Wigzell, H. (1975). "Natural" killer cells in the mouse. II. Cytotoxic cells with specificity for mouse Moloney leukemia cells. Characteristics of the killer cell. European journal of immunology *5*, 117–121.

2. Kiessling, R., Klein, E., and Wigzell, H. (1975). "Natural" killer cells in the mouse. I. Cytotoxic cells with specificity for mouse Moloney leukemia cells. Specificity and distribution according to genotype. European journal of immunology *5*, 112–117.
3. Colucci, F., Caligiuri, M.A., and Di Santo, J.P. (2003). What does it take to make a natural killer? Nature reviews *3*, 413–425.
4. Di Santo, J.P. (2006). Natural killer cell developmental pathways: a question of balance. Annual review of immunology *24*, 257–286.
5. Moretta, A., Bottino, C., Vitale, M., Pende, D., Cantoni, C., Mingari, M.C., Biassoni, R., and Moretta, L. (2001). Activating receptors and coreceptors involved in human natural killer cell-mediated cytolysis. Annual review of immunology *19*, 197–223.
6. Bottino, C., Castriconi, R., Moretta, L., and Moretta, A. (2005). Cellular ligands of activating NK receptors. Trends in immunology *26*, 221–226.
7. Tassi, I., Klesney-Tait, J., and Colonna, M. (2006). Dissecting natural killer cell activation pathways through analysis of genetic mutations in human and mouse. Immunological reviews *214*, 92–105.
8. Kagi, D., Ledermann, B., Burki, K., Seiler, P., Odermatt, B., Olsen, K.J., Podack, E.R., Zinkernagel, R.M., and Hengartner, H. (1994). Cytotoxicity mediated by T cells and natural killer cells is greatly impaired in perforin-deficient mice. Nature *369*, 31–37.
9. Clementi, R., Locatelli, F., Dupre, L., Garaventa, A., Emmi, L., Bregni, M., Cefalo, G., Moretta, A., Danesino, C., Comis, M., et al. (2005). A proportion of patients with lymphoma may harbor mutations of the perforin gene. Blood *105*, 4424–4428.
10. Martin-Fontecha, A., Thomsen, L.L., Brett, S., Gerard, C., Lipp, M., Lanzavecchia, A., and Sallusto, F. (2004). Induced recruitment of NK cells to lymph nodes provides IFN-gamma for T(H)1 priming. Nature immunology *5*, 1260–1265.
11. Hanna, J., Gonen-Gross, T., Fitchett, J., Rowe, T., Daniels, M., Arnon, T.I., Gazit, R., Joseph, A., Schjetne, K.W., Steinle, A., et al. (2004). Novel APC-like properties of human NK cells directly regulate T cell activation. The journal of clinical investigation *114*, 1612–1623.
12. Moretta, L., Bottino, C., Pende, D., Vitale, M., Mingari, M.C., and Moretta, A. (2005). Human natural killer cells: Molecular mechanisms controlling NK cell activation and tumor cell lysis. Immunology letters *100*, 7–13.
13. Nedvetzki, S., Sowinski, S., Eagle, R.A., Harris, J., Vely, F., Pende, D., Trowsdale, J., Vivier, E., Gordon, S., and Davis, D.M. (2007). Reciprocal regulation of natural killer cells and macrophages associated with distinct immune synapses. Blood.
14. Schleypen, J.S., Von Geldern, M., Weiss, E.H., Kotzias, N., Rohrmann, K., Schendel, D.J., Falk, C.S., and Pohla, H. (2003). Renal cell carcinoma-infiltrating natural killer cells express differential repertoires of activating and inhibitory receptors and are inhibited by specific HLA class I allotypes. International journal of cancer *106*, 905–912.
15. Weinberg, R.A. (2006). The Biology of Cancer, 1st edn. (Garland Science).
16. Coca, S., Perez-Piqueras, J., Martinez, D., Colmenarejo, A., Saez, M.A., Vallejo, C., Martos, J.A., and Moreno, M. (1997). The prognostic significance of intratumoral natural killer cells in patients with colorectal carcinoma. Cancer *79*, 2320–2328.
17. Ishigami, S., Natsugoe, S., Tokuda, K., Nakajo, A., Che, X., Iwashige, H., Aridome, K., Hokita, S., and Aikou, T. (2000). Prognostic value of intratumoral natural killer cells in gastric carcinoma. Cancer *88*, 577–583.
18. Villegas, F.R., Coca, S., Villarrubia, V.G., Jimenez, R., Chillon, M.J., Jareno, J., Zuil, M., and Callol, L. (2002). Prognostic significance of tumor infiltrating natural killer cells subset CD57 in patients with squamous cell lung cancer. Lung cancer *35*, 23–28.
19. Lorenzen, J., Lewis, C.E., McCracken, D., Horak, E., Greenall, M., and McGee, J.O. (1991). Human tumour-associated NK cells secrete increased amounts of interferon-gamma and interleukin-4. British journal of cancer *64*, 457–462.
20. Albertsson, P.A., Basse, P.H., Hokland, M., Goldfarb, R.H., Nagelkerke, J.F., Nannmark, U., and Kuppen, P.J. (2003). NK cells and the tumour microenvironment: implications for NK-cell function and anti-tumour activity. Trends in immunology *24*, 603–609.

21. Hokland, M., Kjaergaard, J., Kuppen, P.J., Nannmark, U., Agger, R., Hokland, P., and Basse, P. (1999). Endogenous and adoptively transferred A-NK and T-LAK cells continuously accumulate within murine metastases up to 48 h after inoculation. In vivo *13*, 199–204.
22. Hanna, J., Wald, O., Goldman-Wohl, D., Prus, D., Markel, G., Gazit, R., Katz, G., Haimov-Kochman, R., Fujii, N., Yagel, S., et al. (2003). CXCL12 expression by invasive trophoblasts induces the specific migration of CD16- human natural killer cells. Blood *102*, 1569–1577.
23. Ferlazzo, G., Pack, M., Thomas, D., Paludan, C., Schmid, D., Strowig, T., Bougras, G., Muller, W.A., Moretta, L., and Munz, C. (2004). Distinct roles of IL-12 and IL-15 in human natural killer cell activation by dendritic cells from secondary lymphoid organs. Proceedings of the national academy of sciences of the United States of America *101*, 16606–16611.
24. Romagnani, C., Della Chiesa, M., Kohler, S., Moewes, B., Radbruch, A., Moretta, L., Moretta, A., and Thiel, A. (2005). Activation of human NK cells by plasmacytoid dendritic cells and its modulation by CD4+ T helper cells and CD4+ CD25hi T regulatory cells. European journal of immunology *35*, 2452–2458.
25. Moller, M.J., Kammerer, R., and von Kleist, S. (1998). A distinct distribution of natural killer cell subgroups in human tissues and blood. International journal of cancer *78*, 533–538.
26. Jonges, L.E., Albertsson, P., van Vlierberghe, R.L., Ensink, N.G., Johansson, B.R., van de Velde, C.J., Fleuren, G.J., Nannmark, U., and Kuppen, P.J. (2001). The phenotypic heterogeneity of human natural killer cells: presence of at least 48 different subsets in the peripheral blood. Scandinavian journal of immunology *53*, 103–110.
27. Gazit, R., Gruda, R., Elboim, M., Arnon, T.I., Katz, G., Achdout, H., Hanna, J., Qimron, U., Landau, G., Greenbaum, E., et al. (2006). Lethal influenza infection in the absence of the natural killer cell receptor gene Ncr1. Nature immunology *7*, 517–523.
28. Kim, S., Iizuka, K., Kang, H.S., Dokun, A., French, A.R., Greco, S., and Yokoyama, W.M. (2002). In vivo developmental stages in murine natural killer cell maturation. Nature immunology *3*, 523–528.
29. Tsuchiyama, J., Yoshino, T., Toba, K., Harada, N., Nishiuchi, R., Akagi, T., Furukawa, T., Takahashi, M., Fuse, I., Aizawa, Y., and Harada, M. (2002). Induction and characterization of cutaneous lymphocyte antigen on natural killer cells. British journal of haematology *118*, 654–662.
30. Imai, T., Hieshima, K., Haskell, C., Baba, M., Nagira, M., Nishimura, M., Kakizaki, M., Takagi, S., Nomiyama, H., Schall, T.J., and Yoshie, O. (1997). Identification and molecular characterization of fractalkine receptor CX3CR1, which mediates both leukocyte migration and adhesion. Cell *91*, 521–530.
31. Xin, H., Kikuchi, T., Andarini, S., Ohkouchi, S., Suzuki, T., Nukiwa, T., Huqun, Hagiwara, K., Honjo, T., and Saijo, Y. (2005). Antitumor immune response by CX3CL1 fractalkine gene transfer depends on both NK and T cells. European journal of immunology *35*, 1371–1380.
32. Al-Atrash, G., Shetty, S., Idell, S., Xue, Y., Kitson, R.P., Halady, P.K., and Goldfarb, R.H. (2002). IL-2-mediated upregulation of uPA and uPAR in natural killer cells. Biochemical and biophysical research communications *292*, 184–189.
33. Johansson, B.R. and Nannmark, U. (1996). Ultrastructure of interactions between activated murine natural killer cells and melanoma cells in an extracellular matrix (Matrigel) environment. Natural immunity *15*, 98–106.
34. Johansson, B.R., Unger, M.L., Albertsson, P., Casselbrant, A., Nannmark, U., and Hokland, M. (1996). Infiltration and lysis of tumour cell aggregates by adherent interleukin-2-activated natural killer cells is distinct from specific cytolysis. Natural immunity *15*, 87–97.
35. Palmieri, G., Gismondi, A., Galandrini, R., Milella, M., Serra, A., De Maria, R., and Santoni, A. (1996). Interaction of natural killer cells with extracellular matrix induces early intracellular signalling events and enhances cytotoxic functions. Natural immunity *15*, 147–153.
36. Somasundaram, R., Ruehl, M., Tiling, N., Ackermann, R., Schmid, M., Riecken, E.O., and Schuppan, D. (2000). Collagens serve as an extracellular store of bioactive interleukin 2. The journal of biological chemistry *275*, 38170–38175.
37. Kuppen, P.J., Gorter, A., Hagenaars, M., Jonges, L.E., Giezeman-Smits, K.M., Nagelkerke, J.F., Fleuren, G., and van de Velde, C.J. (2001). Role of NK cells in adoptive immunotherapy

of metastatic colorectal cancer in a syngeneic rat model. Immunological reviews *184*, 236–243.

38. Bajenoff, M., Breart, B., Huang, A.Y., Qi, H., Cazareth, J., Braud, V.M., Germain, R.N., and Glaichenhaus, N. (2006). Natural killer cell behavior in lymph nodes revealed by static and real-time imaging. The journal of experimental medicine *203*, 619–631.
39. Ljunggren, H.G., and Karre, K. (1986). Experimental strategies and interpretations in the analysis of changes in MHC gene expression during tumour progression. Opposing influences of T cell and natural killer mediated resistance? Journal of immunogenetics *13*, 141–151.
40. Karre, K. (2002). NK cells, MHC class I molecules and the missing self. Scandinavian journal of immunology *55*, 221–228.
41. Arnon, T.I., Markel, G., and Mandelboim, O. (2006). Tumor and viral recognition by natural killer cells receptors. Seminars in cancer biology *16*, 348–358.
42. Mandelboim, O., Malik, P., Davis, D.M., Jo, C.H., Boyson, J.E., and Strominger, J.L. (1999). Human CD16 as a lysis receptor mediating direct natural killer cell cytotoxicity. Proceedings of the national academy of sciences of the United States of America *96*, 5640–5644.
43. Raulet, D.H. (2003). Roles of the NKG2D immunoreceptor and its ligands. Nature reviews *3*, 781–790.
44. Welte, S., Kuttruff, S., Waldhauer, I., and Steinle, A. (2006). Mutual activation of natural killer cells and monocytes mediated by NKp80-AICL interaction. Nature immunology *7*, 1334–1342.
45. Bryceson, Y.T., March, M.E., Ljunggren, H.G., and Long, E.O. (2006). Synergy among receptors on resting NK cells for the activation of natural cytotoxicity and cytokine secretion. Blood *107*, 159–166.
46. Geldart, T. and Illidge, T. (2005). Anti-CD 40 monoclonal antibody. Leukemia & lymphoma *46*, 1105–1113.
47. Arnon, T.I., Achdout, H., Levi, O., Markel, G., Saleh, N., Katz, G., Gazit, R., 111. Gonen-Gross, T., Hanna, J., Nahari, E., et al. (2005). Inhibition of the NKp30 activating receptor by pp65 of human cytomegalovirus. Nature immunology *6*, 515–523.
48. Arnon, T.I., Lev, M., Katz, G., Chernobrov, Y., Porgador, A., and Mandelboim, O. (2001). Recognition of viral hemagglutinins by NKp44 but not by NKp30. European journal of immunology *31*, 2680–2689.
49. Mandelboim, O., Lieberman, N., Lev, M., Paul, L., Arnon, T.I., Bushkin, Y., Davis, D.M., Strominger, J.L., Yewdell, J.W., and Porgador, A. (2001). Recognition of haemagglutinins on virus-infected cells by NKp46 activates lysis by human NK cells. Nature *409*, 1055–1060.
50. Arnon, T.I., Achdout, H., Lieberman, N., Gazit, R., Gonen-Gross, T., Katz, G., Bar-Ilan, A., Bloushtain, N., Lev, M., Joseph, A., et al. (2004). The mechanisms controlling the recognition of tumor- and virus-infected cells by NKp46. Blood *103*, 664–672.
51. Pende, D., Parolini, S., Pessino, A., Sivori, S., Augugliaro, R., Morelli, L., Marcenaro, E., Accame, L., Malaspina, A., Biassoni, R., et al. (1999). Identification and molecular characterization of NKp30, a novel triggering receptor involved in natural cytotoxicity mediated by human natural killer cells. The journal of experimental medicine *190*, 1505–1516.
52. Pessino, A., Sivori, S., Bottino, C., Malaspina, A., Morelli, L., Moretta, L., Biassoni, R., and Moretta, A. (1998). Molecular cloning of NKp46: a novel member of the immunoglobulin superfamily involved in triggering of natural cytotoxicity. The journal of experimental medicine *188*, 953–960.
53. Vitale, M., Bottino, C., Sivori, S., Sanseverino, L., Castriconi, R., Marcenaro, E., Augugliaro, R., Moretta, L., and Moretta, A. (1998). NKp44, a novel triggering surface molecule specifically expressed by activated natural killer cells, is involved in non-major histocompatibility complex-restricted tumor cell lysis. The journal of experimental medicine *187*, 2065–2072.
54. Bloushtain, N., Qimron, U., Bar-Ilan, A., Hershkovitz, O., Gazit, R., Fima, E., Korc, M., Vlodavsky, I., Bovin, N.V., and Porgador, A. (2004). Membrane-associated heparan sulfate proteoglycans are involved in the recognition of cellular targets by NKp30 and NKp46. Journal of immunology *173*, 2392–2401.

55. Costello, R.T., Sivori, S., Marcenaro, E., Lafage-Pochitaloff, M., Mozziconacci, M.J., Reviron, D., Gastaut, J.A., Pende, D., Olive, D., and Moretta, A. (2002). Defective expression and function of natural killer cell-triggering receptors in patients with acute myeloid leukemia. Blood *99*, 3661–3667.
56. Moretta, A. (2005). The dialogue between human natural killer cells and dendritic cells. Current opinion in immunology *17*, 306–311.
56. Sivori, S., Pende, D., Bottino, C., Marcenaro, E., Pessino, A., Biassoni, R., Moretta, L., and Moretta, A. (1999). NKp46 is the major triggering receptor involved in the natural cytotoxicity of fresh or cultured human NK cells. Correlation between surface density of NKp46 and natural cytotoxicity against autologous, allogeneic or xenogeneic target cells. European journal of immunology *29*, 1656–1666.
57. Carlsten, M., Bjorkstrom, N.K., Norell, H., Bryceson, Y., van Hall, T., Baumann, B.C., Hanson, M., Schedvins, K., Kiessling, R., Ljunggren, H.G., and Malmberg, K.J. (2007). DNAX accessory molecule-1 mediated recognition of freshly isolated ovarian carcinoma by resting natural killer cells. Cancer research *67*, 1317–1325.
58. Katz, G., Gazit, R., Arnon, T.I., Gonen-Gross, T., Tarcic, G., Markel, G., Gruda, R., Achdout, H., Drize, O., Merims, S., and Mandelboim, O. (2004). MHC class I-independent recognition of NK-activating receptor KIR2DS4. Journal of immunology *173*, 1819–1825.
59. Parolini, S., Bottino, C., Falco, M., Augugliaro, R., Giliani, S., Franceschini, R., Ochs, H.D., Wolf, H., Bonnefoy, J.Y., Biassoni, R., et al. (2000). X-linked lymphoproliferative disease. 2B4 molecules displaying inhibitory rather than activating function are responsible for the inability of natural killer cells to kill Epstein-Barr virus-infected cells. The journal of experimental medicine *192*, 337–346.
60. Fujii, H., Trudeau, J.D., Teachey, D., Fish, J.D., Grupp, S.A., Schultz, K.R., and Reid, G.S. (2006). In vivo control of acute lymphoblastic leukemia by immunostimulatory CpG oligonucleotides. Blood.
61. Chang, C.C. and Ferrone, S. (2006). NK cell activating ligands on human malignant cells: molecular and functional defects and potential clinical relevance. Seminars in cancer biology *16*, 383–392.
62. Gasser, S., Orsulic, S., Brown, E.J., and Raulet, D.H. (2005). The DNA damage pathway regulates innate immune system ligands of the NKG2D receptor. Nature *436*, 1186–1190.
63. Boissel, N., Rea, D., Tieng, V., Dulphy, N., Brun, M., Cayuela, J.M., Rousselot, P., Tamouza, R., Le Bouteiller, P., Mahon, F.X., et al. (2006). BCR/ABL oncogene directly controls MHC class I chain-related molecule an expression in chronic myelogenous leukemia. Journal of immunology *176*, 5108–5116.
64. Coudert, J.D., Zimmer, J., Tomasello, E., Cebecauer, M., Colonna, M., Vivier, E., and Held, W. (2005). Altered NKG2D function in NK cells induced by chronic exposure to NKG2D ligand-expressing tumor cells. Blood *106*, 1711–1717.
65. Li, Z., Groh, V., Strong, R.K., and Spies, T. (2000). A single amino acid substitution causes loss of expression of a MICA allele. Immunogenetics *51*, 246–248.
66. Salih, H.R., Rammensee, H.G., and Steinle, A. (2002). Cutting edge: down-regulation of MICA on human tumors by proteolytic shedding. Journal of immunology *169*, 4098–4102.
67. Wu, J.D., Higgins, L.M., Steinle, A., Cosman, D., Haugk, K., and Plymate, S.R. (2004). Prevalent expression of the immunostimulatory MHC class I chain-related molecule is counteracted by shedding in prostate cancer. The journal of clinical investigation *114*, 560–568.
68. Holdenrieder, S., Stieber, P., Peterfi, A., Nagel, D., Steinle, A., and Salih, H.R. (2006). Soluble MICA in malignant diseases. International journal of cancer *118*, 684–687.
69. Holdenrieder, S., Stieber, P., Peterfi, A., Nagel, D., Steinle, A., and Salih, H.R. (2006). Soluble MICB in malignant diseases: analysis of diagnostic significance and correlation with soluble MICA. Cancer immunology immunotherapy *55*, 1584–1589.
70. Waldhauer, I. and Steinle, A. (2006). Proteolytic release of soluble UL16-binding protein 2 from tumor cells. Cancer research *66*, 2520–2526.
71. Song, H., Kim, J., Cosman, D., and Choi, I. (2006). Soluble ULBP suppresses natural killer cell activity via down-regulating NKG2D expression. Cellular immunology *239*, 22–30.

72. Oppenheim, D.E., Roberts, S.J., Clarke, S.L., Filler, R., Lewis, J.M., Tigelaar, R.E., Girardi, M., and Hayday, A.C. (2005). Sustained localized expression of ligand for the activating NKG2D receptor impairs natural cytotoxicity in vivo and reduces tumor immunosurveillance. Nature immunology *6*, 928–937.
73. Wiemann, K., Mittrucker, H.W., Feger, U., Welte, S.A., Yokoyama, W.M., Spies, T., Rammensee, H.G., and Steinle, A. (2005). Systemic NKG2D down-regulation impairs NK and CD8 T cell responses in vivo. Journal of immunology *175*, 720–729.
74. Fauriat, C., Just-Landi, S., Mallet, F., Arnoulet, C., Sainty, D., Olive, D., and Costello, R.T. (2007). Deficient expression of NCR in NK cells from acute myeloid leukemia: Evolution during leukemia treatment and impact of leukemia cells in NCRdull phenotype induction. Blood *109*, 323–330.
75. Contini, P., Ghio, M., Poggi, A., Filaci, G., Indiveri, F., Ferrone, S., and Puppo, F. (2003). Soluble HLA-A,-B,-C and -G molecules induce apoptosis in T and NK CD8+ cells and inhibit cytotoxic T cell activity through CD8 ligation. European journal of immunology *33*, 125–134.
76. Kageshita, T., Yoshii, A., Kimura, T., Kuriya, N., Ono, T., Tsujisaki, M., Imai, K., and Ferrone, S. (1993). Clinical relevance of ICAM-1 expression in primary lesions and serum of patients with malignant melanoma. Cancer research *53*, 4927–4932.
77. Witkowska, A.M. and Borawska, M.H. (2004). Soluble intercellular adhesion molecule-1 (sICAM-1): an overview. European cytokine network *15*, 91–98.
78. Raulet, D.H., Vance, R.E., and McMahon, C.W. (2001). Regulation of the natural killer cell receptor repertoire. Annual review of immunology *19*, 291–330.
79. Yokoyama, W.M., and Kim, S. (2006). How do natural killer cells find self to achieve tolerance? Immunity *24*, 249–257.
80. Ito, M., Maruyama, T., Saito, N., Koganei, S., Yamamoto, K., and Matsumoto, N. (2006). Killer cell lectin-like receptor G1 binds three members of the classical cadherin family to inhibit NK cell cytotoxicity. The journal of experimental medicine *203*, 289–295.
81. Jarahian, M., Watzl, C., Issa, Y., Altevogt, P., and Momburg, F. (2007). Blockade of natural killer cell-mediated lysis by NCAM140 expressed on tumor cells. International journal of cancer.
82. Markel, G., Lieberman, N., Katz, G., Arnon, T.I., Lotem, M., Drize, O., Blumberg, R.S., Bar-Haim, E., Mader, R., Eisenbach, L., and Mandelboim, O. (2002). CD66a interactions between human melanoma and NK cells: a novel class I MHC-independent inhibitory mechanism of cytotoxicity. Journal of immunology *168*, 2803–2810.
83. Thies, A., Moll, I., Berger, J., Wagener, C., Brummer, J., Schulze, H.J., Brunner, G., and Schumacher, U. (2002). CEACAM1 expression in cutaneous malignant melanoma predicts the development of metastatic disease. Journal of clinical oncology *20*, 2530–2536.
84. Stern, N., Markel, G., Arnon, T.I., Gruda, R., Wong, H., Gray-Owen, S.D., and Mandelboim, O. (2005). Carcinoembryonic antigen (CEA) inhibits NK killing via interaction with CEA-related cell adhesion molecule 1. Journal of immunology *174*, 6692–6701.
85. Stern-Ginossar, N., Nedvetzki, S., Markel, G., Gazit, R., Betser-Cohen, G., Achdout, H., Aker, M., Blumberg, R.S., Davis, D.M., Appelmelk, B., and Mandelboim, O. (2007). Intercellular transfer of carcinoembryonic antigen from tumor cells to NK cells. Journal of Immunology *179*, 4424–34.
86. Markel, G., Seidman, R., Stern, N., Cohen-Sinai, T., Izhaki, O., Katz, G., Besser, M., Treves, A.J., Blumberg, R.S., Loewenthal, R., et al. (2006). Inhibition of human tumor-infiltrating lymphocyte effector functions by the homophilic carcinoembryonic cell adhesion molecule 1 interactions. Journal of immunology *177*, 6062–6071.
87. Kim, S., Iizuka, K., Aguila, H.L., Weissman, I.L., and Yokoyama, W.M. (2000). In vivo natural killer cell activities revealed by natural killer cell-deficient mice. Proceedings of the national academy of sciences of the United States of America *97*, 2731–2736.
88. Kim, S., Song, Y.J., Higuchi, D.A., Kang, H.P., Pratt, J.R., Yang, L., Hong, C.M., Poursine-Laurent, J., Iizuka, K., French, A.R., et al. (2006). Arrested natural killer cell development associated with transgene insertion into the Atf2 locus. Blood *107*, 1024–1030.

89. Orange, J.S. (2006). Human natural killer cell deficiencies. Current opinion in allergy and clinical immunology *6*, 399–409.
90. Orange, J.S. and Ballas, Z.K. (2006). Natural killer cells in human health and disease. Clinical Immunology *118*, 1–10.
91. Jeong, W.I., Park, O., Radaeva, S., and Gao, B. (2006). STAT1 inhibits liver fibrosis in mice by inhibiting stellate cell proliferation and stimulating NK cell cytotoxicity. Hepatology *44*, 1441–1451.
92. Kottilil, S., Shin, K., Jackson, J.O., Reitano, K.N., O'Shea, M.A., Yang, J., Hallahan, C.W., Lempicki, R., Arthos, J., and Fauci, A.S. (2006). Innate immune dysfunction in HIV infection: effect of HIV envelope-NK cell interactions. Journal of immunology *176*, 1107–1114.
93. Lucia, B., Jennings, C., Cauda, R., Ortona, L., and Landay, A.L. (1995). Evidence of a selective depletion of a CD16+ CD56+ CD8+ natural killer cell subset during HIV infection. Cytometry *22*, 10–15.
94. Richards, J.O., Chang, X., Blaser, B.W., Caligiuri, M.A., Zheng, P., and Liu, Y. (2006). Tumor growth impedes natural-killer-cell maturation in the bone marrow. Blood *108*, 246–252.
95. Kashii, Y., Giorda, R., Herberman, R.B., Whiteside, T.L., and Vujanovic, N.L. (1999). Constitutive expression and role of the TNF family ligands in apoptotic killing of tumor cells by human NK cells. Journal of immunology *163*, 5358–5366.
96. Yuen, M.F., Hughes, R.D., Heneghan, M.A., Langley, P.G., and Norris, S. (2001). Expression of Fas antigen (CD95) in peripheral blood lymphocytes and in liver-infiltrating, cytotoxic lymphocytes in patients with hepatocellular carcinoma. Cancer *92*, 2136–2141.
97. Kim, R., Emi, M., Tanabe, K., Uchida, Y., and Toge, T. (2004). The role of Fas ligand and transforming growth factor beta in tumor progression: molecular mechanisms of immune privilege via Fas-mediated apoptosis and potential targets for cancer therapy. Cancer *100*, 2281–2291.
98. Zalcenstein, A., Stambolsky, P., Weisz, L., Muller, M., Wallach, D., Goncharov, T.M., Krammer, P.H., Rotter, V., and Oren, M. (2003). Mutant p53 gain of function: repression of CD95(Fas/APO-1) gene expression by tumor-associated p53 mutants. Oncogene *22*, 5667–5676.
99. Ivanov, V.N., Krasilnikov, M., and Ronai, Z. (2002). Regulation of Fas expression by STAT3 and c-Jun is mediated by phosphatidylinositol 3-kinase-AKT signaling. The Journal of biological chemistry *277*, 4932–4944.
100. Pikarsky, E., and Ben-Neriah, Y. (2006). NF-kappaB inhibition: a double-edged sword in cancer? European journal of cancer *42*, 779–784.
101. Gray-Schopfer, V.C., Karasarides, M., Hayward, R., and Marais, R. (2007). Tumor necrosis factor-alpha blocks apoptosis in melanoma cells when BRAF signaling is inhibited. Cancer research *67*, 122–129.
102. Vaupel, P., Kallinowski, F., and Okunieff, P. (1989). Blood flow, oxygen and nutrient supply, and metabolic microenvironment of human tumors: a review. Cancer research *49*, 6449–6465.
103. Lardner, A. (2001). The effects of extracellular pH on immune function. Journal of leukocyte biology *69*, 522–530.
104. Hellstrand, K. (2003). Melanoma immunotherapy: a battle against radicals? Trends in immunology *24*, 232–233; author reply 234.
105. Nakagomi, H., Petersson, M., Magnusson, I., Juhlin, C., Matsuda, M., Mellstedt, H., Taupin, J.L., Vivier, E., Anderson, P., and Kiessling, R. (1993). Decreased expression of the signal-transducing zeta chains in tumor-infiltrating T-cells and NK cells of patients with colorectal carcinoma. Cancer research *53*, 5610–5612.
106. Fink, T., Ebbesen, P., Koppelhus, U., and Zachar, V. (2003). Natural killer cell-mediated basal and interferon-enhanced cytotoxicity against liver cancer cells is significantly impaired under in vivo oxygen conditions. Scandinavian journal of immunology *58*, 607–612.
107. Hanna, J., Goldman-Wohl, D., Hamani, Y., Avraham, I., Greenfield, C., Natanson-Yaron, S., Prus, D., Cohen-Daniel, L., Arnon, T.I., Manaster, I., et al. (2006). Decidual NK cells regu-

late key developmental processes at the human fetal-maternal interface. Nature medicine *12*, 1065–1074.

108. de Visser, K.E., Eichten, A., and Coussens, L.M. (2006). Paradoxical roles of the immune system during cancer development. Nature reviews cancer *6*, 24–37.
109. Vitale, C., Chiossone, L., Cantoni, C., Morreale, G., Cottalasso, F., Moretti, S., Pistorio, A., Haupt, R., Lanino, E., Dini, G., et al. (2004). The corticosteroid-induced inhibitory effect on NK cell function reflects down-regulation and/or dysfunction of triggering receptors involved in natural cytotoxicity. European journal of immunology *34*, 3028–3038.
110. Chiesa, M.D., Carlomagno, S., Frumento, G., Balsamo, M., Cantoni, C., Conte, R., Moretta, L., Moretta, A., and Vitale, M. (2006). The tryptophan catabolite L-kynurenine inhibits the surface expression of NKp46- and NKG2D-activating receptors and regulates NK-cell function. Blood *108*, 4118–4125.
111. Mehrotra, P.T., Donnelly, R.P., Wong, S., Kanegane, H., Geremew, A., Mostowski, H.S., Furuke, K., Siegel, J.P., and Bloom, E.T. (1998). Production of IL-10 by human natural killer cells stimulated with IL-2 and/or IL-12. Journal of immunology *160*, 2637–2644.
112. Moreau, P., Adrian-Cabestre, F., Menier, C., Guiard, V., Gourand, L., Dausset, J., Carosella, E.D., and Paul, P. (1999). IL-10 selectively induces HLA-G expression in human trophoblasts and monocytes. International immunology *11*, 803–811.
113. Gonen-Gross, T., Achdout, H., Gazit, R., Hanna, J., Mizrahi, S., Markel, G., Goldman-Wohl, D., Yagel, S., Horejsi, V., Levy, O., et al. (2003). Complexes of HLA-G protein on the cell surface are important for leukocyte Ig-like receptor-1 function. Journal of immunology *171*, 1343–1351.
114. Fiorentino, D.F., Zlotnik, A., Vieira, P., Mosmann, T.R., Howard, M., Moore, K.W., and O'Garra, A. (1991). IL-10 acts on the antigen-presenting cell to inhibit cytokine production by Th1 cells. Journal of immunology *146*, 3444–3451.
115. Mocellin, S., Ohnmacht, G.A., Wang, E., and Marincola, F.M. (2001). Kinetics of cytokine expression in melanoma metastases classifies immune responsiveness. International journal of cancer *93*, 236–242.
116. Cai, G., Kastelein, R.A., and Hunter, C.A. (1999). IL-10 enhances NK cell proliferation, cytotoxicity and production of IFN-gamma when combined with IL-18. European journal of immunology *29*, 2658–2665.
117. Lauw, F.N., Pajkrt, D., Hack, C.E., Kurimoto, M., van Deventer, S.J., and van der Poll, T. (2000). Proinflammatory effects of IL-10 during human endotoxemia. Journal of immunology *165*, 2783–2789.
118. Salazar-Onfray, F., Petersson, M., Franksson, L., Matsuda, M., Blankenstein, T., Karre, K., and Kiessling, R. (1995). IL-10 converts mouse lymphoma cells to a CTL-resistant, NK-sensitive phenotype with low but peptide-inducible MHC class I expression. Journal of immunology *154*, 6291–6298.
119. Huang, S., Ullrich, S.E., and Bar-Eli, M. (1999). Regulation of tumor growth and metastasis by interleukin-10: the melanoma experience. Journal of interferon & cytokine research *19*, 697–703.
120. Kundu, N., Beaty, T.L., Jackson, M.J., and Fulton, A.M. (1996). Antimetastatic and antitumor activities of interleukin 10 in a murine model of breast cancer. journal of the national cancer institute *88*, 536–541.
121. Zheng, L.M., Ojcius, D.M., Garaud, F., Roth, C., Maxwell, E., Li, Z., Rong, H., Chen, J., Wang, X.Y., Catino, J.J., and King, I. (1996). Interleukin-10 inhibits tumor metastasis through an NK cell-dependent mechanism. The journal of experimental medicine *184*, 579–584.
122. Li, M.O., Wan, Y.Y., Sanjabi, S., Robertson, A.K., and Flavell, R.A. (2006). Transforming growth factor-beta regulation of immune responses. Annual review of immunology *24*, 99–146.
123. Bellone, G., Aste-Amezaga, M., Trinchieri, G., and Rodeck, U. (1995). Regulation of NK cell functions by TGF-beta 1. Journal of immunology *155*, 1066–1073.

124. Ortaldo, J.R., Mason, A.T., O'Shea, J.J., Smyth, M.J., Falk, L.A., Kennedy, I.C., Longo, D.L., and Ruscetti, F.W. (1991). Mechanistic studies of transforming growth factor-beta inhibition of IL-2-dependent activation of CD3- large granular lymphocyte functions. Regulation of IL-2R beta (p75) signal transduction. Journal of immunology *146*, 3791–3798.
125. Rook, A.H., Kehrl, J.H., Wakefield, L.M., Roberts, A.B., Sporn, M.B., Burlington, D.B., Lane, H.C., and Fauci, A.S. (1986). Effects of transforming growth factor beta on the functions of natural killer cells: depressed cytolytic activity and blunting of interferon responsiveness. Journal of immunology *136*, 3916–3920.
126. Torre-Amione, G., Beauchamp, R.D., Koeppen, H., Park, B.H., Schreiber, H., Moses, H.L., and Rowley, D.A. (1990). A highly immunogenic tumor transfected with a murine transforming growth factor type beta 1 cDNA escapes immune surveillance. Proceedings of the national academy of sciences of the United States of America *87*, 1486–1490.
127. Laouar, Y., Sutterwala, F.S., Gorelik, L., and Flavell, R.A. (2005). Transforming growth factor-beta controls T helper type 1 cell development through regulation of natural killer cell interferon-gamma. Nature immunology *6*, 600–607.
128. Yu, J., Wei, M., Becknell, B., Trotta, R., Liu, S., Boyd, Z., Jaung, M.S., Blaser, B.W., Sun, J., Benson, D.M., Jr., et al. (2006). Pro- and antiinflammatory cytokine signaling: reciprocal antagonism regulates interferon-gamma production by human natural killer cells. Immunity *24*, 575–590.
129. Townsend, M.J., Weinmann, A.S., Matsuda, J.L., Salomon, R., Farnham, P.J., Biron, C.A., Gapin, L., and Glimcher, L.H. (2004). T-bet regulates the terminal maturation and homeostasis of NK and Valpha14i NKT cells. Immunity *20*, 477–494.
130. Castriconi, R., Cantoni, C., Della Chiesa, M., Vitale, M., Marcenaro, E., Conte, R., Biassoni, R., Bottino, C., Moretta, L., and Moretta, A. (2003). Transforming growth factor beta 1 inhibits expression of NKp30 and NKG2D receptors: consequences for the NK-mediated killing of dendritic cells. Proceedings of the national academy of sciences of the United States of America *100*, 4120–4125.
131. Fernandez, N.C., Lozier, A., Flament, C., Ricciardi-Castagnoli, P., Bellet, D., Suter, M., Perricaudet, M., Tursz, T., Maraskovsky, E., and Zitvogel, L. (1999). Dendritic cells directly trigger NK cell functions: cross-talk relevant in innate anti-tumor immune responses in vivo. Nature medicine *5*, 405–411.
132. Borg, C., Terme, M., Taieb, J., Menard, C., Flament, C., Robert, C., Maruyama, K., Wakasugi, H., Angevin, E., Thielemans, K., et al. (2004). Novel mode of action of c-kit tyrosine kinase inhibitors leading to NK cell-dependent antitumor effects. The journal of clinical investigation *114*, 379–388.
133. Mailliard, R.B., Alber, S.M., Shen, H., Watkins, S.C., Kirkwood, J.M., Herberman, R.B., and Kalinski, P. (2005). IL-18-induced CD83+ CCR7+ NK helper cells. The journal of experimental medicine *202*, 941–953.
134. Carbone, E., Terrazzano, G., Ruggiero, G., Zanzi, D., Ottaiano, A., Manzo, C., Karre, K., and Zappacosta, S. (1999). Recognition of autologous dendritic cells by human NK cells. European journal of immunology *29*, 4022–4029.
135. Pende, D., Castriconi, R., Romagnani, P., Spaggiari, G.M., Marcenaro, S., Dondero, A., Lazzeri, E., Lasagni, L., Martini, S., Rivera, P., et al. (2006). Expression of the DNAM-1 ligands, Nectin-2 (CD112) and poliovirus receptor (CD155), on dendritic cells: relevance for natural killer-dendritic cell interaction. Blood *107*, 2030–2036.
136. Spaggiari, G.M., Carosio, R., Pende, D., Marcenaro, S., Rivera, P., Zocchi, M.R., Moretta, L., and Poggi, A. (2001). NK cell-mediated lysis of autologous antigen-presenting cells is triggered by the engagement of the phosphatidylinositol 3-kinase upon ligation of the natural cytotoxicity receptors NKp30 and NKp46. European journal of immunology *31*, 1656–1665.
137. Fauriat, C., Moretta, A., Olive, D., and Costello, R.T. (2005). Defective killing of dendritic cells by autologous natural killer cells from acute myeloid leukemia patients. Blood *106*, 2186–2188.
138. Della Chiesa, M., Vitale, M., Carlomagno, S., Ferlazzo, G., Moretta, L., and Moretta, A. (2003). The natural killer cell-mediated killing of autologous dendritic cells is confined to a

cell subset expressing CD94/NKG2A, but lacking inhibitory killer Ig-like receptors. European journal of immunology *33*, 1657–1666.
139. Moretta, L., Ferlazzo, G., Bottino, C., Vitale, M., Pende, D., Mingari, M.C., and Moretta, A. (2006). Effector and regulatory events during natural killer-dendritic cell interactions. Immunological reviews *214*, 219–228.
140. Marcenaro, E., Della Chiesa, M., Bellora, F., Parolini, S., Millo, R., Moretta, L., and Moretta, A. (2005). IL-12 or IL-4 prime human NK cells to mediate functionally divergent interactions with dendritic cells or tumors. Journal of immunology *174*, 3992–3998.
141. Vitale, M., Della Chiesa, M., Carlomagno, S., Pende, D., Arico, M., Moretta, L., and Moretta, A. (2005). NK-dependent DC maturation is mediated by TNFalpha and IFNgamma released upon engagement of the NKp30 triggering receptor. Blood *106*, 566–571.
142. Vitale, M., Della Chiesa, M., Carlomagno, S., Romagnani, C., Thiel, A., Moretta, L., and Moretta, A. (2004). The small subset of CD56brightCD16- natural killer cells is selectively responsible for both cell proliferation and interferon-gamma production upon interaction with dendritic cells. European journal of immunology *34*, 1715–1722.
143. Adam, C., King, S., Allgeier, T., Braumuller, H., Luking, C., Mysliwietz, J., Kriegeskorte, A., Busch, D.H., Rocken, M., and Mocikat, R. (2005). DC-NK cell cross talk as a novel CD4+ T-cell-independent pathway for antitumor CTL induction. Blood *106*, 338–344.
144. Onizuka, S., Tawara, I., Shimizu, J., Sakaguchi, S., Fujita, T., and Nakayama, E. (1999). Tumor rejection by in vivo administration of anti-CD25 (interleukin-2 receptor alpha) monoclonal antibody. Cancer research *59*, 3128–3133.
145. Shimizu, J., Yamazaki, S., and Sakaguchi, S. (1999). Induction of tumor immunity by removing CD25+ CD4+ T cells: a common basis between tumor immunity and autoimmunity. Journl of immunology *163*, 5211–5218.
146. Ghiringhelli, F., Puig, P.E., Roux, S., Parcellier, A., Schmitt, E., Solary, E., Kroemer, G., Martin, F., Chauffert, B., and Zitvogel, L. (2005). Tumor cells convert immature myeloid dendritic cells into TGF-beta-secreting cells inducing CD4+ CD25+ regulatory T cell proliferation. The Journal of experimental medicine *202*, 919–929.
147. Nishikawa, H., Kato, T., Tawara, I., Takemitsu, T., Saito, K., Wang, L., Ikarashi, Y., Wakasugi, H., Nakayama, T., Taniguchi, M., et al. (2005). Accelerated chemically induced tumor development mediated by CD4+ CD25+ regulatory T cells in wild-type hosts. Proceedings of the national academy of sciences of the United States of America *102*, 9253–9257.
148. Trzonkowski, P., Mysliwska, J., Szmit, E., Wieckiewicz, J., Lukaszuk, K., Brydak, L.B., Machala, M., and Mysliwski, A. (2003). Association between cytomegalovirus infection, enhanced proinflammatory response and low level of anti-hemagglutinins during the anti-influenza vaccination–an impact of immunosenescence. Vaccine *21*, 3826–3836.
149. Wolf, A.M., Wolf, D., Steurer, M., Gastl, G., Gunsilius, E., and Grubeck-Loebenstein, B. (2003). Increase of regulatory T cells in the peripheral blood of cancer patients. Clinical cancer research *9*, 606–612.
150. Ghiringhelli, F., Menard, C., Terme, M., Flament, C., Taieb, J., Chaput, N., Puig, P.E., Novault, S., Escudier, B., Vivier, E., et al. (2005). CD4+ CD25+ regulatory T cells inhibit natural killer cell functions in a transforming growth factor-beta-dependent manner. The journal of experimental medicine *202*, 1075–1085.
151. Morse, M.A., Garst, J., Osada, T., Khan, S., Hobeika, A., Clay, T.M., Valente, N., Shreeniwas, R., Sutton, M.A., Delcayre, A., et al. (2005). A phase I study of dexosome immunotherapy in patients with advanced non-small cell lung cancer. Journal of translational medicine *3*, 9.
152. Smyth, M.J., Teng, M.W., Swann, J., Kyparissoudis, K., Godfrey, D.I., and Hayakawa, Y. (2006). CD4+ CD25+ T regulatory cells suppress NK cell-mediated immunotherapy of cancer. Journal of immunology *176*, 1582–1587.
153. Barao, I., Hanash, A.M., Hallett, W., Welniak, L.A., Sun, K., Redelman, D., Blazar, B.R., Levy, R.B., and Murphy, W.J. (2006). Suppression of natural killer cell-mediated bone marrow cell rejection by CD4+ CD25+ regulatory T cells. Proceedings of the national academy of sciences of the United States of America *103*, 5460–5465.

154. Oida, T., Zhang, X., Goto, M., Hachimura, S., Totsuka, M., Kaminogawa, S., and Weiner, H.L. (2003). CD4+ CD25- T cells that express latency-associated peptide on the surface suppress CD4+ CD45RBhigh-induced colitis by a TGF-beta-dependent mechanism. Journal of immunology *170*, 2516–2522.
155. Liu, C., Yu, S., Kappes, J., Wang, J., Grizzle, W.E., Zinn, K.R., and Zhang, H.G. (2007). Expansion of spleen myeloid suppressor cells represses NK cell cytotoxicity in tumor bearing host. Blood.
156. Gorelik, E., Bere, W.W., and Herberman, R.B. (1984). Role of NK cells in the antimetastatic effect of anticoagulant drugs. International journal of cancer *33*, 87–94.
157. Rickles, F.R., and Edwards, R.L. (1983). Activation of blood coagulation in cancer: Trousseau's syndrome revisited. Blood *62*, 14–31.
158. Gunji, Y. and Gorelik, E. (1988). Role of fibrin coagulation in protection of murine tumor cells from destruction by cytotoxic cells. Cancer research *48*, 5216–5221.
159. Nierodzik, M.L., Plotkin, A., Kajumo, F., and Karpatkin, S. (1991). Thrombin stimulates tumor-platelet adhesion in vitro and metastasis in vivo. The journal of clinical investigation *87*, 229–236.

Chapter 10
Contribution of the Microenvironment to the Pathogenesis of EBV-Positive Hodgkin and Nasal NK/T-cell Lymphomas

Eva Klein, Miki Takahara*, and Lóránd Levente Kis

Abstract Depending on the differentiation and maturation of the EBV-carrying cell the virally encoded proteins can be expressed in various assortments. The expression of these proteins determines the fate of the EBV-harboring cell. Expression of the six nuclear and three membrane-associated EBV-encoded proteins, designated Type III latency, induces cell proliferation. This occurs only in B lymphocytes. In spite of the transforming capacity of EBV, humans are virus carriers without the emergence of B-cell malignancy, because the immune system recognizes and eliminates the cells which express the growth-promoting proteins. However, immunosuppressed individuals have a high risk for EBV-induced B-cell malignancies.

In EBV-associated malignancies of other cell types, such as Hodgkin and NK/T-cell lymphomas, the expression of virally proteins is restricted. EBV does not induce the autonomous proliferation of the cells. However, it induces phenotypical changes that influence the fate of the cell. The virally encoded genes function in evasion of apoptosis, induction of cellular interactions, production and response to cytokines. Contribution of the microenvironment to the survival and proliferation of the EBV-carrying malignant cells is thus pivotal in these tumors.

Keywords Epstein-Barr virus, Hodgkin lymphoma, NK/T-cell lymphoma, microenvironment, cytokine

10.1 Background

The evolution of the complex interaction between man and EBV originated with the primate ancestors. During the long history of the coexistence it evolved to be largely harmless, that is mainly determined by the variability of viral gene expression

Department of Microbiology, Tumor and Cell Biology (MTC) Karolinska Institutet, Stockholm, 17177, Sweden regulated by the differentiation of the host cell. Today, man and all Old, but not New World, primates harbor EBV-like viruses [1]. After the primary infection humans become virus carriers for lifetime and develop immunity against the virally encoded antigens. The maintenance of the virus in face of the immune response is secured by viral programs in which the immunogenic viral proteins are not expressed.

EBV shows a high degree of B-cell tropism. It binds to a B lymphocyte-specific surface molecule, CD21 (receptor for the C3d fragment of complement). Binding of the viral envelop to the receptor induces cell activation. The infected cells can enter into the mitotic cycle and maintain proliferating capacity [1–3]. The EBV generated B-cell lines are referred to as lymphoblastoid cell lines (LCL). In these cells the viral genome is maintained in episomal form and expresses nine proteins. The growth transformation is accompanied by phenotypic changes in the cells, including the expression of co-stimulatory cell surface molecules that are pivotal for their recognition by the immune system. Consequently, unless their immune response is compromised, healthy individuals are saved from the emergence of EBV-induced malignancies.

The harmless maintenance of this transforming virus in humans is ensured on one hand by the modulation of the viral gene expression, allowing maintenance of the viral genome in B-cells without imposition of immunogenicity; on the other hand by the strong immunogenicity of the cells that express the growth promoting EBV-encoded proteins and are therefore eliminated [4].

Primary infection occurs usually in adolescents. In about half of the individuals it is followed by infectious mononucleosis (IM), a benign disease with highly variable severity and symptoms. The symptoms reflect the presence of activated B-cells and the developing immune response.

Infection of B-cells *in vitro* has revealed the molecular details of the complexity of virus–host cell interaction and the immunological recognition of the latently infected B-cells. In contrast to B lymphocytes, infection of epithelial-, T-, and NK-cells *in vitro* can be achieved only through manipulation of the virus or the host cell, such as using virus constructs with introduced selection markers and target cells with inserted viral receptors. However, the development of sensitive assays has led to the discovery of EBV-positive malignancies originating in T and NK lymphocytes, epithelial and mesenchymal cells indicating that these cell lineages are occasionally infected *in vivo*.

The EBV-harboring malignant B-cell lymphomas differ considerably from the non-B malignancies. The resident viral genome can induce proliferation autonomously only in the former, while additional factors, among them ones contributed by the microenvironment, contribute to the latter.

For the discussion of the role of EBV in the pathogenesis of the various lymphomas and of the contribution of the microenvironment it is necessary to consider the various strategies of EBV gene expression and their impact on the target cell.

10.2 Expression of EBV-encoded Proteins in Lymphocytes with Latent Infection

The EBV-encoded genes expressed in latent infection were characterized in the *in vitro* generated lymphoblastoid cell lines, LCLs, and the pattern was designated as Type III or growth program [5, 6]. In these cells the EBV genomes reside as covalently closed circles, episomes. Only rare cells in the LCLs produce virus. The expression of six virally encoded proteins localized in the nucleus, EBNAs, is regulated by one of the alternative viral promoters, W and C [1]. The spliced products of a giant message are translated to EBNA-1 to - 6 or EBNA -1, -2, -3a, -3b, -LP, -3c. In addition the virus encodes three cell membrane-associated proteins (LMP-1, LMP-2a, and LMP-2b), and two non-translated small RNAs, EBERs. For the induction and maintenance of the transformed phenotype the EBV-encoded proteins interact with each other and with cellular transcription factors and co-activators. The expression of EBNA-2 and LMP-1 are essential and play a pivotal role with multiple functions in the establishment of the proliferating phenotype [7, 8].

LMP-1 induces conspicuous phenotypical changes, expression of several adhesion and activation related molecules and by this it imposes strong immunogenicity of the Type III cell. Consequently such B lymphocytes survive and proliferate only in patients with impaired immune functions, such as transplant recipients who receive immunosuppressive treatments [9]. These patients have therefore a high risk for EBV-driven lymphoproliferations (post-transplant lymphoproliferative disorders, PTLD).

EBV was discovered in a B-cell lymphoma, Burkitt lymphoma, BL [10]. The transforming capacity of the virus for B lymphocytes to yield LCLs *in vitro* [2] was regarded as the *in vitro* correlate for the generation of this malignancy. However, it was soon shown that BL cells and LCLs differ [3] in that the former resemble resting cells, whereas the latter are similar to activated lymphoblasts, and the two cell populations are driven to proliferate by different, unrelated mechanisms. While the proliferation of the LCL is regulated by the EBV-encoded proteins, the proliferation of the BL cell is independent of EBV. In the BL cells constitutive activation of the myc gene, deregulated by the Ig/myc translocation (occurring also in EBV-negative BLs,), is the driving force.

Among the EBV-encoded proteins that were discovered in the LCLs only one, the EBNA-1, is present in the BL cells [11]. This viral expression pattern is designated as Type I. This program has two important characteristics. It does not induce proliferation and it does not change the phenotype of the cells. Due to the absence of activation makers the cells are not "seen" by the EBV-specific immunity. This EBV program is exhibited by the virus genome carrier normal B-cells in the memory compartment [12, 13]. They secure the maintenance of EBV in humans firstly through the undisturbed presence of the viral genome carrying cell, and secondly by not endangering the life of the virus carrying individual.

Table 10.1 Variation of EBV gene expression in EBV-associated lymphoid malignancies

Latency designation[a]	EBV-encoded proteins[b]	Malignancies
I	EBNA-1	Burkitt lymphoma
IIa	EBNA-1, LMP-1, LMP-2	Hodgkin lymphoma and nasal NK/T-cell lymphoma
IIb	EBNA-1 to EBNA-6	PTLD
III	EBNA-1 to EBNA-6 and LMP-1, LMP-2	PTLD and lymphoma in AIDS

[a]cells with these latency types have been detected in lymphoid tissues of patients with IM
[b]two non-coding RNAs, EBER1–2, are expressed in all EBV-carrying cells

The common characteristic of EBV-carrying malignancies of the hemopoetic system, originating from cells other than the mature B lymphocyte, such as the Hodgkin-, T-, and NK-lymphomas, is the expression of EBNA-1 and LMP-1 [14, 15]. Due to the lack of B lymphocyte-specific transcription factors, the C and W-promoters are inactive and EBNA-2 (along with EBNA-3, -4, -5, -6) is not expressed in these malignancies. This viral program does not induce cell proliferation autonomously because this requires expression of EBNA-2. This restricted program was first seen in the epithelial cells of nasopharyngeal carcinoma (NPC) [16], and it was designated later as Type II EBV latency [5]. We denoted it as Type IIa in order to distinguish it from a differently restricted program, Type IIb, that expresses the nuclear proteins, EBNAs but not LMP-1 [17]. The latter was first seen in B-CLL cells infected *in vitro* [18, 19].

On the basis of the behavior of cells the Type IIa and Type IIb restricted EBV gene expression represent a group in that they lack one or the other proteins pivotal for transformation: EBNA-2 in the former and LMP-1 in the latter [17]. They differ in the regulation of EBV gene expression and in the characteristics of the cells. Their common property is however, that cell survival and proliferation can lead to malignancies with contribution of factors provided by the microenvironment.

The list of the known EBV expression patterns in lymphoid malignancies is given in Table 10.1. All of them can be found in the lymphoid tissue of IM patients.

10.3 The Changes of Viral Expression Concomitant with B-cell Maturation

In healthy carriers EBV resides in the memory B-cell compartment, in which the growth program of the virus is not expressed [12, 13]. The origin of these cells is not yet clarified. The *in vitro* B-cell transformation system has emphasized the establishment of type III pattern upon infection. This is partly due to the growth capacity of Type III cells that readily proliferate in the infected culture, overgrowing thus other EBV-carrier cells. It was proposed that the infection event leads to

Type III protein expression and the various virus–cell phenotypes may represent successive changes with switches occurring concomitantly with maturation stages of the B lymphocyte [6]. However the viral expression in the latent infection of B lymphocytes can be determined directly by the differentiation of the target B-cell at the occasion of infection as well.

According to the model that employs changes in the expression program of the resident viral genome, naïve B-cells are infected with EBV and than they express the type III pattern [6]. In healthy individuals such cells are eliminated by the immune response. Escaping cells may enter and differentiate in the germinal centers (GC) of secondary follicles and switch to type I viral expression. We proposed that the expression of LMP-1 in the EBV-carrying GC B-cells is induced by cytokines [20, 21] and is not a consequence of cell differentiation, as proposed by Babcock et al. [22]. When these cells leave the GC, the EBV-encoded proteins are turned off. In the resting memory B-cells the viral genome is retained, but only the non-coding RNAs, EBERs, are detectable, unless the cells become stimulated and enter the mitotic cycle, when EBNA-1 is also expressed [23].

Indeed, cell phenotype-associated changes in the expression of the resident EBV genome have been shown in LCLs, in BL-derived lines and in somatic cell hybrids [11, 24]. Our results with B-CLL cells infected *in vitro* indicates however that the property of the target cells at the occasion of the infection can also determine the expression of the viral genes [19, 25].

10.4 EBV-carrying Hematopoetic Malignancies with Type IIa Restricted Viral Gene Expression

The EBV-carrying hematopoetic malignancies with Type IIa restricted viral gene expression, Hodgkin- and nasal NK/T-lymphomas express LMP-1 [14, 15, 26]. This molecule has a multitude of functions and interacts with viral and cellular genes. Accordingly its role is different in the two tumor types. Their shared characteristics are that LMP-1 is instrumental in cooperation with the normal tissue environment. However, while the presence of EBV is not obligatory to the development of the HL pathology, all nasal NK lymphomas carry EBV.

LMP-1 is an integral membrane protein composed of a short cytoplasmic N-terminal, six transmembrane domains and a cytoplasmic C-terminal tail. By self aggregating and engaging signalling intermediates of the TNF-receptors (so called TRAFs) it acts like a constitutively active receptor independently of ligand binding. Through the interaction of its C-terminal cytoplasmic tail with the signaling intermediates TRAFs and TRADD, LMP-1 activates the NF-kappa B (both canonical and non-canonical), the p38/MAPK, the JNK, and the phosphatidylinositol 3 kinase (PI3-K) pathways that in turn results in the expression of multiple cellular genes. These are members of the anti-apoptotic family (A20, Bfl-1, Mcl-1, c-IAP2), cytokines (IL-6, IL-8, IL-10, IL-18), cell surface activation or adhesion molecules

(ICAM-1, SLAM, CD23, CD25, CD40, EGFR) and many others. *In vitro* overexpression studies have shown that LMP-1 has oncogenic effects in non-lymphoid cells. When expressed in rodent fibroblast cell lines, LMP-1 induced cellular transformation. Transgenic expression of LMP-1 in the B-cells of CD40-deficient mice has provided *in vivo* proof that LMP-1 mimics partially CD40 signals, since it could induce extrafollicular B-cell differentiation, but not germinal center formation.

The contribution of EBV in the development of HL is indicated by the elevated risk for EBV-positive cases in children with a history of IM [27]. However, the characteristic molecular mechanisms were shown to be shared by the virus positive and negative HL. Significance has been attributed to the activation of the NF-kappa B pathway that occurs in both, though induced by different mechanisms.

For the mechanism of HL development, disturbance of B-cell differentiation is proposed to be decisive [28]. According to their scheme, in the process of somatic hypermutation the faulty B-cells that arise in the germinal center are the progenitor of H-RS cells. While the correctly differentiated cells exit the GC, these "crippled" surface Ig-negative B lymphocytes are eliminated by apoptosis [28]. They may be rescued however by EBV infection for which the functions of LMP-1 and LMP-2a are held responsible.

10.4.1 Hodgkin lymphoma, HL

Depending on the geography and histological type the frequency of EBV-positive HLs vary. The classical HL tissue contains only a few (about 1–10%) Hodgkin-Reed Sternberg (H-RS) cells [29]. Whether these H-RS cells carry or not the viral genome, the lymphoma tissues are made up by T-, B-cells, macrophages, eosinophils, and plasma cells, thus they have an inflammatory character. The EBV gene expression pattern in the H-RS cells is regularly Type IIa with abundant LMP-1. While commonly the H-RS cells belong to the B-lymphoid lineage [30], they do not express all B-cell-specific genes, such as transcription factors and B-cell-associated markers, like CD19, CD20 and immunoglobulins [31, 32]. Detailed analysis of the gene expression profile of the HL lines combined with immunohistochemical analysis of tumor samples showed considerable deviations from the B-lymphocyte pattern, showing that the differentiation program of the H-RS cell is impaired. Since the regular EBV latent program in B lymphocytes is Type III or Type I, this reflects also that the cells deviate from normal B-cells. EBNA-2 expression requires B-cell-specific transcription factors such as Oct-2 and BSAP (PAX-5) [33, 34]. In their absence the Wp and Cp promoters that regulate EBNA-1, -2, -3, -4, -5, -6 expression during type III latency is inactive.

HL has been studied in great detail and its literature is abundant. As mentioned above the characteristics of the HL tissues are similar in the EBV-negative and -positive cases. Special properties that could be ascribed to the presence of the virus in the H-RS cells have not been discovered yet; however some studies concluded that the prognosis of EBV-positive cases is more favorable for certain age groups and histologic subtypes.

LMP-1 may not only function in the rescue of the faulty differentiated B lymphocytes, but could contribute to the establishment of the granulomatous tissue as well. For the phenotype and survival of the H-RS cell activation of the nuclear factor (NF)-kappa B, that is the hallmark of these cells, is decisive [35]. In the EBV-carrying cases LMP-1, in the EBV-negative cases mutations in the NF-kappa B inhibitors (IkB-alpha) [36], amplification of the REL gene (encodes an NF-kB family member) [37] or signals acting on surface receptors CD30 and CD40 [38, 39], may activate NF-kappa B.

The HL granuloma is a complex interactive environment to which a multitude of cellular and soluble factors contribute. Secretion of the CCL17/ TARC and CCL22/ MDC chemokines by the H-RS cells can attract CCR4-expressing Th2 and regulatory T-cells [40, 41]. The H-RS cells may get survival signals from the infiltrating Th-2 cells, while the regulatory T-cells may protect them from the EBV-specific immune response [42, 43]. Furthermore the H-RS cells secrete immunosuppressive Cytokines, such as TGF-beta and IL-10. Though LMP-1 could impose immunogenicity on the cell, the same molecule released from cells has been shown to be immunosuppressive [44].

The difficulty to establish cell lines from EBV-carrying HL is in line with the complex structure of the lymphoma tissue. In spite of considerable effort, HL-derived cell lines with type IIa EBV latency have not been established [45]. With one exception – the Type III L591 line, that however does not represent the *in vivo* H-RS cell – the few available HL-derived cell lines are all EBV-negative [46]. This shows that the Type IIa EBV expression does not induce cell proliferation, unless additional cellular changes or signals from the microenvironment contribute.

The phenotype of the H-RS cells deviates thus from the mature B lymphocytes. Its particular phenotype and the lack of *in vitro* type IIa latency model motivated our studies for the cell–virus interaction. With the help of a recombinant EBV strain containing the neomycin resistance gene, we established a virus-carrying subline of the EBV-negative KM-H2 line [47]. For maintenance of the virus, the cells had to be carried in the selective medium. Importantly, only the EBNA-1 protein was expressed in these cells, thus they were Type I. Similar results were reported by Baumforth and co-workers [48]. Thus, the *in vitro*-established EBV carrier state of the KM-H2 cells did not correspond to the *in vivo* expression of viral proteins in HRS cells. However, LMP-1 became induced when IL-4 and contact with CD40-ligand (CD40L) were provided to the cells, and this was not accompanied by EBNA-2 expression [47]. We showed thus for the first time that the EBV gene expression pattern can be influenced by cytokines and LMP-1 can be induced in HL-derived cells in the absence of EBNA-2 expression [47]. This condition can prevail *in vivo* because in the lymphogranuloma the H-RS are surrounded by activated CD4-positive T-cells that produce a variety of cytokines and express CD40L [38, 49].

Similarly to the experimentally created KMH2-EBV cells, EBNA-2 could not be induced in the EBV-carrying cell lines derived from body cavity lymphoma (pleural effusion lymphoma, PEL) [50] that have also lost several B-cell characteristics [51].

On the basis of our results showing cytokine-induced expression of LMP-1 we modify the scheme proposed by Thorley-Lawson and discussed above. We introduce the contribution of the microenvironment for the generation of Type IIa cells, in that we propose that LMP-1 expression is induced by extracellular signals in the GC, provided by cytokines and cell contacts. (Fig. 10.1). This is in line also with the observations that EBV-positive memory B-cells isolated from tonsils express

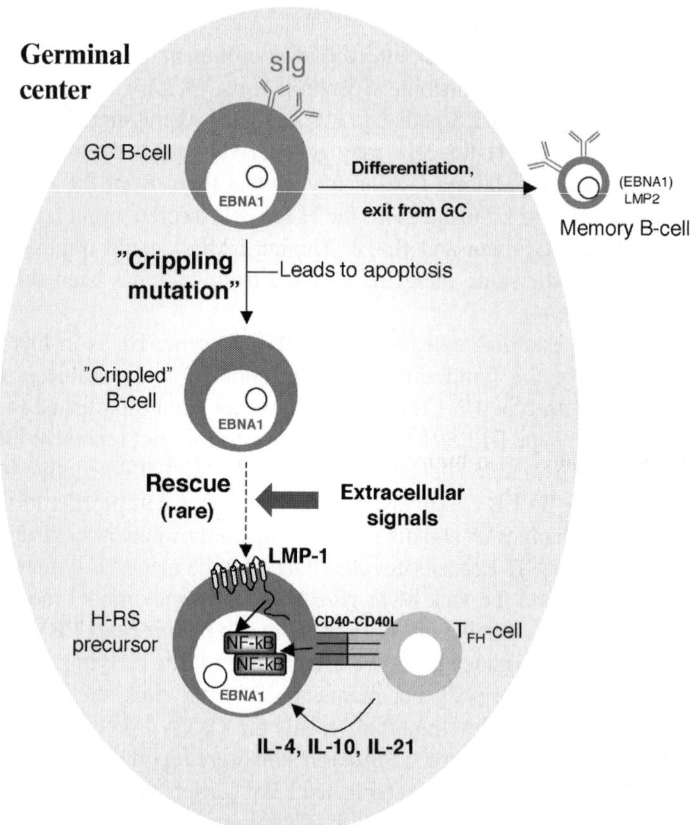

Fig. 10.1 Development of the EBV-positive H-RS cell in the germinal center, emphasizing the development of Type IIa EBV latency under the influence of the microenvironment. EBV carrier normal B-cells in GC express the Type I viral pattern. When they leave the GC, the virally encoded proteins are turned off. During the process of somatic hypermutation occasional cells acquire non-functional mutations, they do not express sIg, and succumb to apoptosis. EBV can rescue these cells. Signals provided by T-cells through the CD40L and cytokines, particularly IL-4, IL-10 and IL-21, induce the expression of LMP-1. Interaction of these cells with the environment, cell–cell contacts and cytokines leads to their survival and entry to the cell cycle. Constitutive activation of NF-kB, induced by LMP-1 has a key role in the malignant property of the H-RS cell. The circles in the nucleus of the cells represent the viral episome. sIg denotes surface immunoglobulin

LMP-1, while the ones in blood are LMP-1 negative [52]. Upon exit from the GC, the long-lived, resting memory B-cells do not express the viral proteins. The H-RS cells originate from "crippled "B-cells that are normally eliminated [28]. When the cells acquire EBV infection, their fate changes. LMP-1 (and LMP-2a) is induced and they escape apoptosis. Occasional cells enter in a mutual stimulatory interaction with the normal cellular components of the lymph-node, persist, and can enter the mitotic cycle.

Elimination of the EBV-carrying type III latent B-cells is critical for the infected individual since these cells have a proliferative potential. LMP-1 and LMP-2a expression together with the co-stimulatory molecules and intact antigen-presenting machinery would be expected to elicit an immune response in HL as well. In the HL granuloma, however, the immune response seems to be locally inhibited [53] by immunosuppressive cytokines (IL10, TGF-beta) and by regulatory T cells [42, 43].

10.4.2 Extranodal, Nasal NK/T-cell Lymphoma

This rare and highly malignant lymphoma occurs mainly in Asia. The tumor cells are intermixed with inflammatory cells in a necrotic tissue with obvious vascular damage [54]. These lymphomas are regularly EBV-positive, with some variation in the expression of the virally encoded genes, the tissue contains both Type I and Type II cells, the latter with high variation in the level of LMP-1 [55, 56]. EBV-negative and EBV-positive cell lines have been established, and, except for one, that is Type I, the EBV carrier ones maintained the Type II pattern. In contrast to the *in vivo* tumor the level of LMP-1 in the cell lines is homogeneous, though it differs in the various cell lines. Importantly, these tumor-derived lines with the NK phenotype require IL-2 for *in vitro* proliferation, thus in this respect they resemble normal NK cells. LMP-1 expression is coupled to cell proliferation, it is dependent on the supply of IL-2 [57]. We demonstrated that in absence of IL-2 cells treated with the cytokines IL-10 and IFN-gamma expressed LMP-1, but cell proliferation was not sustained [57]. Thus the LMP-1 protein in the NK tumor cells was not sufficient for induction of proliferation. However it modified the growth potential of the cells indirectly because the requirements for IL-2 were changed in the cytokine-treated cells. Concomitantly with the induction of LMP-1 the expression of CD25, the high affinity component of the IL-2 receptor, increased and this led to lower IL-2 concentration requirement for growth [57]. We concluded therefore that while the EBV-encoded proteins expressed in the NK-lymphoma do not induce proliferation, LMP-1 can contribute to the pathogenesis of the NK-lymphoma, by potentiating the response of the malignant NK-cells to growth-promoting cytokines.

The following sequence of events may thus contribute to the development of NK/T lymphomas. This malignancy is confined to the nasal area, a site where inflammation is frequent. In the inflammatory tissue B, T and NK lymphocytes are activated. For the part of B lymphocytes, the ones carrying the EBV genome may be induced for virus production. Through direct contact, the otherwise rare event, NK-cells may be infected with the virus and the cells exhibit the Type IIa protein pattern (Fig. 10.2) Activated T-cells in the tissue produce IL-2, and IL-10. Activated NK-cells express IL-2 receptors and respond with proliferation. The sensitivity to the effect of IL-2 is potentiated by the influence of IL-10 that is also provided by the T lymphocytes, and by macrophages as well. In addition the tumor cells are induced to produce IL-10 exhibiting thus autocrine upregulation of LMP-1. By concomitantly up-regulating the expression of CD25 (IL-2R-alpha), IL-10 may have a key role in the proliferation of the malignant cells because it may be responsible for efficient exploitation of the available amount of IL-2.

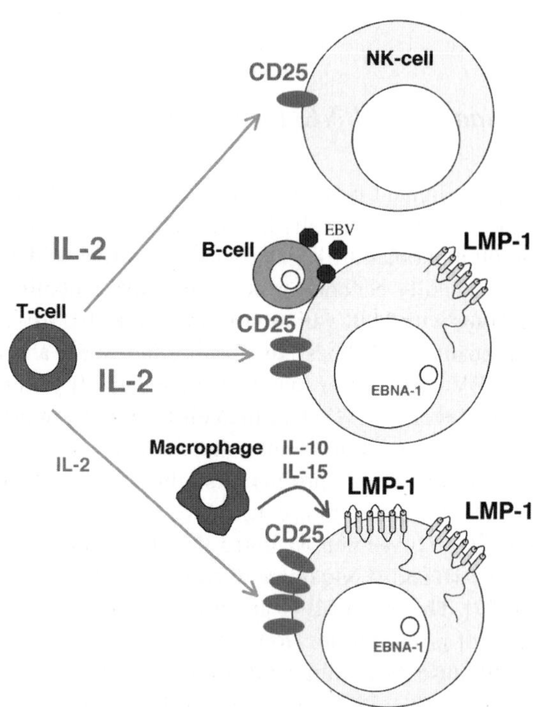

Fig. 10.2 Contributing factors to the development of nasal NK-lymphoma, emphasizing the role of EBV. In the inflammatory tissue NK-, B-, T-cells and macrophages are activated. Occasional EBV-carrying B-cells produce viral particles and infect NK-cells through direct contact. In these, the EBV-encoded protein expression is Type IIa. Activated NK-cells express the IL-2-receptor, consequently they can respond to the growth-promoting stimulus of IL-2 that is produced by T-cells. IL-10 and IL-15 is secreted by macrophages that elevate the level of LMP-1 that may directly increase CD25 expression. Consequently these cells can be stimulated for division by relatively lower level of IL-2. The assignment of cytokines to a particular cell type is not exclusive, e.g. IL-10 can be produced by T-cells and by the tumor cells as well. The circles in the nucleus of the cells represent the viral episome

Table 10.2 The role of EBV in lymphomagenesis

Malignancy	Key factors for in vitro proliferation	The role of EBV
Burkitt lymphoma	Ig-myc translocation	Anti-apoptosis?
Hodgkin lymphoma	Constitutive NF-kappa B activity	Rescue of faulty GC B-cells from apoptosis
Nasal NK/T-cell lymphoma	Exogeneous IL-2	IL-2R-alpha (CD25) upregulation
AIDS-lymphomas and PTLD	EBV-encoded, growth promoting genes	Induction of proliferation

Thus unlike B lymphocytes in which LMP-1 together with other EBV-encoded proteins (particularly EBNA-2) drives cell proliferation, the role of EBV in the NK/T lymphomas is confined to potentiate the sensitivity to growth promoting cytokines received from the components of the inflammatory tissue. Whether LMP-1 is directly involved in the expression of the IL-2R alpha (CD25), or its regulation of expression is concomitant, remains to be seen.

10.5 Conclusion

EBV contributes to tumorigenesis of the cells of the hemopoetic system (Table 10.2). Except for the nasal lymphomas these malignancies have alternative pathogenesis as well, because their association with EBV is not obligatory. In variable proportions, lymphoid proliferations with similar pathological and clinical features occur without the presence of EBV.

All the EBV-encoded proteins that act in concert for growth promotion are expressed only in B lymphocytes. Among these the LMP-1 molecule has a key role. In accordance with the multitude of LMP-1 functions, depending on the type of the infected cell it can act in various ways. The impact of LMP-1 on the cell phenotype encompasses the enhancement of interactions with the microenvironment both through direct cell-contacts and through soluble factors.

Though the cell type of origin differs, Hodgkin and NK/T-cell lymphomas have common characteristics, such as: (1) abundancy of normal cells, mainly those belonging to the immune system, (2) type II expression of the resident EBV genome, (3) important role of the EBV-encoded membrane-associated molecule LMP-1, and (4) mutual interactive signals with the cellular components of the inflammatory tissue that contribute to the survival and proliferation of the malignant cells. While the role of LMP-1 differs in the two malignancies cytokines are involved in the generation and maintenance of both tumors. This implies the possibility for inhibitory interventions with therapeutic potential by neutralization of growth promoting soluble factors and/or blockade of the relevant receptors.

Acknowledgements Supported by funds from the Swedish Cancer Society, Sweden. L.L.K. and M.T. are recipients of Cancer Research Fellowship of Cancer Research Institute (New York)/ Concern Foundation (Los Angeles).

References

1. Rickinson AB, Kieff E (2001). *Epstein-Barr virus*. In: Knipe DM, Howley PM (eds). *Fields virology*, vol. 2, 4th edn. Lippincott Williams and Wilkins: Philadelphia, PA, pp 2575–2628.
2. Pope JH, Horne MK, Scott W (1968). *Transformation of foetal human keukocytes in vitro by filtrates of a human leukaemic cell line containing herpes-like virus*. Int J Cancer 3(6): 857–866.
3. Nilsson K, Ponten J (1975). *Classification and biological nature of established human hematopoietic cell lines*. Int J Cancer 15(2): 321–341.
4. Klein G (1994). *Epstein-Barr virus strategy in normal and neoplastic B cells*. Cell 77(6): 791–793.
5. Rowe M, Lear AL, Croom-Carter D, Davies AH, Rickinson AB (1992). *Three pathways of Epstein-Barr virus gene activation from EBNA1-positive latency in B lymphocytes*. J Virol 66(1): 122–131.
6. Thorley-Lawson DA (2001). *Epstein-Barr virus: exploiting the immune system*. Nat Rev Immunol 1(1):75–82.
7. Hammerschmidt W, Sugden B (1989). *Genetic analysis of immortalizing functions of Epstein-Barr virus in human B lymphocytes*. Nature 340(6232): 393–397.
8. Kaye KM, Izumi KM, Kieff E (1993). *Epstein-Barr virus latent membrane protein 1 is essential for B-lymphocyte growth transformation*. Proc Natl Acad Sci USA 90(19): 9150–9154.
9. Young L, Alfieri C, Hennessy K, Evans H, O'Hara C, Anderson KC, Ritz J, Shapiro RS, Rickinson A, Kieff E, Cohen JI (1989). *Expression of Epstein-Barr virus transformation-associated genes in tissues of patients with EBV lymphoproliferative disease*. N Engl J Med 321(16): 1080–1085.
10. Epstein MA, Achong BG, Barr YM (1964). *Virus particles in cultured lymphoblasts from Burkitt's lymphoma*. Lancet 15: 702–703.
11. Rowe M, Rowe DT, Gregory CD, Young LS, Farrell PJ, Rupani H, Rickinson AB (1987). *Differences in B cell growth phenotype reflect novel patterns of Epstein-Barr virus latent gene expression in Burkitt's lymphoma cells*. EMBO J 6(9): 2743–2751.
12. Chen F, Zou JZ, di Renzo L, Winberg G, Hu LF, Klein E, Klein G, Ernberg I (1995). *A subpopulation of normal B cells latently infected with Epstein-Barr virus resembles Burkitt lymphoma cells in expressing EBNA-1 but not EBNA-2 or LMP1*. J Virol 69(6): 3752–3758.
13. Babcock GJ, Decker LL, Volk M, Thorley-Lawson DA (1998). *EBV persistence in memory B cells in vivo*. Immunity 9(3): 395–404.
14. Deacon EM, Pallesen G, Niedobitek G, Crocker J, Brooks L, Rickinson AB, Young LS (1993). *Epstein-Barr virus and Hodgkin's disease: transcriptional analysis of virus latency in the malignant cells*. J Exp Med 177(2): 339–349.
15. Chiang AK, Tao Q, Srivastava G, Ho FC (1996). *Nasal NK- and T-cell lymphomas share the same type of Epstein-Barr virus latency as nasopharyngeal carcinoma and Hodgkin's disease*. Int J Cancer 68(3): 285–290.
16. Fahraeus R, Fu HL, Ernberg I, Finke J, Rowe M, Klein G, Falk K, Nilsson E, Yadav M, Busson P, Tursz T, Kallin B (1988). *Expression of Epstein-Barr virus-encoded proteins in nasopharyngeal carcinoma*. Int J Cancer 42(3): 329–338.
17. Klein E, Kis LL, Klein G (2007). *Epstein-Barr virus infektion in humans: from harmless to life endangering virus-lymphocyte interactions*. Oncogene, In Press.
18. Doyle MG, Catovsky D, Crawford DH (1993). *Infection of leukaemic B lymphocytes by Epstein Barr virus*. Leukemia 7(11): 1858–1864.
19. Teramoto N, Gogolak P, Nagy N, Maeda A, Kvarnung K, Bjorkholm M, Klein E (2000). *Epstein-Barr virus-infected B-chronic lymphocyte leukemia cells express the virally encoded nuclear proteins but they do not enter the cell cycle*. J Hum Virol 3(3): 125–136.
20. Kis LL, Takahara M, Nagy N, Klein G, Klein E (2006a). *IL-10 can induce the expression of EBV-encoded latent membrane protein-1 (LMP-1) in the absence of EBNA-2 in B lymphocytes and in Burkitt lymphoma- and NK lymphoma-derived cell lines*. Blood 107(7): 2928–2935.
21. Kis LL, Takahara M, Nagy N, Klein G, Klein E (2006b). *Cytokine mediated induction of the major Epstein-Barr virus (EBV)-encoded transforming protein, LMP-1*. Immunol Lett 104(1–2): 83–88.

22. Babcock GJ, Hochberg D, Thorley-Lawson DA (2000). *The expression pattern of Epstein-Barr virus latent genes in vivo is dependent upon the differentiation stage of the infected B cell.* Immunity 13(3): 497–506.
23. Hochberg D, Middeldorp JM, Catalina M, Sullivan JL, Luzuriaga K, Thorley-Lawson DA (2004). *Demonstration of the Burkitt's lymphoma Epstein-Barr virus phenotype in dividing latently infected memory cells in vivo.* Proc Natl Acad Sci USA 101(1): 239–244.
24. Contreras-Brodin BA, Anvret M, Imreh S, Altiok E, Klein G, Masucci MG (1991). *B cell phenotype-dependent expression of the Epstein-Barr virus nuclear antigens EBNA-2 to EBNA-6: studies with somatic cell hybrids.* J Gen Virol 72(Pt 12): 3025–3033.
25. Maeda A, Bandobashi K, Nagy N, Teramoto N, Gogolak P, Pokrovskaja K, Szekely L, Bjorkholm M, Klein G, Klein E (2001). *Epstein-barr virus can infect B-chronic lymphocytic leukemia cells but it does not orchestrate the cell cycle regulatory proteins.* J Hum Virol 4(5): 227–237.
26. Pallesen G, Hamilton-Dutoit SJ, Rowe M, Young LS (1991). *Expression of Epstein-Barr virus latent gene products in tumour cells of Hodgkin's disease.* Lancet 337(8737): 320–322.
27. Hjalgrim H, Askling J, Rostgaard K, Hamilton-Dutoit S, Frisch M, Zhang JS, Madsen M, Rosdahl N, Konradsen HB, Storm HH, Melbye M (2003). *Characteristics of Hodgkin's lymphoma after infectious mononucleosis.* N Engl J Med 349(14): 1324–1332.
28. Kanzler H, Kuppers R, Hansmann ML, Rajewsky K (1996). *Hodgkin and Reed-Sternberg cells in Hodgkin's disease represent the outgrowth of a dominant tumor clone derived from (crippled) germinal center B cells.* J Exp Med 184(4): 1495–1505.
29. Kuppers R (2002). *Molecular biology of Hodgkin's lymphoma.* Adv Cancer Res 84: 277–312.
30. Kuppers R, Rajewsky K, Zhao M, Simons G, Laumann R, Fischer R, Hansmann ML (1994). *Hodgkin disease: Hodgkin and Reed-Sternberg cells picked from histological sections show clonal immunoglobulin gene rearrangements and appear to be derived from B cells at various stages of development.* Proc Natl Acad Sci USA 91(23): 10962–10966.
31. Hertel CB, Zhou XG, Hamilton-Dutoit SJ, Junker S (2002). *Loss of B cell identity correlates with loss of B cell-specific transcription factors in Hodgkin/Reed-Sternberg cells of classical Hodgkin lymphoma.* Oncogene 21(32): 4908–4920.
32. Schwering I, Brauninger A, Klein U, Jungnickel B, Tinguely M, Diehl V, Hansmann ML, Dalla-Favera R, Rajewsky K, Kuppers R (2003). *Loss of the B-lineage-specific gene expression program in Hodgkin and Reed-Sternberg cells of Hodgkin lymphoma.* Blood 101(4): 1505–1512.
33. Contreras-Brodin B, Karlsson A, Nilsson T, Rymo L, Klein G (1996). *B cell-specific activation of the Epstein-Barr virus-encoded C promoter compared with the wide-range activation of the W promoter.* J Gen Virol 77(Pt 6): 1159–1162.
34. Tierney R, Kirby H, Nagra J, Rickinson A, Bell A (2000). *The Epstein-Barr virus promoter initiating B-cell transformation is activated by RFX proteins and the B-cell-specific activator protein BSAP/Pax5.* J Virol 74(22): 10458–10467.
35. Bargou RC, Emmerich F, Krappmann D, Bommert K, Mapara MY, Arnold W, Royer HD, Grinstein E, Greiner A, Scheidereit C, Dorken B (1997). *Constitutive nuclear factor-kappaB-RelA activation is required for proliferation and survival of Hodgkin's disease tumor cells.* J Clin Invest 100(12): 2961–2969.
36. Cabannes E, Khan G, Aillet F, Jarrett RF, Hay RT (1999). *Mutations in the IkBa gene in Hodgkin's disease suggest a tumour suppressor role for IkappaBalpha.* Oncogene 18(20): 3063–3070.
37. Martin-Subero JI, Gesk S, Harder L, Sonoki T, Tucker PW, Schlegelberger B, Grote W, Novo FJ, Calasanz MJ, Hansmann ML, Dyer MJ, Siebert R (2002). *Recurrent involvement of the REL and BCL11A loci in classical Hodgkin lymphoma.* Blood 99(4): 1474–1477.
38. Carbone A, Gloghini A, Gruss HJ, Pinto A (1995). *CD40 ligand is constitutively expressed in a subset of T cell lymphomas and on the microenvironmental reactive T cells of follicular lymphomas and Hodgkin's disease.* Am J Pathol 147(4): 912–922.
39. Horie R, Watanabe T, Morishita Y, Ito K, Ishida T, Kanegae Y, Saito I, Higashihara M, Mori S, Kadin ME, Watanabe T (2002). *Ligand-independent signaling by overexpressed CD30 drives NF-kappaB activation in Hodgkin-Reed-Sternberg cells.* Oncogene 21(16): 2493–2503.

40. van den Berg A, Visser L, Poppema S (1999). *High expression of the CC chemokine TARC in Reed-Sternberg cells. A possible explanation for the characteristic T-cell infiltratein Hodgkin's lymphoma.* Am J Pathol 154(6): 1685–1691.
41. Nakayama T, Hieshima K, Nagakubo D, Sato E, Nakayama M, Kawa K, Yoshie O (2004). *Selective induction of Th2-attracting chemokines CCL17 and CCL22 in human B cells by latent membrane protein 1 of Epstein-Barr virus.* J Virol 78(4): 1665–1674.
42. Marshall NA, Christie LE, Munro LR, Culligan DJ, Johnston PW, Barker RN, Vickers MA (2004). *Immunosuppressive regulatory T cells are abundant in the reactive lymphocytes of Hodgkin lymphoma.* Blood 103(5): 1755–1762.
43. Ishida T, Ishii T, Inagaki A, Yano H, Komatsu H, Iida S, Inagaki H, Ueda R (2006). *Specific recruitment of CC chemokine receptor 4-positive regulatory T cells in Hodgkin lymphoma fosters immune privilege.* Cancer Res 66(11): 5716–5722.
44. Dukers DF, Meij P, Vervoort MB, Vos W, Scheper RJ, Meijer CJ, Bloemena E, Middeldorp JM (2000). *Direct immunosuppressive effects of EBV-encoded latent membrane protein 1.* J Immunol 165(2): 663–670.
45. Staratschek-Jox A, Wolf J, Diehl V (2000). *Hodgkin's disease.* In: Masters JRW, Palsson BO (eds). *Human cell culture*, vol. 3. Kluwer: Dordrecht, pp 339–353.
46. Vockerodt M, Belge G, Kube D, Irsch J, Siebert R, Tesch H, Diehl V, Wolf J, Bullerdiek J, Staratschek-Jox A (2002). *An unbalanced translocation involving chromosome 14 is the probable cause for loss of potentially functional rearranged immunoglobulin heavy chain genes in the Epstein-Barr virus-positive Hodgkin's lymphoma-derived cell line L591.* Br J Haematol 119(3): 640–646.
47. Kis LL, Nishikawa J, Takahara M, Nagy N, Matskova L, Takada K, Elmberger PG, Ohlsson A, Klein G, Klein E (2005). *In vitro EBV-infected subline of KMH2, derived from Hodgkin lymphoma, expresses only EBNA-1, while CD40 ligand and IL-4 induce LMP-1 but not EBNA-2.* Int J Cancer 113(6): 937–945.
48. Baumforth KR, Flavell JR, Reynolds GM, Davies G, Pettit TR, Wei W, Morgan S, Stankovic T, Kishi Y, Arai H, Nowakova M, Pratt G, Aoki J, Wakelam MJ, Young LS, Murray PG (2005). *Induction of autotaxin by the Epstein-Barr virus promotes the growth and survival of Hodgkin lymphoma cells.* Blood 106(6): 2138–2146.
49. Atayar C, Poppema S, Visser L, van den Berg A (2006). *Cytokine gene expression profile distinguishes CD4+/CD57+ T cells of the nodular lymphocyte predominance type of Hodgkin's lymphoma from their tonsillar counterparts.* J Pathol 208(3): 423–430.
50. Anastasiadou E, Boccellato F, Cirone M, Kis LL, Klein E, Frati L, Faggioni A, Trivedi P (2005). *Epigenetic mechanisms do not control viral latency III in primary effusion lymphoma cells infected with a recombinant Epstein-Barr virus.* Leukemia 19: 1854–1856.
51. Arguello M, Sgarbanti M, Hernandez E, Mamane Y, Sharma S, Servant M, Lin R, Hiscott J (2003). *Disruption of the B-cell specific transcriptional program in HHV-8 associated primary effusion lymphoma cell lines.* Oncogene 22: 964–973.
52. Babcock GJ, Thorley-Lawson DA (2000). *Tonsillar memory B cells, latently infected with Epstein-Barr virus, express the restricted pattern of latent genes previously found only in Epstein-Barr virus-associated tumors.* Proc Natl Acad Sci USA 97: 12250–12255.
53. Frisan T, Sjoberg J, Dolcetti R, Boiocchi M, De Re V, Carbone A, Brautbar C, Battat S, Biberfeld P, Eckman M, Öst A, Christensson B, Sundström C, Björkholm M, Pisa P, Masucci MG (1995). *Local suppression of Epstein-Barr virus (EBV)-specific cytotoxicity in biopsies of EBV-positive Hodgkin's disease.* Blood 86(4): 1493–1501.
54. Harabuchi Y, Yamanaka N, Kataura A, Imai S, Kinoshita T, Mizuno F, Osato T (1990). *Epstein-Barr virus in nasal T-cell lymphomas in patients with lethal midline granuloma.* Lancet 335(8682): 128–130.
55. Tsuge I, Morishima T, Morita M, Kimura H, Kuzushima K, Matsuoka H (1999). *Characterization of Epstein-Barr virus (EBV)-infected natural killer (NK) cell proliferation in patients with severe mosquito allergy; establishment of an IL-2-dependent NK-like cell line.* Clin Exp Immunol 115(3): 385–392.

56. Zhang Y, Nagata H, Ikeuchi T, Mukai H, Oyoshi MK, Demachi A, Morio T, Wakiguchi H, Kimura N, Shimizu N, Yamamoto K (2003). *Common cytological and cytogenetic features of Epstein-Barr virus (EBV)-positive natural killer (NK) cells and cell lines derived from patients with nasal T/NK-cell lymphomas, chronic active EBV infection and hydroa vacciniforme-like eruptions*. Br J Haematol 121(5): 805–814.
57. Takahara M, Kis LL, Nagy N, Liu A, Harabuchi Y, Klein G, Klein E (2006). *Concomitant increase of LMP1 and CD25 (IL-2-receptor alpha) expression induced by IL-10 in the EBV-positive NK lines SNK6 and KAI3*. Int J Cancer 119(12): 2775–2783.

Index

M

N

T

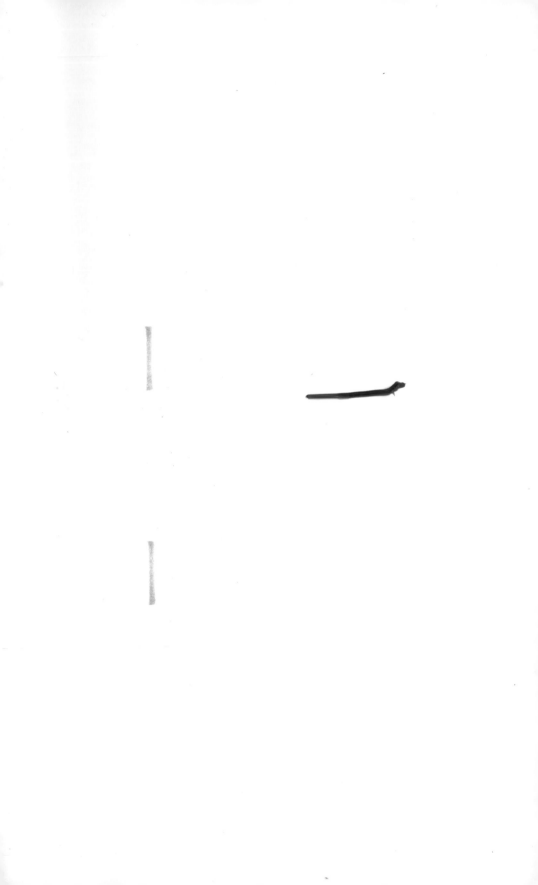